David A. White

Derrida on Being as Presence

David A. White

Derrida on Being as Presence

Questions and Quests

Managing Editor: Anna Michalska

DE GRUYTER
OPEN

ISBN 978-3-11-054013-0
e- ISBN 978-3-11-054033-8
ISBN EPUB 978-3-11-054014-7

Library of Congress Cataloging-in-Publication Data
A CIP catalog record for this book has been applied for at the Library of Congress.

Published by De Gruyter Open Ltd, Warsaw/Berlin
Part of Walter de Gruyter GmbH, Berlin/Boston
The book is published with open access at www.degruyter.com.

Managing Editor: Anna Michalska

www.degruyteropen.com
Cover illustration: © firebrandphotography

There are few things that we phrase properly; most things we phrase badly: but what we are trying to say is understood.

Augustine, *The Confessions*, Book XI, Section XX

...the sentiment that words were acts and acts were steps in life....

Henry James, *The American*

For last year's words belong to last year's language
And next year's words await another voice.

T. S. Eliot, "Little Gidding" from *Four Quartets*

Contents

Part II: The Aggregations of Language

Preface

Presence. Being as presence. Metaphysics as a play of presence. These references appear in many of Derrida's writings and almost all his earlier work. They embody a notion of singular importance for deconstruction and, when deconstruction is considered as a narrative whole, they stand as a locus of intersection for all other major concepts of that narrative.

Invoking the phrase "being as presence" rings with privileged authority for many advocates of deconstruction as an epitome of the foundational vision that underlies and guides its practice. The implication is that in some uniquely incisive way, Derrida discerned a feature permeating the history of western thought, a feature heretofore unrecognized in its full implications by any of the great figures defining that tradition–including Martin Heidegger, who made the presence of being a cardinal point of ontological identification.

In this essay, I examine in detail a series of pivotal passages showing how Derrida introduces and develops his notion of being as presence. If we then approach Derrida's position by questioning the structure and implications of being as presence, basing this inquiry on certain traditional concepts and distinctions, it becomes possible, if not essential, to surround the phrase in question with an aura of incisive critical concern. The ramifications of such investigation will challenge the scope and effects of Derrida's proposed insight and, along with coordinated argument, disrupt the foundational elements of deconstruction. John Dewey's elegant trope, "eulogistic predicate," underscores an approach to a philosopher's basic vocabulary which justifies this procedure. Without the kind of scrutiny and critical attention argued in this essay, being as presence remains a prime candidate for Dewey's tersely elitist description, regardless how often "being as presence" animates deconstructive inquiries and regardless of the extent to which advocates of deconstruction believe that it displays self-evident and decisive relevance in a myriad of theoretical contexts.

The essay proceeds as follows: the exposition and critical analysis draw primarily on key passages from Derrida's early work. There are two reasons for this limited approach. First, these texts lay the foundation for his position on being as presence. Second, the texts are often pitched recognizably as arguments, i.e., in the classic deployment of premisses yielding asserted conclusions, this in contrast to much of Derrida's later work which tends toward the oracular in tone and presentation. This is not to maintain, as some readers have done on occasion (e.g., Jürgen Habermas), that Derrida does not argue his positions but it is to emphasize the advantage for the student of Derrida to confront texts arranged according to such classic protocols. These discursive contexts, tautly interpreted, facilitate recognizing the kinds of evidence Derrida employed in formulating this seminal principle as well as the sequence of reasons introduced to elicit its effects.

Derrida cast his philosophical net wide. It may therefore be observed from the standpoint of interpretive practicality that this essay presupposes general familiarity with Derrida's projects concerning language as such and deconstruction as a programmatic "gesture" for interpreting linguistic events. I also have assumed that the reader has some experience interacting with the various stylistic modalities Derrida employed to verbalize those projects. In addition, spare consideration has been given to secondary work on Derrida. As far as I am aware, writers on Derrida have either not pursued this topic with much critical interest or, when they have mentioned being as presence, analyzed it with the depth and interpretive rigor which I have applied in this essay.

In an earlier work, *Derrida on Formal Logic An Interpretive Essay* (see Bibliography), I discussed Derrida's various treatments of certain basic concepts and principles in formal logic. This interpretive approach, both expository and critical, concluded with a series of speculative suggestions regarding the deconstructive thrust of Derrida's thinking in relation to the regulative stance typically ascribed to formal logic. In basic structure, that work and this essay complement one another, especially in contexts marked by high conceptual generality. I want to emphasize, however, that other than a series of correlated themes referenced in footnotes, the present essay stands by itself in content and as a prospective contribution to the secondary literature on Derrida–especially with regard to the seminal importance of being as presence–and may be so judged by the reader.

The essay is in three parts. In Part I, the discussions primarily pursue Derrida's various configurations of being as presence, with critical analysis typically generated from the areas in philosophy traditionally referred to as metaphysics. Part I concludes with a set of conclusions and questions, tentatively posed, intended to sketch ways to evaluate the cogency of Derrida's texts on being as presence and to outline areas of problematic concern, especially given that Derrida's inquiry encompasses the entirety of western metaphysics.

In Part II, I have integrated inferences drawn from the analyses of Part I into a series of contexts which focus on an intersection between metaphysics in a broad sense and metaphysically controlled inquiries into language. The analyses concentrate on more circumscribed and purely theoretical description of structural components traditionally associated with language. Derrida had much to say concerning this intersection and Part II raises questions about a set of his most important investigations–reference, context, negation, iterability–all of which include abstract considerations which animate their structure.

The final section, Part III, brings together the far-ranging themes of Parts I and II by showing how Derrida's approach to the history of metaphysics played into the formulation–seriously problematic to its core–of concepts and principles now typically referred to collectively as *deconstruction*, a name for a methodology advanced for a disclosive reading of texts. The essay concludes with a set of speculative developments–quests, in a word–concentrating on a series of basic concepts often appearing in metaphysical accounts, but concepts redrawn with the intention of incorporating the general movement of Derrida's thinking within these more traditional frameworks. The footnotes cite and discuss selected commentary on themes advanced in the essay and also include parallel and contrasting passages on these themes drawn from Derrida's later work.

The essay's conclusions are established by argument in the usual sense, i.e., premisses yielding conclusions. This may seem an odd feature of the essay to emphasize in a Preface, since presumably such reasoning is hardly an atypical procedure for a work intended to be philosophical. I wish to call attention to it, however, for the sake of readers who may be committed advocates of deconstruction in general and Derrida in particular. That there are many individuals so inspired is evident from the extensive secondary work on Derrida. But also noteworthy, I believe, is the hagiographical aura surrounding much (but not all) of this commentary. For those readers of Derrida and prospective students of deconstruction, the conclusions in this essay may appear distressingly antithetical to their understanding of Derrida's teaching–or if this word exudes programmatic mustiness, to the textual interpretations he set as examples. If so, then one possible response to the lines of argument developed below is to adopt the same procedure as that of the essay–that is, producing counter positions by embodying reasons and arguments, ordered into recognizable premisses and conclusions. Response so developed will allow discussion of Derrida's thought to proceed in a manner promising greater clarification for the benefit of the philosophical community at large as well as for all who, whatever their ideological bent in philosophy and related disciplines, have taken the time and effort to read him. If this essay contributes a measured advancement of such discussion, it will have achieved its end.

Abbreviations

Works by Jacques Derrida cited in this essay follow these abbreviations. Publication information for these sources appears in the Bibliography.

LI *Limited Inc.*, 1988
MP *Margins of Philosophy*, 1982
N *Negotiations*, 2002
OG *Of Grammatology*, 1976
P *Positions*, 1981
PS *Psyche Inventions of the Other*, 2007
SP *Speech and Phenomena*, 1973
WD *Writing and Difference*, 1978

Part I: **Presence and the History of Metaphysics**

1 Being as Presence: Systemic Considerations

Toward the conclusion of the brief preface Derrida included for his early work *Of Grammatology*, we read: "...my interpretation of Rousseau's text [which appears in Part III of OG] follows implicitly the propositions ventured in Part I, propositions that demand that reading should free itself, at least in its axis, from the classical categories of history–not only from the categories of the history of ideas and the history of literature, but also, and perhaps above all, from the categories of the history of philosophy."[1] Then, several pages later at the end of the equally brief "Exergue," the following passage appears: "The future can only be anticipated in the form of an absolute danger. It is that which breaks absolutely with constituted normality and can only be proclaimed, *presented*, as a sort of monstrosity" (OG, 5–italics in text).

Strong, dire, even apocalyptic claims. *Of Grammatology* is intended to show that we need to jettison "the categories of the history of philosophy" when reading that selfsame history (as well as, presumably, anything else of comparable weight and import), with such categorical withdrawal thrusting us into a future which can be described as "a sort of monstrosity."

A reader approaching *Of Grammatology* must do so with sympathy toward its project–interpretive politeness requires such neutral receptivity. But questions immediately arise in the face of its announced goal. Broadly stated: (a) why is it necessary to practice the art of reading apart from the traditional categories of the history of philosophy? And (b) why would the results of philosophical activity pursued after embracing this radically revamped procedure be monstrous in content, form or perhaps both? The answers to these question revolve around the notion of being as presence and a set of diverse implications which Derrida drew from that notion.

1.1 Principles of Interpretation: Procedures and Scope

Derrida deploys presence (*présence*) primarily in two distinct but related contexts: the history of metaphysics and the philosophical analysis of language. This essay is organized to reflect these allied concerns. Chapter 1 of Part I develops passages from Derrida which detail the relation between being as presence and the history of metaphysics; this chapter and all of Part I are devoted to abstract considerations befitting the tenor of metaphysical thinking regarding such matters. Chapter 4 of Part II focuses on passages concerning the relation between being as presence and its function as one of the foundational elements in Derrida's thoughts on language; this

1 OG, lxxxix; cf. OG, 163, where Derrida asserts that the "classical categories" proper to reading are also "the founding categories of metaphysics." Subsequent passages from *Of Grammatology* will be cited in the body of the text. See note 5 for the system employed in citing passages for commentary.

account initiates a four-chapter analysis of a set of fundamental concepts in Derrida's approach to understanding linguistic elements from a deconstructive perspective.

This separation of themes will clarify the identification of systemic elements pertaining to being as presence (Part I) and application of those elements to language (Part II). I have named these results "systemic" rather than "structural" to underline the fact that their overall organization is relatively loose, reflective of Derrida's discursive style, but that in the present context they are also as tight as is feasible given the apparent content of the texts analyzed. The basic assumption is that Derrida remained a product of his place in the history of metaphysics to such an extent that it is necessary to read him from that perspective, if only provisionally, in order to be in a position to appreciate and to evaluate what he might have foreseen as the future of philosophical activity.[2]

The passages on being as presence discussed in Chapter 1 have been selected from Derrida's earlier works. They represent the core and central ramifications of Derrida's position concerning this crucial component of his teaching. Since Derrida often wrote in modes distant from the lines of thought developed and narrated in traditionally systematic protocols, I have coordinated the passages so that their sequence in this chapter forms a linear progression of topics and illustrates, in Derrida's distinctive style of presentation, justifications for conclusions he maintained. Discussion of these passages is linked so that the reader can appreciate the continuity and progression in Derrida's thinking; in addition, the arrangement and examination of individual passages allow more concentrated individual critiques of claims and arguments maintained in each of these passages.

Finally, a point worth repeating (and a theme developed in the essay proper): the questions applying to the strictly metaphysical considerations pertaining to being as presence (Part I) also apply to Derrida's systemic approach to language (Part II). If problems arise and, perhaps, proliferate during the process of Derrida's thinking on matters of high generality, then the cumulative weight of those difficulties should have, or at least may have, a bearing on the final legitimacy attributed to Derrida's project. For if Derrida cannot adequately establish a context of inquiry as philosophically problematic because of ineffective parameters intended to localize, define and justify that problem, then it is not obvious why that context remains problematic merely because Derrida's texts proclaim it to be so.

---

2 Derrida insists that any substantive position emerging from the application of deconstructive analysis must to some degree remain within the bounds of metaphysical discourse: OG, 19; WD, 36, 281-2, 284; MP, 177, 219, 237, 263; Jacques Derrida 1992, 50; Richard Kearney 2004, 143. Commentators have noted this insistence: Irene Harvey 1986, 44, 96, 100, 124; Michael Naas 2003, 100; Barry Stocker 2006, 184; Russell Daylight 2011, 9. Salmi Haddad 2013 says, 141: "...in his own inheritance of past thinking, Derrida can only ever destabilize certain notions by stabilizing others." See also Ch8n95. Chapter 10 of this essay develops implications from Haddad's perceptive observation.

1.2 Derrida's Narrational Strategy

Jacques Derrida published three separate works in the same year (1967), but upon investigation it seems obvious that the intention was to advance a concerted and integrated position laid out separately, in tripartite configuration.[3] Thus most of his technical terms, or at least terms which have solidified themselves as staples in the ranks of commentaries on Derrida, appear in *Of Grammatology*. It may be assumed then that the more critical attention paid to the content displaying the highest degree of abstraction as developed in these works–all three yielding passages analyzed in this essay–the more informed will be the educible connections between these technical terms and the foundational elements in Derrida's approach to the underlying characteristics of deconstruction taken as a method or, expressed in a mode which the advocate of Derrida might find more congenial, a chain of gestures involved in the reading of texts.

The other pertinent ramification concerns the doctrinal links between positions Derrida worked out on being as presence in the earlier texts and the themes and topics which occupied him in his middle and late periods–e.g., hospitality, the concept of gift, death, various political themes and issues, etc. Whether Derrida derived the substance of his accounts of these topics from direct consideration of factors pertaining to being as presence or whether these accounts originated primarily or even solely from Derrida's own creativity is a question of provenance which can be addressed and, perhaps, resolved only if Derrida's early thought on being as presence is fully explored. The present essay concentrates on analyzing Derrida on this abstract theme with the hope that its results may, by extension, be useful to readers of Derrida who might wish to pursue the structure, cogency and implications of his later thinking.[4]

3 See Daylight, 6-7 for comments on the publication history of Derrida's first three books.

4 Two recent sources explore a wide range of Derrida's later work: Michael Naas' *Miracle and Machine Jacques Derrida and the Two Sources of Religion, Science and the Media* and Samir Haddad's *Derrida and the Inheritance of Democracy* (see Bibliography). See also Jacques Derrida 2001, 101, for Derrida's own list of topics pursued in his later writings. Whether the theoretical positions embodied in Derrida's early works can be consistently developed in such substantive directions is a question which may be posed based on the argument of this essay. Existing commentary on being as presence does not attempt to develop systematically its structure and implications. See, for example, Thomas Baldwin 2008, 108-9, 110-11; Jonathan Culler 1983, 93-4; Richard Rorty 1985, 135.

1.3 The Deployment of Presence

1.3.1 Truth and Language

Early in *Of Grammatology,* Derrida makes the following axiomatic claim:

> **Passage 1A** *The history of* (the only) *metaphysics*, which has, in spite of all differences, not only from Plato to Hegel (even including Leibniz) but also, beyond these apparent limits, from the pre-Socratics to Heidegger, always assigned the origin of truth in general to the logos: the history of truth, of the truth of truth, has always been–except for a metaphysical diversion that we shall have to explain–the debasement of writing, and its repression outside "full" speech (OG, 3–italics in text).[5]

This passage emphasizes that metaphysics is characterized throughout its history as originating its stance toward truth in terms of logos–or "logocentrism," that is, the location of truth within the "privilege of an interior, self-present voice."[6] The immediate confines of "logos" in this voiced sense. i.e., word, account, argument, etc.,

5 Passages analyzed in this essay are sequentially identified by chapter (in which the passage first appears) and number. This is **Passage 1A**; all references to **Passage 1A** refer to this passage. Subsequent passages in Chapter 1 are identified **Passage 1B, 1C** etc. In addition to **Passage 1A**, Derrida also emphasizes the unity of the history of metaphysics at WD, 193. The passages quoted from Derrida vary in length from a single sentence to entire paragraphs. This approach maximizes the factor of context as much as possible (against the strictures Derrida imposed on assigning limits to given contexts, a theoretical position discussed in detail in Chapter 6 below); it also addresses and presumably avoids what Christopher Norris 2012, 185n102, labels "highly selective or snippety reading," a practice which Norris claims is used by commentators (such as Richard Rorty) "who seek to recruit Derrida...to the cause of their own wholesale anti-foundationalist, anti-realist or neo-pragmatist crusade."

6 Daylight, 6, has offered this definitional sketch of "logocentrism." Derrida uses "logocentrism" on the first page of OG's Exergue but leaves the term undefined at that point (see **Passage 1B**). Daylight adds later, 20, that logocentrism is the: "self-presence of mental impressions prior to or without recourse to language"; for Stocker, 51, logocentrism is "...an approach at the heart of metaphysics according to which truth, knowledge or being are present at some particular moment"; for Culler, 92, logocentrism represents "...the orientation of philosophy toward an order of meaning–thought, truth, reason, logic, the Word–conceived as existing in itself, as foundation." These three accounts display considerable variance in emphasis, content, and the extent to which logocentrism may include, as part of its structure, experience of a philosophically-inclined investigator. Derrida invites this diffuseness of interpretation when in *Positions* he says: "In *Of Grammatology* I simultaneously proposed everything that can be reassembled under the rubric of *logocentrism*–and I cannot pursue this any further here–along with the project of *deconstruction*" (P, 51–italics in text). Cf. MP, 64-5: LI, 20, 104; Kearney, 143, 146, 148-9, 154. Derrida is surely justified in refusing to repeat himself but the "simultaneously" in this claim is perhaps the reason why his commentators diverge as widely as they do regarding the import of the term in question. Derrida's inclusive approach to logocentrism also underlines the importance of the current study, i.e., analyzing and evaluating as clearly as possible how the groupings of philosophers in **Passage 1A** exhibit the singular term "logocentrism" in the context of being as presence. For additional comment on logocentrism, see Art Berman 1988, 279; Hans-Georg Gadamer 1989, 95; Rorty 1989; Stocker, 52; Norris 2012, 134; Daylight, 22.

all presuppose the determinate existence of fixed elements of significance which Derrida calls the "transcendental signified." We will clarify both "logocentrism" and "transcendental signified" as technical terms as the essay unfolds. For now, we observe that this passage underscores Derrida's insistence that the emphasis under scrutiny has debased writing and repressed its function relative to what is typically held as "full" speech, i.e., the spoken word. Derrida contends that one and only one history of metaphysics exists, implying that whatever can be construed as metaphysical in tone or structure (or both) must be located within the sweep of this single historical, panoramic phenomenon. The one exception Derrida mentions does not affect the emphatic sense of unity conferred on this history.[7] It will be observed that Derrida marks the differences which punctuate the history of metaphysics–an element in this context which we will examine closely–but these differences do not affect the possibility of generalizing about the entire history of metaphysics.

The immediate scope of this generalization concerns language, in particular the relation between written and spoken discourse. Thus "logos" refers to discourse taken in a very broad sense. For Derrida, logos pertains to "the origin of" truth, which leaves open whether truth–granting for now its feasibility in a public or justified sense–should be understood just as the event of language, whether spoken or written, or whether it includes a relation between spoken or written language and something other than language. If the latter, language arises necessarily from human origin but its full character when viewed philosophically is not limited to that factor taken by itself. This possibility becomes increasingly significant once we elicit the implications of Derrida's position on being as presence, especially when these implications enter the sphere of language.

1.3.2 Presence and the History of Metaphysics

A few pages later, Derrida again correlates in greater detail the history of metaphysics with being as presence:

> **Passage 1B** We already have a foreboding that phonocentrism merges with the historical determination of the meaning of being in general as *presence* [*présence*], with all the subdetermina-

7 The notion of unity is crucial for Derrida's various agendas–e.g., MP, 259 on the unity of metaphysics as a branch of western thought; see also WD, 167, 281-2; MP, 121; OG, 3, 13; P, 51; PS, 301. Commentators differ concerning the status of unity in Derrida–e.g., John Caputo 1997, 31, denies that unity applies to the history of metaphysics while Harvey supports the relevance of unity in several contexts, 96-102, 106; see also in this regard, Christopher Fynsk 2001, 156; Daylight, 58; Richard Shusterman 1992, 75-6; Robert Bernasconi 1989, 247; and Leslie Hill 2007, 97, on the unity of a given text. Additional commentary on the status of unity appears in David A. White 2011 (hereafter cited DFL), 59-62. See also below, Ch3n40. Unity will be analyzed in a series of fundamental contexts in Chapters 4, 5 and 6 below.

> tions which depend on this general form and which organize within it their system and their historical sequence (presence of the thing to the sight as *eidos*, presence as substance/essence/existence [*ousia*], temporal presence as point [*stigmè*] of the now or of the moment [*nun*], the self-presence of the cogito, consciousness, subjectivity, the co-presence of the other and of the self, intersubjectivity as the intentional phenomenon of the ego, and so forth). Logocentrism would thus support the determination of the being of the entity as presence (OG, 12–italics in text).

This passage from *Of Grammatology* may be read with the following passage from "Structure, Sign and Play in the Discourse of the Human Sciences," a lecture given in 1966 and published as the concluding essay in *Writing and Difference* in 1967, the same year as *Of Grammatology*.

> **Passage 1C** The history of metaphysics, like the history of the West, is the history of these metaphors and metonymies. Its matrix–if you will pardon me for demonstrating so little and for being so elliptical in order to come more quickly to my principal theme–is the determination of Being [*de l'être*] as *presence* [*présence*] in all senses [*les sens*] of this word. It could be shown that all the names related to fundamentals, to principles, or to the center have always designated an invariable [*l'invariant*] presence–*eidos, archē, telos, energeia, ousia* (essence, existence, substance, subject) *alētheia*, transcendentality, consciousness, God, man, and so forth (WD, 279-80).

For Derrida, phonocentrism names an attitude toward language which privileges spoken over written discourse given that the former necessarily occurs in the present, thereby reinforcing presence as the most important medium of expression. The specification of this privilege represents a crucial theme since Derrida infers that this preference, the product of millenia of philosophical thought, has made it essential to revitalize our attitude toward the writing of philosophy and writing in general. But the dominant concern of this passage is the identification "of the meaning of being in general as *presence*," along with "all the subdeterminations" which derive from the traditional approach to being as understood in this way.

In **Passages 1B** and **1C**, Derrida generalizes by collecting pivotal moments occurring throughout the history of metaphysics without individuating these moments by the name of the originating philosopher but rather by the principal element in that philosopher's thinking. These elements, listed in historical order, indicate that the philosophers in question include Plato, Aristotle, Descartes and Husserl. Derrida does not mention (although he does write "and so forth"), either by doctrine or name, the metaphysical positions developed by, for example, Jean-Paul Sartre or Alfred North Whitehead, or by the large number of metaphysicians in the Anglo-American tradition, from F. H. Bradley through Paul Weiss. And if we postdate the authorship referred to in the passages cited above, additional instances of metaphysics may be cited: Nelson Goodman, David Lewis, and even more recently in the sector of analytic philosophy a group of thinkers pursuing what they have named "metametaphysics."[8]

---

8 For example, the 2009 anthology *Metametaphysics New Essays on the Foundations of Ontology*. See David Chalmers, et. al. in the Bibliography.

The fact that such modes of metaphysical thinking are not identified–whether they exist before, during or after the time of Derrida's pronouncements–cannot imply that some or all of them fall outside the effective scope of the "being as presence" designation. Such putative exclusion, if it were permissible within Derrida's generalizations, would allow significant areas and individual figures throughout the history of philosophy, up to the present (and, presumably, into the future), to remain apart from and impervious to whatever limitations Derrida will institute into the structure of being as presence. In this hypothetical eventuality, the "other" metaphysical systems, those promulgated and enduring beyond the boundaries of being as presence, would have at least the possibility of open access to, for example, whatever may have been (and, for Derrida, definitely has been) concealed by the collective agency of philosophical thought insofar as that thought was and continues to be marked by subservience to the factor named "being as presence." We conclude therefore that for Derrida (save for the exception he cites at OG, 2), each and every one of these moments, whether named or unnamed, constitute instances of being as presence.

Passage 1B refers to what Derrida calls "the meaning of being in general as presence." This phrase may be considered as a unity, a whole of parts. As it stands, the phrase as a whole is ambivalent between the following readings. If (a) the phrase is divided into "the meaning of" and "being as presence," then "being as presence" exhausts the meaning of being; the key point then is to recognize that "the meaning of" being is *nothing but* presence. However, if (b) the phrase is taken as an indivisible unity, then "the meaning of being as presence" can be restated as: "the meaning of being as presence *when being is in fact presence*." And this reading leaves open the possibility that being could be something *other* than presence; if so, then being may have a very different meaning.

Since Derrida writes "being in general," context suggests that the first of these two possibilities is the intended sense, that being as presence is all and only what being can be, that is, from the perspective Derrida has assumed on the history of metaphysics. The fact that "presence" is italicized also supports this reading so we shall take it in this sense. However, when he writes "the meaning of being in general as presence," it appears as if Derrida thinks that he has, in fact, clearly and unambiguously stated the meaning of being when being is qualified "as presence." In other words, at this point in his analysis of the history of metaphysics, nothing more need be said about being as such other than that it is "presence." Thus when Aristotle famously asserts (*Metaphysics* VII 1028^{a}10) that "being can be said in many ways,"[9] Derrida's position implies that whatever metaphysical diversifications Aristotle discerned in being, *all* such manifestations may, indeed must, be reduced to "being as presence."

9 All quotations from Aristotle are from translations in the *Basic Works of Aristotle* (see Bibliography).

In Chapter 2, we will demonstrate that Derrida's uniform designation of being "as presence" becomes seriously problematic and in fact runs counter to at least one of the texts in early classical western metaphysics. It will then follow that "being as presence," if taken as an epithet duly summarizing the entirety of western metaphysics, conceals–and arguably distorts–more than it may reveal about being. Furthermore, in Part III of the essay we argue that the second reading, (b) above, is in fact more congruent with implications derivable from Derrida's texts on being as presence. If so, then Aristotle's metaphysical sensibilities were correct after all. Thus Derrida's reformulation of being insofar as it is allowed to function within the entire history of metaphysics is best construed "in many ways" rather than according to the single strict and rigid rubric of "being as presence." But this line of interpretation remains only conjectural until it has been justified in what follows.

In **Passage 1B**, phonocentrism and logocentrism, both of which include references to language, "merge" and "support" respectively the uniformity of presence. It may safely be inferred then, first, that at least some of the considerations derived from the development of the position on being as presence will pertain to Derrida's position on language. As we noted earlier in this chapter, these considerations will be examined in Part II of this essay.

For now, second, consider the important word "merge" (*se confond*) especially when used in close proximity to Derrida's phrase "the meaning of being as presence." Is the merger simple conjunction–(a) the fact of spoken language coexisting with (b) the fact of being construed as presence with each element in this conjunction maintaining its own distinct identity, or more organic and penetrating such that "being as presence" is defined, at least in part, by the fact of discourse itself? If the latter, then a necessary but not sufficient condition for being as presence is that it includes the fact of language articulating the character of being. It would then follow that clarifying the structure of being as presence would include analyzing how language contributes to determining this structure. It would also follow, apparently, that the burden is on the advocate of being as presence to explain why, as an opening and obvious question, it should not be construed as a species of idealism. The point is not to demean idealism as a grounding metaphysical vision; only that it now becomes urgent to determine, if possible, whether idealism does indeed belong to being as presence on condition that "being as presence," as a single and seemingly unified referring phrase endowed with a fixed, referential object, is characterized at least potentially by a uniform structure.[10]

10 Stocker, 29-30, denies that Derrida is an idealist. For Derrida's impression of such a descriptor applied to his thought, see P, 50; cf. LI, 93.

1.3.3 Presence and Difference

In **Passage 1A**, truth was located in logos throughout the history of metaphysics "in spite of all differences" marking that history. In **Passage 1B**, Derrida refers to "the subdeterminations" which "depend on this general form," i.e. being as presence. Derrida's "subdeterminations" has a more positive connotation than the word "differences," which tends to evoke a more passive reliance on an underlying wholeness. But terminological nuances aside, these differences and subdeterminations do not affect and therefore do not diminish the force of the generalization Derrida wants to make, i.e., that the system and sequence of the history of metaphysics are strictly determined as variations on being as presence.

Although the differences between the spare extant texts of Heraclitus and Parmenides are stark, the far more articulated and subtle differences between Plato and Aristotle are, in their own way, even more dramatic. However, it must be presumed that Derrida's factor of being as presence construed as "invariable" (*invariant*) can and will encompass all these positions, at least insofar as they exhibit allegiance to at least some of the basic elements of metaphysics as exhibited throughout its history. If this invariability does not obtain, then "being as presence" becomes, in John Dewey's evocative phrase, a eulogistic predicate,[11] applicable only to a certain strain of thinking as that strain has emerged from the history of thought but without its having been established, as Derrida so maintains, as the controlling factor in all western metaphysical deliberations. It may then be assumed that the specific doctrinal content of all the epochal entries cited in **Passages 1B** and **1C** as well as all other moments in the history of metaphysics not explicitly mentioned must share something in common in order to justify ascribing the term "presence" as the single predicate proper to the one and only history of metaphysics.

1.4 Presence: Linearity and Process

Being as presence spans the history of metaphysics but to construe the character of the historical factor itself as an element in this ultimate philosophical datum requires careful consideration:

> **Passage 1D** This pluri-dimensionality does not paralyze history within simultaneity, it corresponds to another level of historical experience, and one may just as well consider, conversely, linear thought as a reduction of history. It is true that another word ought perhaps to be used; the word history has no doubt always been associated with a linear scheme of the unfolding of presence, where the line relates the final presence to the originary presence according to the

11 See John Dewey 1958, 28, for a concise explanation of "eulogistic predicate." For discussion concentrating on affinities between deconstruction and pragmatism, see Chantal Mouffe 1996.

> straight line or the circle. For the same reason, the pluri-dimensional symbolic structure is not given within the category of the simultaneous. Simultaneity coordinates two absolute presents, two points or instances of presence, and it remains a linear concept (OG, 85).

If we think of the history of metaphysics represented as a straight line joining, e.g., certain pre-Socratics at one end and Heidegger at the other, then Derrida objects that this configuration is reductive and that history so conceived amounts to an omnipresent simultaneity. His explanation appeals to what he calls "the originary presence" and its relation to "the final presence." However, it is possible that this distinction between originary and final is not canonic Derrida but rather an interpretive result tied to the hypothetically linear approach to history. If this approach to the conceptualization of history errs by dint of reductiveness, then one may reason that any designation falling along an inappropriately envisioned historical line would also be equivalently suspect.

The following conjectures may nonetheless be considered as possibilities worth introducing here, with a final sense of plausibility dependent upon examination and interpretation of additional texts. Presence can be a process, i.e., between an "originary presence" and a "final presence" although it would be premature to conclude that presence as such is always and necessarily a process between a starting point and some sort of conclusion. A subsidiary question would be whether or not this process, if indeed such a concept is pertinent, has any characteristic typically predicated of processes (as in Aristotle–e.g., the continuousness of division[12]). A more expansive surmise will hold that presence can be (a) both a process while the process itself is in motion as well as (b) the "end points" of such a process. What would Derrida's use of "unfolding" then refer to in this context? Is it limited to sheer temporality, as in the time that passes between an event's origin and its conclusion? Or is it both the time component as well as the elements of the event itself insofar as these elements are in motion from a certain point of origin to a certain point of cessation or fulfillment? The texts introduced at this juncture do not answer these questions, but a thoughtful quest for the content of being as presence will find such issues useful to consider.

More directly relevant given future lines of argument in this essay is the concluding analysis of simultaneity. Derrida refers to "two absolute presents," then adds "two points or instances of presence," concluding that simultaneity linking two "presents" in this way remains a "linear concept" and therefore presumably inadequate to the proper understanding of being as presence.

This description is suggestive in a number of respects: first, if simultaneity "coordinates" two "instances of presence" then it is implied that presence as such is not reducible to simultaneity since if it were, simultaneity would *constitute* presence, not merely coordinate it.[13] Thus presence presumably includes temporality–note

---

12 For the consequences of division applied to a continuum, see Aristotle, *Physics*, $207^{a}8$; $231^{b}15$.

13 Derrida registers concern about the apparent cogency of the notion of simultaneity–see WD, 14,

Derrida's appeal to the relation between "two absolute presents"–insofar as temporality occupies but is not equivalent to the present. The question then becomes: what constitutes presence over and above the temporality which conditions presence to emerge in the present? The implication is that presence must include a reference to *what is present* and not simply isolate the temporal factor of an already, pre-established existent "present" entity in order to posit that factor *as* being as presence. This inference, if sound, is crucial since it entails investigation into the organizational makeup of entities or events as a prerequisite to identifying more precisely the content of being as presence.

In that vein, second, what are the referents of "two absolute presents"? The immediate contexts suggest events and entities, since simultaneity is often applied to both types of reality and in a variety of contexts (although event and entity are clearly quite different from one another). But other possible referents include (a) two distinct metaphysical systems or (b) two distinct elements within one given metaphysical system. In both cases, "absolute present" could readily apply to either (a) a complete metaphysical system taken as a unity and thus as an absolute present or (b) to two (or more) elements of a single given metaphysical system. The texts from Derrida examined so far are not decisive in these areas of inquiry but, again, the topics broached are well worth additional consideration.

1.5 Presence and the Metaphysics of Form

The series of conjectures offered in the previous section addresses a fundamental issue–what *is* being as presence? The following passage appears in Derrida's article "Form and Meaning: a Note on the Phenomenology of Language," which was originally published in *Speech and Phenomena* in 1967, the same year as *Of Grammatology*. This passage opens a new dimension for achieving a more precise determination of how being as presence should be understood:

> **Passage 1E** All the concepts by which *eidos* or *morphē* could be translated and determined refer back to the theme of presence in general. Form is presence itself. Formality is what is presented, visible, and conceivable of the thing in general (SP, 108).

When Derrida asserts with dramatically discursive force that "form [*forme*] is presence itself," does he intend "is" to be identity or predication? Although this distinction, marked with such clean-cut opposition, is itself perhaps a candidate for deconstructive analysis, we preserve it here in order to explore alternative readings of the claim.

24; MP, 55, 56, 58. For discussion, see DFL, 173-6.

If the "is" in this context is of identity, then it follows that form must be functional and operative in each and every determination of being in all the lists Derrida provides. Derrida's following claim, that formality, the character of being formal, is "what is presented, visible, and conceivable of the thing in general," then becomes seminal. We are told that what is presented is "visible and conceivable." But should the "and" linking visible and conceivable be taken literally, as a logical conjunction? If so, then something existing as presence is *both* visible to the sense of sight *and* capable of being understood and conceptualized by the mind–i.e., both properties apply to *all* instances of being as presence. However, another reading is possible. In this case, each entry in the list of three properties applies to one, perhaps some, instances of being but not to all instances. Thus some instances of presence are visible, other instances of presence are capable of being conceived, but no single instance of presence is *both* visible *and* conceivable. The pair of properties, visible and conceivable, is therefore intended as a summary description applicable to the entire sequence of instances but aligned with the interpretive restrictions just introduced.

In the latter case, the conjunction "and" is misleading since its effective logical sense is *disjunctive*, i.e., Derrida is claiming that being as presence is in some cases visible and in other cases conceivable, but he is not claiming that any one instance of being as presence is both visible and conceivable. However, note what follows immediately for both alternative readings: if being as presence is in some respect (or respects) the same and "invariable" throughout the history of metaphysics, then being as presence when it is visible is and must be *the same as* being as presence when it is conceivable.[14] The question then becomes: can we identify characteristics common to both visible and conceivable instances of being as presence? If the answer is no, then there is reason to suspect that *either* Derrida has not accurately characterized being as presence with respect to this conjunction of properties and has done so at an especially crucial point since the context purports to articulate a generalized, comprehensive description of being as presence *or* that being as presence is fundamentally and intractably ineffable. Either disjunct of this conclusion has important implications, as we shall see.

Assume now that the intended sense of "is" should be read not as an identity indicator but as predicational. In this case, presence becomes broader in import than form, implying that presence, if its nature is amenable to description, could not be reduced by identity to the nature of form. If so, presence includes elements other than and in addition to form–however form is construed. It may then be assumed that these elements can be articulated according to designations similar to "form" and all

---

14 Derrida often refers to sameness; also, what could be taken as an apparent synonym of sameness–invariability–is equally important to the structural characteristics assigned to being as presence. Sameness will be introduced and discussed below (in the section "Presence and *Différance*"). Also, the factor of invariability and its role in the structure of being as presence will become increasingly important to the argument of this essay.

that resonates with the term "form" throughout its variegated appearances during the history of metaphysics. Without such articulation, the import of presence, other than its animation of form as specified by Derrida, will remain ineffable and presumably inaccessible to any mode of investigation, deconstructive or otherwise.

1.6 Being as Presence–the "Subjective" Turn

Derrida connected being as presence with the notion of form and did so with a high degree of intimacy. We have sketched some preliminary reflection on the relevance of this connection with respect to clarifying the content of being as presence. However, in an essay on Hegel based on a lecture delivered in 1968, one year after the three works already explored, we read a general claim which considerably broadens the context of inquiry:

> **Passage 1F** In determining being as presence (presence in the form of the object, or self-presence under the rubric of consciousness), metaphysics could treat the *sign* only as a *transition*. Metaphysics is even indistinguishable from such a treatment of the sign. And neither has such a treatment somehow overtaken the concept of the sign: it has constituted it (MP, 71–italics in text).

This passage–the first words in the essay (excluding a series of quotations from Hegel)–is important for detailing the connection between metaphysics and the way in which Derrida confronts the concept of "sign." We will devote considerable attention to the concept of "sign" and its relevance to metaphysics in Chapter 4. For present purposes, however, the parenthesis in the opening sentence commands our attention since it reveals a new approach to determining the character of being as presence. I suggest that we may interpret as complementary parts of one theoretical whole the texts appearing early in *Of Grammatology* (and the other, contemporary sources cited above), which emphasize the content of a number of systems throughout the history of metaphysics, and the passage from the essay on Hegel just introduced.

1.6.1 Presence, Form, Consciousness

We begin with yet another decision on an ostensibly small logical factor but one with crucial interpretive implications. If the "or" in this passage indicates strict disjunction, then being as presence can be *either* "presence in the form of the object" *or* "self-presence under the rubric of consciousness;" also, neither disjunct can be either reduced to or co-exist with the other. We may then interpret the distinction as contrasting being as presence when it arises from "the form of the object" and being as presence when it emerges as "self-presence" within, as Derrida puts it, "the rubric of consciousness."

These characterizations of being as presence are in some respect (or respects) dissimilar. Being as presence applies to identifications of basic metaphysical components (via the concept of form) drawn from key figures in the history of metaphysics as well as to consciousness itself insofar as it functions as a foundational reality for one of those figures (Hegel–and others, e.g., in the phenomenological tradition). The use of "rubric" in this context suggests that for Derrida, consciousness is a name standing as a traditional place marker encompassing a welter of possible sub-determinations rather than as a clearly established concept worth employing as a point of departure for additional and putatively firm analysis. It may be inferred then that "consciousness" appearing in contexts such as those considered at this point conceals more than it reveals with respect to ulterior reaches of metaphysical insight.[15]

If the two designations of being as presence are indeed intended to differ significantly from one another and if the second designation indicates that the defining characteristic that justifies the ascription of "being as presence" is necessarily limited to the sphere of consciousness, then it seems necessary to construe the parallel defining characteristic of "presence in the form of the object" to be something *other than* consciousness. If this otherness were not the case, then both evocations of being as presence would include some dimension of consciousness; but if that were so, then at least from this perspective no difference would mark the two designations of being as presence, apparently nullifying the text's parenthetical point.

Derrida's position on the invariable character of being as presence nonetheless requires that the presence emergent from the systems referred to in the first set must be the same at least in some respects as the presence that emerges from the exemplar of being as presence in the second identification; if not, then the single phrase "being as presence" does not apply uniformly to the history of metaphysics. We have suggested that systems in the first set do not include, as an essential factor in instantiating being as presence, consciousness or some specific function or element derivable from consciousness. But if so, then the sameness underlying the uniformity of being as presence must be based on a feature of presence generated from characteristics pertaining to being as such, i.e., being construed *apart from* consciousness as the locus of any experience of being. In sum, Derrida's strategy in the opening assertion of the Hegel essay points toward the necessity to derive some meaning for "presence in" which simultaneously (a) does not include whatever aspect of presence could be pointed to as a feature of consciousness–e.g., functioning in the present and (b) instantiates the intended sense of form.

Here is a passage from the Introduction to *Speech and Phenomena* which may initially appear to pursue this allied goal:

15 See Derrida's observation on the relation between consciousness and the "I," MP, 283. For additional approaches to the phenomenon of consciousness, see SP, 147; Derrida 1981, 254.

> **Passage 1G** The factor of *presence*...is itself modified, without being lost, each time there is a question of the presence (in the two related senses, of the proximity of what is set forth as an object of an intuition, and the proximity of the temporal present which gives the clear and present intuition of the object its form) of any object whatever to consciousness, in the clear evidence of a fulfilled intuition (SP, 9–italics in text).

Derrida refers to the "factor of presence" and then prefaces his parenthesis with "the two related senses," suggesting that these are the *only* two senses that characterize presence and that they are related to one another. The first sense emphasizes that something is "set forth as an object of intuition." Apparently anything existing in any sense whatsoever could instantiate this kind of intuition as long as the setting forth of that existence was in "proximity." But proximity to what or whom? Presumably to an agent equipped to experience the intuition Derrida identifies as an element in the first sense of presence. The second sense of presence given in this passage appeals to the "proximity of the temporal present" in which the object of intuition receives its "form." This sense stresses that presence is a concentrated experience that, apparently, occurs *in* the temporal present. But if there is a temporal present, does it follow that the temporality proper to presence also exists in more encompassing temporal dimensions, e.g., the past and the future? Is presence therefore a kind of omnipresence pervading all standard dimensions of temporality? Or is presence necessarily restricted to experience occurring *only* in the present, thus to the exclusion of past and future? And how should the introduction of "form" in this extremely abstract statement be understood? As Derrida has himself noted, "form" exists in many guises throughout the history of metaphysics.

Although this passage may appear initially to be a straightforward statement clarifying the meaning of "being as presence," I suggest that if this was indeed Derrida's intention, the statement is arguably incomplete. If we take the passage and its dual related senses to function in this capacity, then being as presence becomes essentially *relational*, and the relation in question is constituted by a subject-object connection. But much has been left unspecified: is the "proximity" that produces formality defined *solely* by temporality (whether in the present or, as noted above, in other dimensions of time) or does this proximity include other factors of a traditional metaphysical cast (with space the obvious instance)? Is the temporality generated by necessity from the cognitive agent–in, e.g., a quasi-Kantian fashion–or is it purely "objective" temporality in the manner of Newton? And even more crucial is the indeterminate status of "object" in "object of intuition." Will any elementary object as identified conceptually in any systematic metaphysics satisfy this sense of "object"? If so, then if the objects are characterized within two (or, of course, more) metaphysical systems, which is clearly Derrida's intent when he so often surveys the climactic points in western metaphysical thought, the resulting depictions of "object" will differ vastly from one another.

The question again arises: even if the "two related senses" in this passage give the student of Derrida a direction for determining the "presence" factor of being as presence (i.e., the subject-pole), no guidance whatsoever appears as far as identifying what remains the same, what is "invariable," given that being as presence penetrates and defines the entire history of western metaphysics. In other words, the "being" factor (the object-pole) receives only the most cursory treatment in this passage, although what Derrida wrote at this point clearly exhibits the realist dimension of being as presence. The object-pole of being as presence as well as its relational aspect (which factor will become more and more pervasive as our quest to clarify the import of being as presence continues) remain areas which require additional analysis.

1.6.2 Modulated Possibilities

The following is a series of provisional inferences based on this reading of the opening of Derrida's Hegel essay; it will become evident that this complex Hegelian context necessitates a number of important interpretive decisions if the strategy is to confront Derrida on Hegel for purposes of seeing more clearly the nature of being as presence.

A. Derrida refers to "self-presence" in the second section of the parenthesis. Derrida also used the locution "self-presence" in **Passage 1B**, but at that point he wrote "self-presence of the cogito," suggesting strongly that he had Descartes in mind. If so, then the question concerns whether Cartesian "self-presence" is equivalent to Hegelian "self-presence." If we deny this equivalence given the vast differences in notional content between the final positions in the respective systems of Descartes and Hegel, then how do the instances of self-presence differ from one another and yet both constitute proper exemplifications of being as presence? Presumably some facet of consciousness "as such" will distinguish the two modes of self-presence; otherwise, "self-presence" would depend on something other than its "selfhood" in order to complete the sense in which self-presence instantiates being as presence.[16]

B. The appeal to "self-presence" suggests that consciousness carries its own source of verification with respect to being as presence. Thus the mere presence, so to speak, of consciousness, especially in a Hegelian context, establishes that being as presence is indeed present at hand. A passage from the Introduction to Derrida's commentary on Husserl supports this possibility; here Derrida characterizes (**Passage 1H**) "self-presence in consciousness" as "nothing other than the possibility of the self-presence of the present in the living present" (SP, 9). I suggest that what Derrida contends regarding consciousness in Husserl also applies, perhaps even more broadly, to the self-presence of consciousness in Hegel and ultimately to the distinctive character of this mode of being as presence.

16 For additional sources on the notion of self-presence, see MP, 52, 125-6; Kearney, 148-9; Derrida 1992, 46. In the early essays contained in *Writing and Difference*, Derrida occasionally uses "full presence." As the discussion in Section B indicates, whether "full presence" and "self-presence" are synonyms is not clear. See also WD, 194, 279, 292, 296, 333n20.

If this reading is accurate, the implications are crucial. Derrida is not claiming that consciousness must be intentional, i.e., related to something "outside" consciousness, in order to justify the conclusion that consciousness illustrates "self-presence." Rather, consciousness itself, the activity of consciousness, generates its own constitutive instantiation of being as presence. Consciousness itself is self-presence. Consciousness does not require something other than itself (e.g., an object insofar as it is experienced in any respect by consciousness) to establish a relation so that it generates what Derrida has in mind by being as presence for a metaphysician such as Hegel, with his concerted and systematic integration of consciousness in relation to a myriad of objects. Therefore, the term "subjective" as used in the title of this section emphasizes the fact that for Derrida, the content of being as presence in Hegel is located exclusively within the field of consciousness itself rather than in an extra-subjective range of being.

Of course, prudence suggests that this reading of "self-presence" is tentative especially given that in the first three sentences of the Hegel essay's second paragraph, Derrida uses these locutions: "two moments of full presence," "one presence to another," "original presence," "final presence," and "self-presence;" then, a few lines later, we read a reference to "the movement of lost presence." It is fair to say that clarification of these locutions as employed within the Hegel essay is textually spare; thus the reader concerned to have the most adequate awareness of how Derrida is developing these contexts of presence must either search for assistance in other Derrida texts or attempt to glean some measure of understanding from reflection defined by a narrowly circumscribed immediacy of context.

However, even if all the distinctions introduced in this very abbreviated section of the prefatory remarks (MP, 71-2) are unique to this essay because of its subject matter, i.e., Hegelian dialectic and metaphysics, it is still incumbent on an interpreter advancing a comprehensive reading of Derrida's overall position to determine a sense for all the distinctions so introduced given that the opening of the essay suggests that being as presence is no less prevalent in Hegel as it is in every other figure in the history of metaphysics. Even if all the distinctions and configurations of presence itemized at this juncture in the essay cannot be transposed to other thinkers in the history of metaphysics, they are as important to the stature of Derrida's notion of being as presence as Hegel is to the history of metaphysics, since it may be assumed that as a cardinal figure in that history, he is very important indeed.

Self-presence appears to have a global function in Derrida's reading of Hegel but Derrida introduces additional dimensions to the ways in which presence functions, at least with respect to the inner workings of Hegel's systematic concerns. Here is one such function, although as we shall see, its interpretation can be only conjectured. Thus (**Passage 1I**):

> As the site of the transition, the bridge between two moments of full presence, the sign can function only as a *provisional* reference of one presence to another (MP, 71–italics in text).

The usage of "sign" in this passage will be analyzed in the following section of this chapter. For now, consider the elaborations of presence in terms of "moments."

The expression "two moments of full presence" appears ambivalent, at least as it stands. When sign functions provisionally by referring "one presence to another," the usage of "one presence" could refer to a given entity in relation to another given entity, i.e., two distinct "presences" linked by the functioning of a sign. However, the phrase could also refer to a *single instance* of "full presence" defined as a type of unity which is "full" with respect to instantiation of presence precisely by virtue of a link or connection between two moments. In this reading, Derrida is not saying that each moment by itself has full presence and that the conjunction of the two moments yields (a) a first instance of full presence followed by (b) a second and completely different instance of full presence. Rather, only the linkage provided by the sign of one present to the other present generates full presence. An example: if a word can function as a sign, then when a word refers to an object in one context and the same word refers to the same object in a different context, then the sign so used has generated "two moments of full presence." The ambivalence rests on specifying what "one present" means, especially with respect to its becoming involved in relations to a sign, and also what "full presence" means given that it can be discriminated according to "moments." If precision in these regions of inquiry is available, it is fair to say that considerably more analysis is required in order to achieve it.

1.7 The "Transcendental Signified"

Being as presence becomes manifest in various modes, as the essay on Hegel has shown. One of these modes is consciousness itself. Precisely how consciousness is or becomes "self-presence" is complex and, as argued in the previous section, not readily amenable to discursive analysis. Apparently consciousness has its own justification, a unique type of self-characterization. If, however, we take consciousness as a given and also as a foundation for being as presence, we are prepared for the following additional considerations pertaining to the structure of presence.

> **Passage 1J** [C. S.] Peirce goes very far in the direction that I have called the de-construction of the transcendental signified, which, at one time or another, would place a reassuring end to the reference from sign to sign. I have identified logocentrism and the metaphysics of presence as the exigent, powerful, systematic, and irrepressible desire for such a signified (OG, 49).

A preliminary point: There is no textual warrant for assuming that "being as presence" and "the metaphysics of presence" do not refer to the same conceptual state of affairs. If so, Derrida has approached its structure by dividing his account into distinct phases. The context within which being as presence is now analyzed focuses on the "transcendental signified." The following discussion offers introductory exegesis of this important notion.[17]

First, the characterization of "the metaphysics of presence" in **Passage 1J** certainly appears to be sharply, even radically, distinct from the descriptions of being as presence which appear earlier in *Of Grammatology*. Derrida now emphasizes, with four qualifications, the "desire" for a "transcendental signified." Thus an affective response or interest arising from the subject is a necessary but presumably not sufficient condition for establishing "the metaphysics of presence" in relation to being as presence. But the more detailed phase of the subjective turn taken at this point in *Of Grammatology* does not end there. The object of the stipulated desire–the "transcendental signified"–exists as a "reference from sign to sign," that is, insofar as this phenomenon occurs within the internal constitution of the cognitive agent and speaker of language. The key interpretive questions are: (a) what does Derrida mean by "signified" and (b) how is this item signified in a way which justifies being described as "transcendental"?

Passage 1F introduced the notion of sign and its relation to metaphysics. As noted, we shall attempt to clarify the structure of sign later in this essay. However, as a preamble I submit that an adequate explanation will include attention to three distinct elements: word, concept, referent. Here is one straightforward account which is germane to Derrida's usage of "sign." Consider a glass of water and the phrase "glass of water." In this state of affairs, "glass of water" is a set of words, the concept *glass of water* exists in the mind of a speaker of those words (and, in some contexts, in the mind of a listener), and the object referred to is the actually existing container of liquid, a complex entity the contents of which are suitable for quenching thirst (as well as a variety of other purposes if this entity is viewed imaginatively). What "sign" means for Derrida will be constituted by a configuration of one or more of these three elements. For present purposes, I will take Derrida's use of sign to refer to the supposed mental correlate of a word; thus a word is a necessary condition for a sign but a sign, as such, cannot be reduced to the word which initiates the existence of the sign in the mind of a cognitive agent.[18]

Derrida generalizes in **Passage 1F** that "metaphysics" treats the sign only as a *transition*, with the question of identifying the elements of a transition–between what and what–not addressed at this juncture. However, this reference to "metaphysics" presupposes that it is not germane to the point concerning a sign's transitional character which system in the history of metaphysics one chooses to consider. The entire history of metaphysics can be taken as uniformly consistent with respect to Derrida's interpretation of a sign. Presumably then all metaphysical systems employ

---

17 Derrida's use of "transcendental signified" is analyzed in Chapter 4.

18 For Derrida's assertion of an apparently necessary link between word and notion, see WD, 236. Daylight's comments on complexities in Derrida's employment of sign are useful–see 54, 57, 82, 110, 133. The structure of the components of signification is explored in Chapter 4.

signs in the same way. The connection any one moment within that history might have assumed toward the interpretation of a sign is indistinguishable *qua* metaphysics from the position toward the interpretation of a sign taken by any other moment within that history–a generalization of considerable theoretical amplitude.

The emphasis at hand is on language in all its guises and functions since the claim about the desire for a transcendental signified holds whenever a sign event, as it might be called, occurs within language in any of its settings, not only in the rarified instances of discourse when metaphysicians are plying their art and assigning language to encompass and articulate the nature and structure of reality. This comprehensive environment should be noted but our concern in the essay is with strictly philosophical matters (although the relevance of what David Hume called "common life" in this regard will emerge in both Parts II and III of the essay).

The correlation between being as presence and the "transcendental signified" is multi-dimensional. The tone of **Passage 1J** suggests that although strong reasons and justifications exist for seeking to command the existence and relevance of the "transcendental signified," for Derrida no such element can be secured, at least not in the "pure" sense in which the protagonists of traditional approaches to both the analysis of language as well as to epistemology have believed. We will continue to reconsider and critically evaluate Derrida's approach to this element, the transcendental signified, in Parts II and III below, where the context of scrutiny is more immediately concerned with language. For now, consider why the introduction of the notion of the transcendental signified coupled with the statement just provisionally offered concerning its nature is so crucial for determining as rigorously as possible the sense in which being as presence characterizes the entire history of western metaphysics, especially the appeal to the commonality of being as presence throughout that history.

The key nexus is the (putative) transcendental signified pointing to System A (e.g., a Platonic Form as an element in that system) in relation to the (putative) transcendental signified pointing to system B (e.g., an Aristotelian form in that system). Given the commonality established by being as presence, the relation between the sign and entity-as-referent-in-system A must be the same as the relation between the sign and entity-as-referent-in-system B in order to justify the uniformity of being evoked by "being as presence." What is present within the mode of existence of the sign in the two instances of relation must be rendered and justified as *the same* in order to maintain that the single notion, being as presence, accurately evokes being as represented by the two relations. But here a fundamental difficulty emerges. If some attribute or property is in fact educed from the conjunction of systems A and B (omitting for the moment that this attribute or property must be constant throughout the history of metaphysics) and shown to be the same, then a pivotal distortion must have been wrought on the content of the respective systems given that the texts of those systems, i.e., Plato and Aristotle, describe fundamentally distinct types of being.

In sum, the presence-element in the relation of sign to what the sign is *of* when this relation occurs within system A must be the same as the presence-element in the relation of sign to what the sign is *of* when this relation occurs within system B. But the referents of system A and system B differ from one another in, apparently, irreducible ways. Thus Form for Plato is very different from Aristotle's use of form. How then can being as presence establish and preserve the same elements from such diverse positions? This question will grow in importance as this essay proceeds. But we may observe that the denial of a stable "transcendental signified" appears to jeopardize the cogency of Derrida's position regarding the comprehensiveness of sameness resident in the notion of being as presence (we will devote more attention to this implication in Parts II and III).

1.8 Particularity

Derrida has referred to "two points or instances of presence" (**Passage 1D**) and discussion of that passage introduced aspects of particularity relative to the implied plurality regarding the reference to "two" presents. And in the Hegel essay discussed above, Derrida refers to "one presence and another," implying a multiplicity of instances exemplifying presence. These ways of detailing being as presence suggest a significant contrast relevant to this phase of our investigation. For if being as presence refers to what in traditional terminology would be called a principle and "two points or instances of presence" refer to particulars defined by that principle, how should the contrast between the two usages of presence be understood?

The contrast appears to indicate at least a pair of ways to look at being as presence. If so, the next question becomes how to deploy that distinction given what Derrida assures us is the fundamental character of being as presence. However, any demarcation attempting to clarify the distinction by way of amplification risks distorting it if the language used is borrowed from the schemas of western metaphysics. We recall Derrida's alert early in *Of Grammatology* regarding the necessity to revamp the traditional vocabularies employed in reading philosophical texts. With that cautionary note in view, I suggest that the contrast in the phrase evokes a part-whole distinction. Thus "presence" represents the whole or a principle with (as we have assumed) at least one determinable property whereas "presents" represents instances of that whole.

This reading once reintroduced into the history of western metaphysics raises a series of basic questions: first, how to isolate and describe presence as such if it can be separated, if only logically, from its instances. Does it have a unity which differs from the unity of any single instance of presence? And regarding the latter, does "presents" refer to entities understood without any qualification? Thus does each and every individual entity also exist as an instance of being as presence? Another area of concern: if each entity is an instance of presence, does it follow that entities must be subject to human experience in order to exist as a "present"? This seems to be a yes/no question. If no, then a given present exists on its own, autonomously, as an instance of being as presence, without the active or even passive agency of human experience. But if yes, then the significance of any given entity defined with respect to being as presence as necessarily existing in relation to human experience can be determined only by taking into account the indwelling of that entity within a sphere or arena of human experience.

The nature of an entity as a particular instance of presence–the texts assure us that this ascription is appropriate to a full articulation of being as presence–stands as a question of considerable importance and complexity.

1.9 Presence and the Principle of Identity

One could make the case that as the contexts in which Derrida pursues being as presence increase in number and also in diversity of content, so does the inherent difficulty of identifying the shared elements proper to being as presence. We now see that this complexity become even more intricate.

In *Speech and Phenomena*, Derrida opens his analysis of Husserl with a broad and blunt challenge:

> **Passage 1K** ...being interested in language only within the compass of rationality, determining the logos from logic, Husserl had, in a most traditional manner, determined the essence of language by taking the logical as its telos or norm. That this telos is that of being as presence is what we here wish to suggest (SP, 8).

The relevant context does not involve whether this reading of Husserl is accurate or even defensible. The point is what Derrida took from Husserl as a conclusion, and also as a challenge–in this case, showing that the "telos or norm" of "the logical" is "being as presence." Derrida renders the connection between logic and presence more precise in the following passage (part of a parenthesis), taken from his 1966 essay on Freud:

> **Passage 1L** Logic obeys consciousness, or preconsciousness, the site of verbal images, as well as the principle of identity, the founding expression of a philosophy of presence...) (WD, 207).

This passage is noteworthy in at least two respects: first, in rendering logic subservient or obedient to something other than itself. Thus "verbal images" appear in consciousness (or preconsciousness) and these images command the apparently apodictic character of logic. It is tempting to reply that a verbal image is itself possible only on condition that the principles of logic obtain, in effect reversing the priority Derrida maintains. Consider the principle of contradiction. The explanation for the impossibility of producing a verbal image of a contradiction is based on the character of contradiction itself, not on the supposition that verbal images take a certain shape and, as a direct result of this shape, contradiction emerges as a fundamental principle of logic. The second point is the elevation of the principle of identity as the most basic principle of logic on which, apparently, all other logical principles rest and a principle which, according to Derrida, is "the founding expression of a philosophy of presence."

The principle of identity, stated formally, is A = A. Derrida's "the founding expression of a philosophy of presence" is intended to obviate taking A = A as a logical axiom which somehow *contributes to* the formal conceptualizing and articulation of "being as presence" as a phenomenon characterizing the history of metaphysics. Thus to construe A = A as "purely formal" is to denature it, that is, to locate identity in a privileged realm which, given its apparently necessary function, allows identity to remain apart from substantive considerations. But if so, then the principle can be restated in order to indicate this foundational status relative to the history of metaphysics. Therefore, we will write the principle: A is present to A. What then does the relation "is present to" convey?

This interpretive gambit suggests that the relation "is present to" must add something to the pure formality of the identity sign. A = A, the standard expression of the principle of identity, indicates nothing about A over and above its identity to itself. Therefore, unless "is present to" adds something to the identity sign, "is present to" and "=" become indistinguishable from one another. However, if what is added to the equality sign by virtue of "is present to" imports any substantive content to A, then this importation must be justified. But how?

Justification is possible if it could be shown that "A is present to A" can exist only on condition that its status as *a*–or, if Derrida is correct, *the*–principle of logic presupposes at least some characteristics accruing to being as presence. This stipulation is warranted on condition that the principle of identity is indeed "the founding expression of a philosophy of presence." If this analysis is sound, then it becomes all the more crucial to investigate what Derrida means by being as presence, since he claims that the most intimate connection possible obtains between this principle, A = A, and the history of metaphysics understood as being as presence. For Derrida, the standard interpretation of the principle of identity as a purely formal axiom is metaphysically naive if not presumptuous. For in fact, adoption of this principle motivated from an exigent desire deep within the site of verbal images resident in consciousness has contoured and controlled the subsequent history of logic and, by extension, the history of metaphysics.

Here is an attempt at such justification: "A is present to A" can stand as a fundamental principle of logic only if properly stated–thus "A is present to A" if and only if this principle connects consciousness, the seat of its conception, with whatever emerges "in" consciousness. For this appearance in consciousness allows consciousness to depict this A, whatever it may be, to exist so that it becomes present to itself through the relation of equality. Again, this justification is only a sample analysis; if this formulation and development are sound, then it follows that A = A is dependent on something other than itself, i.e., consciousness. For only if such a thing as consciousness existed would it be possible to engender this kind of purely "formal" approach to existence proper. (This alternative approach to the supposedly "formal" character of a principle as abstract as self-identity is intriguing and bears additional scrutiny, quite apart from its relevance to the question of determining the nature of being as presence.)[19]

1.10 Presence and *Différance*

As we have seen, for Derrida the entire history of western metaphysics can be meaningfully encompassed as a unity constituted by a multitude of metaphysical differences. Furthermore, a structural feature of that history is open to thoughtful scrutiny, i.e., through the notion of *différance*. *Différance* appears in various early texts of Derrida, sometimes with accompanying comments, sometimes as a "given" in Derrida's technical vocabulary with the understanding that the interested reader will secure an appropriate grasp of its definition and description. The account and application of *différance* for Derrida's overall project occupies a level of significance even more fundamental than that of his revisionist approach to the foundational status of the principle of identity. We offer here an introductory survey of this notion, emphasizing its relevance for determining being as presence. More concentrated expository and critical analysis of *différance* appear later in the essay.

1.10.1 Sameness and Invariance

Derrida goes to great lengths to emphasize what *différance* is *not*, so it is a decided challenge in interpretation to isolate what *différance* in fact *is*; the following passages contextualize its development:

> **Passage 1M** Indeed one must understand the *incompetence* of science which is also the incompetence of philosophy, the *closure* of the *epistémè*. Above all it does not invoke a return to a prescientific or infra-philosophic form of discourse. Quite the contrary. This common root, which

19 At WD, 260, Derrida asserts axiomatically that logic is itself only "an interpretation." At MP, xx, Derrida describes logic as a "hierarchy of regional beings." See also SP, 148; MP, 39; LI, 92-3, 104; and cf. **Passage 1L**. For discussion of Derrida's position on the principle of identity, see DFL, 154-61. The notion of self-identity is also analyzed from the perspective of its importance to being as presence in Chapters 3, 5 and 7 below. The concept of "consciousness" used here must be distinguished from the consciousness Derrida cites in **Passage 1B** as one of many instances of being as presence found in the history of metaphysics.

> is not a root but the concealment of the origin and which is not common because it does not amount to the same thing except with the unmonotonous insistence of difference, this unnameable movement of *difference-itself*, that I have strategically nicknamed *trace*, *reserve*, or *differance*, could be called writing only within the *historical*-closure, that is to say within the limits of science and philosophy (OG, 93–italics in text).

The common root which underlies both science and philosophy has a concealed origin. But Derrida indicates that what is common to science and philosophy "does not amount to the same thing," from which one may infer that whatever makes philosophy philosophical and science scientific is not the same, even though Derrida appears to assert some type of commonality at this level of supreme generality. Although Derrida's initial reference to a "common root" orients the dimensions of *différance* towards an apparently accessible uniformity, his development of this descriptor takes an unexpected turn. Derrida insists that this "common root" is neither a root nor common. Thus the commonality does "not amount to the same thing" except as a signal to the "unmonotonous insistence of difference." Derrida's use of "unmonotonous" (*si peu monotone*) suggests that difference is essential to the sweep of the history of metaphysics and thus is not at all "monotonous" with respect to the unity of that history; however, the negative prefix alerts the reader that difference as such is not the most fundamental feature of metaphysics taken as a unity. This approach to difference within the totality of metaphysics is an "unnameable movement" which Derrida has variously categorized, including by means of the neologism–"*différance*."[20]

The "incompetence" regarding the "common root" to which Derrida refers toward the beginning of **Passage 1K** is not a condition which could be remedied by greater measures of genius on the part of the most perceptive and penetrating protagonists of philosophy or, less dramatically, on the possibility of increasing the work ethic of philosophers generally. The incompetence in question results from the gamut of being as presence and its overriding dominance throughout the entire epoch of western metaphysics.

What then is "unnameable"? It is apparently a feature or features of the "movement" of "difference itself," or *différance*. Derrida emphasizes in various ways, e.g., "unmonotonous" as a qualifier of difference, the dimensions of negativity which surround what he is attempting to convey to the reader with regard to *différance*. The following passage introduces a more substantive approach to the "unnameable movement" embodied in *différance*.

---

20 Passage 1M apparently suggests that "trace" is a synonym for *différance* but discussion in Chapter 9 contends that such an equivalence does not seem possible. See the critical assessments of *différance* in Chapters 8 and 9 below. For additional analysis and critical evaluation of *différance*, see DFL, Chapter 4 ("*Différance* and Difference"). See also Harvey 203-5, 212; Shusterman, 79-80; Fynsk, 153; David Wood 1988, 64; Vincent Descombes 2002, 30-43.

Passage 1N Reflection on this last determination [i.e., Heidegger's Being/beings] will lead us to consider differance as the *strategic* note or connection–relatively or provisionally *privileged*–which indicates the closure of presence, together with the closure of the conceptual order and denomination, a closure that is effected in the functioning of traces (SP, 130-1–italics in text).

Derrida cautiously refers to *différance* as "provisionally privileged," although how "provisionally privileged" should be understood is not obvious. Thus "provisionally" with respect to what other possibilities or alternatives? Is the provisional character intended with respect to (a) whether being as presence can admit "closure" at all or (b) whether the appropriate vehicle of closure is something other than *différance* as described in the present essay? These qualifications indicate that Derrida is fully aware that he has elevated a single term to a status which, given the leveling tendencies of deconstructive analysis, invites close critical scrutiny in its own right. The appeals to qualifications of the privileged status of *différance* as "relative" or "provisional" are apparently intended to remove *différance* from traditionally deployed philosophical evaluation. However, it is difficult to see how Derrida can be so reserved given all the work that *différance* does for his overall position. Does the fact that *différance* is only "provisionally" privileged hold open the possibility that *différance* may either not exist at all or exist but in a sense even more qualified than that asserted by Derrrida in this passage? Surely this possibility is untenable, at least in Derrida's comprehensive frame of theoretical and speculative reference. These explicit qualifications will become important later, when we pay more concentrated critical attention to *différance* (Chapters 7 and 8).

For Derrida, Heidegger constituted the effective termination of western thought, i.e., the advent of the "closure of presence." The fact that Derrida announces this closure is, of course, not accidental to the advanced position. In this regard, the notion of *différance* becomes crucial since *différance* names the most appropriate way to conceive and articulate the history of western metaphysics from its inception with the pre-Socrates to its apparent completion in Heidegger.[21]

It now becomes all the more pressing to determine the meaning of "being as presence" as precisely as possible since the inevitable conclusion of *différance* is the "closure of presence." Thus the import of Derrida's notion of "closure" can be fully appreciated if and only if we can determine what has been subjected to closure–i.e., being as presence. The following passage indicates, or at least implies, that this quest is not completely quixotic:

21 Derrida's use of "closure" is highlighted in the central argument of Chapter 9; Heidegger is of signal importance as the "last metaphysician," i.e., the "closing figure" in Derrida's reading of the history of metaphysics. Thus, MP, 63: "The extraordinary trembling to which classical ontology is subjected in *Sein und Zeit* still remains within the grammar and lexicon of metaphysics." See "*Ousia* and *Grammē*: Note on a Note from *Being and Time*," esp. MP, 33-4n6; see also MP, 127-8, 131; Kearney, 141.

> **Passage 1O** We provisionally give the name *differance* to this *sameness* which is not *identical*... (SP, 129–italics in text).

In this passage, Derrida provides, albeit "provisionally," a positive element of *différance*–it functions as a *sameness*. In **Passage 1M**, Derrida writes of a "common root, which is not a root but the concealment of the origin and which is not common because it does not amount to the same thing except with the unmonotonous insistence of difference." But **Passage 1O** stipulates that precisely such a sameness is at hand when *différance* is viewed synoptically, as the organizing notion of western metaphysics. It is also important to appreciate Derrida's insistence that sameness understood at this uniquely fundamental level is distinct from *identity*. The denial of that which is "not common" and the assertion of a primordial "sameness" may appear initially to be at odds logically. But the fact that being as presence is "invariable" (**Passage 1C**) must be interpreted and I submit that this invariability is in fact apposite with the sameness Derrida recognizes in **Passage 1O**. This sameness, if it were in any sense accessible to inquiry, would reveal at least presentiments of the currently concealed "origin" referred to in **Passage 1M**, presumably the ultimate "monstrous" goal of all deconstructive inquiry.[22]

Even a cursory glance at Derrida's lists of instances of being as presence readily shows that these instances differ among themselves. To claim that sameness–but not identity–obtains among members of a class is to appeal to at least one (or more) characteristics simultaneously shared by all members of that class. This simultaneity holds even if in all other respects no member of this class has a property identical to properties of all other members (taken distributively) of that class. Also, some properties may be identical to all members of the class but not all properties, for in this case no distinction would obtain between sameness and identity. In sum, A and B will be the same in (at least) one respect but will not be identical in all other respects. Furthermore, whatever core belongs to sameness as sameness can be, indeed must be, stated as itself a self-identity, thus formally allowing the differences marking the history of philosophy to be gathered under the single notion of sameness. This necessity presupposes in turn the reliability of the principle of identity, a point worth mentioning since Derrida has denied any special privilege to this principle–indeed, as we saw above, Derrida contends that identity is the paramount principle of being as presence.

22 At one point in the essay on *différance*, Derrida asserts: "there is no essence of *différance*..." (SP, 158). The conjoint (a) affirmation of sameness as invariability and (b) denial of essence to *différance* may appear logically incompatible. However, if "essence" is taken in a sense (or senses) defined exclusively by existent metaphysical accounts employing this concept (e.g., Aristotle–see Ch2n30), then no incompatibility arises on condition that the invariability in question is not determinable according to "essentialist" parameters. This implication reinforces the urgency of determining the content of this invariable commonality as clearly and rigorously as Derrida's texts permit.

In these respects, I suggest that as already noted, the sameness in question should be connected to the appeal to an "invariable" presence (**Passage 1C**), that is, to the definition of being "in all senses of this word," for it is precisely at this point in **Passage 1C** that Derrida provides one of his extended lists of fundamental concepts drawn from the history of metaphysics.

To what extent this invariability is–or can be–defined by determinable properties remains to be seen. However, although the perspective Derrida emphasizes consistently when he refers to *différance* is its promulgation of differences as they define the gamut of metaphysical systems, it is crucial to the intelligibility of *différance* that it incorporate, even if only in the most minimal sense, some measure of positive content. The reasoning: if presence is invariable, then when Derrida asserts that *différance* produces an ineluctable and apparently unique kind of sameness, whatever properties open to articulation from presence which are indeed invariable will constitute the sameness providing the core of Derrida's notion of *différance*.

In fact, this appeal to invariability justifies recalling that Derrida offered a clue concerning the content of this commonality. Since in **Passage 1B,** Derrida writes "temporal presence" referring to "the now" (probably Aristotle, in his *Physics*), it may be inferred that presence as such is broader than temporality in its typical divisions (e.g., "now" as present where past and future are coordinate specifications). But it cannot be determined from this passage alone whether this breadth of reference includes (i) both a past and future component, (ii) one of these two temporal divisions and not the other, or (iii) additional characteristics which may not be usually associated with temporality. Textual evidence from the passages cited in this chapter thus suggests that (a) presence is not limited to temporality in the sense of the now, or the present as the mid-segment between past and future; also that (b) presence is discernible as a process, or at least that being as presence has process-elements as a component feature of its structure. It would be premature to put too much weight on these inferences, drawn as they are from isolated passages. However, such assertions suggest that being as presence is at least not entirely ineffable. As a result, we may ask whether it is possible to characterize being as presence in more specific terms even if these terms are themselves drawn, necessarily at this point but nonetheless provisionally, from the vocabulary of western metaphysics.

Two concluding points. First, can we clarify what Derrida offers as evidence for his position regarding the fundamental character of being as presence? As stated, the claim about being as presence is an assertion, not the conclusion of an argument. It may of course be a *true* assertion but it is legitimate to ask whether, if true, it is the sort of claim which can be justified by discursive means. Is it possible to establish criteria for evaluating this position, especially given the fact that for Derrida, being as presence spans the entire history of western metaphysics?

And second, as a phase in evaluating the ultimate justification of being as presence as epochally definitive for the history of metaphysics, is it possible to identify concepts or principles (or both) which Derrida must assume as formal prerequisites in order to introduce and develop positions *vis-à-vis* the history of philosophy, positions inspired by deconstructive ideals? These formal prerequisites would presumably be public and thus accessible to readers of Derrida's texts. Again, more will be said below on this important question.

1.10.2 Form and Opposition

For Derrida, *différance* has articulated and diversified being as presence by means of a specific metaphysical generality–distinctive and commonly employed–in conjunction with a dominant type of relation. The following passage clarifies both elements:

> **Passage 1P** As soon as we use the concept of form–even to criticize another concept of form–we must appeal to the evidence of a certain source of sense. And the medium of this evidence can only be the language of metaphysics. For that language we know what "form" means, how the possibility of its variations is ordered, what its limits are, and the field of all conceivable disputes concerning it. The system of oppositions in which something like form can be considered, the formality of form, is a finite system (SP, 108).

Derrida confidently states how crucial the "concept of form" is to envisioning the structure and limitations of the history of metaphysics. Since the language of metaphysics has been determined–both for individual systems as well as for the sequence of the history in its entirety–by the revelatory possibilities inherent in *différance*, it follows that form will be the most dominant single notion contouring that history. Thus the "language of metaphysics," properly shaped per deconstructive protocols, will provide the evidence to critique the notion of form while concurrently funding that very history through the history of its texts. In effect then, metaphysics is self-referential but sufficiently supple in its possibilities to be critically incisive in its self-referentiality.

Examination of the details of metaphysical systems reveals that form has played itself out according to a certain set conceptual deployment. Note then the intimacy between form in this global and determinative sense and the "system of oppositions," a duadic structure permeating much of the history of metaphysics and typically marked by a rigidity of contrast which Derrida's deconstructive practices are capable of dismantling as fully and as completely as possible. We will investigate and assess Derrida's appeals to form (in Chapters 2 and 3) and opposition (in Chapters 4 and 8) as perhaps the primary notional target of deconstruction.

1.11 Presence, the History of Metaphysics, Closure

Passage 1L illustrated the close connection between being as presence and closure; and, as we saw earlier, **Passage 1K** generalized concerning the closure of the epistemological phase of western metaphysical inquiry. Thus closure encompasses two distinct phenomena–knowledge (*epistémè*) and presence–in turn engendering two principal themes: the limits of being as presence and the suggestion that something beyond those limits may be available to some type of inquiry as well as some modes of discourse.

If "closure" refers to a point or axis of termination or completion as Derrida views the history of metaphysics (indeed, history in general–cf. the opening of **Passage1C** with its undeveloped appeal to "the history of the West"), then the obvious question becomes: what will happen to knowledge and reality, or to the knowledge *of* reality, on the other side of this closure? I employ the blunt locator "on the other side of" by intent in order to emphasize that if the history of metaphysics has exhausted its possibilities and if philosophers or thinkers of any stripe will continue their work in interrogating the world around them, is there any way to determine what will happen next?

Derrida explicitly referred to the "limits" of philosophy (as well as science) and also, again, to the "incompetence" of philosophy, as if something worthy of thoughtful consideration existed "outside" the limits of the entire history of western metaphysics. The inside/outside configuration is a favorite target of deconstructive treatment but the contrast so stated seems appropriate to introduce at this juncture. Think of "closure" as a door that has been closed; on one side of the door lies what has happened in the history of the west, on the other side of the door we may find... what? If these comments and the resultant questions strike the reader as too empirical and rough-edged, they at least have the virtue of sharpening the point of inquiry so that a more sophisticated, layered response might be forthcoming. One additional thought: if presence has reached a point of saturation within the gamut of western thought so that it can be said to have reached "closure," does this event sanction the possibility that philosophizing can be articulated in ways which are in some sense "outside" the domain of concepts and language affected by being as presence? *Of Grammatology* opens with Derrida alerting his readers to the real possibility that following his thinking will culminate in a sort of "monstrous" philosophizing. If so, then such monstrosity may be only the articulation of what currently lies "outside" the realm of traditional metaphysical discourse.

Derrida used three terms in **Passage 1M** as apparent synonyms for the "unnamable movement of *difference-itself*." This terminological generosity suggests that it is possible to reduce the ineffability pertaining to *différance* by tracing through, so to speak, those places in Derrida's texts where these synonyms are discussed with more confidence and at least a modicum of development. However, even if "difference-itself" is ineffable, at least at this level of abstraction, its functions can be described. And most importantly for present purposes, difference itself when it has been codified under "*différance*" establishes the power of closure for the entire history of western metaphysics.

Derrida concludes this passage by asserting that closure in his intended sense "is effected in the functioning of traces." The notion of "trace" is clearly pivotal in this context and will be the subject of extended discussion in Chapter 9. For now, we note that if Derrida's use of "trace" includes even a minimal sense of its typical meaning, then "trace" presupposes (a) something leaving a trace and (b) that the evidence of its passage can be somehow discerned. Thus if a given trace emerges from the history of metaphysics, indicates in some way the closure of that history, and also traverses the totality of western metaphysics, then due attention devoted to such phenomena–assume for the moment that "trace" admits of plurality–will provide an avenue toward insight into what has been concealed "under," so to speak, the effective quality of a given trace. We shall see then that trace refers to elements in Derrida's reading of the history of metaphysics which occur both *within* that history and also *beyond* that history. If so, then following traces read in this sense will allow us to anticipate, perhaps to articulate, what the limited and "incompetent" history of metaphysics has systematically overlooked. We return to this line of thought in Chapters 9 and 10.

A final word on writing. At the end of **Passage 1M**, "writing" becomes a technical term of restricted import relative to Derrida's systematic approach to the history of metaphysics. For Derrida, "writing" names everything that conditions the possibility of reference in language and also how everything was, has been and could be referred to throughout the history of metaphysics. Thus "writing" becomes a virtual synonym of totality insofar as any existing thing can enter the sphere of discourse, whether philosophical or non-philosophical (e.g., what is traditionally referred to as "literature"). Derrida's oft-quoted motto, "There is nothing outside the text," becomes more intuitively accessible once "writing" is seen in the sense just outlined. The intimate thematic links between writing (in Derrida's technical usage) and extra-linguistic reality will be examined in detail in Part II, especially Chapter 4.

1.12 Summary: In Quest of Being as Presence

Being. Being as presence. The account in Chapter 1 of variously diverse settings in which Derrida describes being as presence should make it clear that this notion is both crucial to Derrida's general approach to traditional philosophical issues as well as decidedly complex in its formulation. The context for determining the effective scope and content of being as presence may be stated thus: Being as presence (i) encompasses all explicitly asserted systematic accounts of metaphysics; (ii) straddles the differences which define each of those systems relative to all the other systems constituting the history of metaphysics; (iii) occupies a level of abstraction which controls the formal breadth of logical concepts and principles, such as identity; (iv) is represented as an "invariable" determinant emerging from the commonality encompassing the totality of all these different systems; (v) has reached a limit or closure of sorts, suggesting–and the following formulation is avowedly metaphysical–that at least in one respect, being as presence has realized whatever potential it may have had during its history.

But a context, even one as broadly characterized as that just described, evokes only a set of underlying circumstances. What does Derrida *mean* by being as presence? A review of the five conditions enumerated in the previous paragraph might accurately surround the content of being as presence but this enumeration does not, by itself alone, disclose the meaning of being as presence. The extensive variety of positions characterizing the history of metaphysics coupled with the stipulated condition that being as presence is invariable and, as the product of *différance*, describes a common sameness in the midst of all the differences animating that history severely complicates specifying how exactly being as presence should be understood. Is presence in Derrida's intended technical sense something that belongs to being as such, to being in relation to beings, or to the relation between human beings and being insofar as this relation establishes the ultimate ground of "the meaning" of being?

A related but distinct question: Does being as presence have a determinable structure or, more loosely stated, an assemblage of constituent elements? This is a different question since the meaning of being as presence might be advanced in synonyms, or in phraseology which remains to a certain degree apposite but yet is also undeveloped and unarticulated. If, however, a structural account of being as presence exists and is accessible to some form of language, then the more this structure is brought into the open, the greater will be the opportunity to subject what is arguably the most fundamental term in Derrida's technical vocabulary to informed and rigorous critical scrutiny.

Finally, in view of the privileged function Derrida assigns to being as presence within the history of metaphysics in tandem with what appears to be the inherent opacity of being as presence, is it possible (a) to establish criteria for specifying the content of being as presence as well as pursuing its critical evaluation? Or should Derrida's position in this context be taken as (b) an attempt executed by fiat or through some direct intuitive vision, standing or falling on the basis of how much apparent insight can be derived from its narration and application to the history of metaphysics? The former possibility should be pursued as a matter of sustained inquiry before the latter–i.e., a *laissez-faire* philosophizing without any obvious dimensions of evaluative control–is adopted. Chapter 2 develops an instance of such pursuit.

2 Presence and the Question of Evidence

The quest to identify or at least to approximate with some measure of precision what Derrida refers to by the phrase "being as presence" has led to the following realizations. We know that being as presence spans the entire history of metaphysics and we know a number of classic positions within that history which exemplify being as presence because Derrida has explicitly identified them in this context, on occasion mentioning by name their philosophical authors. There is also strong textual evidence that being as presence includes consciousness, although the precise sense in which consciousness should be incorporated under the heading "being as presence" is, at the moment, indeterminate (more on this question later in Chapter 2). And we know that formal logic in the contemporary sense is, for Derrida, "an interpretation" and in this respect remains subservient to being as presence in terms of what stands as philosophically ultimate.

The single most important aspect of what we do not yet know is how to collect the differences marking the individual members of the sequence known as western metaphysics in such a way as to identify how a single characteristic–presence–can remain the same, or "invariable," throughout this sequence, be recognized *as* the same, and yet concurrently preserve the "unmonotonous" differences punctuating the history of metaphysics. Derrida's notion of *différance* is pivotal in this context since *différance* grounds the sameness in question and deploys the range of differences which individually and collectively define metaphysics. However, determining the meaning of this sameness appears to be a quest of considerable difficulty, involving an intersected diversity of contexts encompassing a broad and detailed spectrum. How then should the inquiry proceed?

2.1 Being as Presence: The Possibility of Demonstration

Consider a passage already cited and discussed in Chapter 1 which we now examine from a different perspective:

> **Passage 2A [cf. Passage 1C]** The history of metaphysics, like the history of the West, is the history of these metaphors and metonymies. Its matrix–if you will pardon me for demonstrating so little and for being so elliptical in order to come more quickly to my principal theme–is the determination of Being [*de l'être*] as *presence* [*présence*]in all senses [*les sens*] of this word. It could be shown that all the names related to fundamentals, to principles, or to the center have always designated an invariable [*l'invariant*] presence–*eidos, archē, telos, energeia, ousia* (essence, existence, substance, subject) *alētheia*, transcendentality, consciousness, God, man, and so forth (WD, 279-80).

Our concern is with Derrida's concentrated regrets for "demonstrating so little" and "for being so elliptical," asides prior to another list of examples illustrating "the determination of Being as presence." What does Derrida mean by asking the reader to accept his lack of demonstration–has he failed to demonstrate (1) *that* his list of positions does in fact comprise instances of being as presence or (2) *how* these examples so instantiate that especially privileged fundamentum? Logically, demonstrating (2) presupposes that (1) is in fact the case. Thus if what Derrida claims he has not demonstrated is *how* the designated instances from the history of metaphysics illustrate and constitute being as presence, then it will become necessary at some point to describe what each instance shares with–or is the same as–all other instances in order to justify the predication of the single descriptor, "being as presence."

This reading reinforces the decisive appeal to sameness as the operative factor in Derrida's notion of *différance*. Furthermore, (1) could be granted without addressing (2) on condition that we can establish by inspection or some sort of direct insight that the instances listed are indeed proper to being as presence. However, the analysis in Chapter 1 has shown that (2) is the problematic and crucial question, since only if different metaphysical positions yield an underlying sameness is it possible to recognize that "being as presence" exhibits a measure of determinate content. Derrida's phrasing suggests that demonstration of this sameness is indeed available to interested parties–but he does not have time to execute it, at least not in the work from which this claim was taken. The quest for the invariable quality of being as presence must therefore continue.

The following passage from *Of Grammatology* (also examined in Chapter 1 from a different perspective) provides a series of foundational assumptions for this phase of our inquiry:

> **Passage 2B** Indeed one must understand the *incompetence* of science which is also the incompetence of philosophy, the *closure* of the *epistémè*. Above all it does not invoke a return to a prescientific or infra-philosophic form of discourse. Quite the contrary. This common root, which is not a root but the concealment of the origin and which is not common because it does not amount to the same thing except with the unmonotonous insistence of difference, this unnameable movement of *difference-itself*, that I have strategically nicknamed *trace*, *reserve*, or *differance*, could be called writing only within the *historical*-closure, that is to say within the limits of science and philosophy (OG, 93–italics in text).

Derrida's expository *via negativa* is dramatically prominent throughout the terminology in this passage–"incompetence," "does not invoke," "not a root," "not common," "not amount to," "unmonotonous," "unnameable".... And yet he inflects all the negative locutions with an appeal to a "common root" which undergoes an immediate "concealment of the origin." This concerted stylistic negativity reflects "the *incompetence* of science which is also the incompetence of philosophy." And the goal, once being as presence has been identified as the dominant veil of concentrated excess, is to revivify and if possible to approach at least an approximation of whatever has been concealed at the origin by the advance of western metaphysics and its unconsidered–until Derrida made it known–and uncontested reliance on being as presence.[23]

For Derrida, "this unnameable movement of *difference-itself*" appropriately describes *différance*, the theoretical instigator of metaphysics. But one of the primary assumptions of this essay is that the sameness which rests at the core of *différance* can at least be approximated with appropriate names as a precursor to more precise delimitation by means of some sort of apt vocabulary. Derrida then mentions a number of "nicknames" for the "unnameable movement of *difference-itself*" as if they are interchangeable in terms of what each of them denote. If so, then the nicknames should perhaps be taken as a set of terms by means of which Derrida will pursue what both science and philosophy have concealed. Closure represents the "limits" of what philosophy, particularly metaphysics, has accomplished through its millennial history, and it is reasonable to read Derrida's texts as if his strategy included giving at least some indication how to move beyond those limits, if not actually bridging them.

2.2 Form, Presence and Evidence

In Chapter 1, we cited a number of passages in which Derrida cataloged seminal instances of being as presence with respect to their defining the history of metaphysics when that history is taken as a serial unity. But we require a narrower focus in order to address the problem at hand, that is, specifying in more precise detail the nature of being as presence. In **Passage 2B**, Derrida emphasizes his primary technical term *différance*, and we recall again from **Passage 1O** that the defining core of *différance* is "this *sameness* which is not *identical*...." A pivotal argument in Chapter 1 regarding the content of being as presence was that the element of sameness is the principal agent through which being as presence, understood as a unity spanning the entirety of metaphysics, functions in this epochally definitive fashion. We asked in the previous section: *how* should being as presence be understood when Derrida directly confronts the major texts of the western philosophical tradition? The following passage (the first section of which, not included here, appeared in Chapter 1 as **Passage 1P**) provides an answer to this question:

> **Passage 2C** The system of oppositions in which something like form can be considered, the formality of form, is a finite system. Furthermore, it is not enough to say that "form" has a sense for us, a center of evidence, or that its essence is given to us as such: indeed, this concept is, and always has been, indissociable from the concepts of appearance, sense, evidence, or essence.

23 Derrida's interest in negation has implications for negative theology (noted at SP, 134). Negation in its more neutral and seminal senses in western metaphysics is crucial for Derrida and will be discussed in Chapter 6 and in various additional contexts below.

> Only a form is evident, only a form has or is an essence, only a form presents itself as such. This is a point of certainty that no interpretation of the Platonic or Aristotelian conceptual system can dislodge (SP, 108).

Since this passage is involuted and important, its analysis spans several sections. Derrida begins by deploying the "formality of form" according to what he calls the "system of oppositions," implying that it is possible to determine the limits of the effects of "form" by studying the regnant series of oppositions as they have occurred throughout the career of philosophy in general and metaphysics in particular. I suggest that "formality of form" not only includes this specification but that it also be taken as a generic term naming whatever has been dubbed "form" and its cognates. Thus an inquiry into the formality of form, when restricted to figures such as Plato and Aristotle, will attempt to discover what makes a form to be what it is for Plato in contrast to what makes a form what it is for Aristotle. These accounts, if divergent (and of course they do diverge in content), will both fall under the notion of formality as will any account of form other than its Platonic and Aristotelian versions insofar as these accounts define the history of metaphysics. This grouping function is legitimate given the emphasis Derrida puts on form and its intimacy with the realization of being as presence. We will subsequently enumerate examples of the kinds of opposition Derrida intends, but however numerous these oppositions will be, they are, according to Derrida, finite in number and arranged systematically.[24]

The conclusion of the passage states an inference concerning interpretation of both Plato and Aristotle as systematic thinkers with respect to their use of "form." Derrrida's claim is strong and invites critical assessment (which will be provided below). Of course, the more generalized point, and the dictum which concerns us in this chapter, is that "only a form presents itself as such." Thus if X is a form, then X presents itself. But the qualifier "as such" suggests that presence may be associated, perhaps to a lesser degree, with something other than a form. We will not pursue that possibility here; however, if the connection between form and *how* form "presents itself as such" can be elucidated, then we may reasonably assume that this account will contribute to our understanding of being as presence. This expectation is why Derrida's assertion about the "point of certainty" concerning interpretation of the respective conceptual systems of Plato and Aristotle becomes crucial. For if the claim about Plato and Aristotle as systematic thinkers is indeed a "point of certainty" and inquiry can appropriately analyze these conceptual systems, we may legitimately anticipate that such inquiry will clarify the parameters of being as presence–especially with respect to the attribution of apparently distinct types of evidence, that is, of "how" form presents itself.

---

24 The notion of opposition stands as the primary target of deconstructive analysis and the notion itself, as well as its exemplifications, will be critically discussed in Chapter 8.

2.3 Form and Presence: Along the Gamut of Exhibition

We direct the initial phase of this inquiry toward interpreting "appearance, sense, evidence, or essence" as this set of elements pertains to form, since for Derrida it is axiomatic that "only a form presents itself as such" and that form "is, and always has been, indissociable" from the set of elements just enumerated. We might reason that being as presence includes, necessarily, *various ways in which an entity exhibits itself as presence.* The logical connective "or" in this quartet then becomes as important as the "and" analyzed in the comment on **Passage 1E** in Chapter 1. For if the "or" is intended as a strict disjunction, then Derrida's claim is that a form is connected *either* to appearance *or* to sense *or* to evidence *or* to essence–i.e., presence exists in its fullest metaphysical specification just as long as *any one* of these four elements is in play (we grant in context that this list of modes of display is exhaustive).

Consider Derrida's sequence: "appearance, sense, evidence, or essence." Upon examination, a number of differences in categorical status emerge in this list. The first three terms–appearance, sense, evidence–can be read as pertaining to the subjective side of a relation between a cognitive agent and the object of this agent's concern. Furthermore, appearance and sense are approximately circumscribable as species of the same genus whereas evidence in this context is significantly broader; indeed, evidence subsumes appearance and sense, as well as a number of other possibilities. This logically more inclusive aspect of evidence may be part of what Derrida had in mind with the phrase "center of evidence," that is, that evidence proper is a genus term with species as types subsumable under it. Finally, essence differs from the other three elements in at least two ways: first, it is a substantively metaphysical concept whereas the other three are primarily said of epistemological experiences; second, essence is almost exclusively a concept applied to Aristotelian metaphysics and is seldom said of Platonic metaphysical concerns. We will discuss these aspects of essence later in this chapter.

If this disjunctive interpretation holds, then one way to approach this passage is to assume that each of the four elements–appearance, sense, evidence, or essence–displays presence *differently*, since otherwise there would be no reason to distinguish the elements in the first place. We proceed with that assumption in hand.

Finally, since Derrida has emphasized that "conceptual systems" of Plato and Aristotle are exemplars of this reading of form and its import with respect to presence, we assess this claim by doing what Derrida says in **Passage 2A** he does not have time to do–examine a text drawn from this pair of preeminent metaphysicians in order to show how it exemplifies being as presence. This interpretation is directly relevant to the final sentence of **Passage 2C**. Derrida insisted there that the connection between form and being as presence deployed according to this broad dimension of exhibition is a "point of certainty" which cannot be dislodged regardless how either the Platonic or the Aristotelian system is interpreted. It follows that being as presence holds with equal sway for both Platonic and Aristotelian metaphysics, in particular with respect to each philosopher's notion of form as a pivotal element in their systems. In order to evaluate this inference, we shall reflect on one of the positions Derrida explicitly indicates in **Passage 2A**, thereby pursuing the possibility of at least glimpsing more definite content for being as presence.

2.4 Plato's *Phaedo*: Form and Experience

We examine several closely grouped texts from Plato's *Phaedo*. This dialogue is seminal in the history of metaphysics for announcing an early version of what is often referred to as the theory of Forms. The dialogue says a great deal about formality in what are presumably relevant Platonic contexts.

At *Phaedo* 65d-66a, Socrates has asked Simmias whether he, Simmias, agrees that there is such a thing as Just itself, Beautiful itself, Good itself. Simmias assents. Socrates then asks whether Simmias has ever seen any of these realities with his eyes. Simmias says that he has not. Here is Socrates' response (Tredennick's translation):

> **Passage 2D** Well, have you ever apprehended them with any other bodily sense? By 'them' I mean not only absolute tallness or health or strength, but the real nature of any given thing–what it actually is. Is it through the body that we get the truest perception of them? Isn't it true that in any inquiry you are likely to attain more nearly to knowledge of your object in proportion to the care and accuracy with which you have prepared yourself to understand that object in itself?[25]

After Simmias agrees with this conclusion, Socrates continues:

> **Passage 2E** Don't you think that the person who is likely to succeed in this attempt most perfectly is the one who approaches each object, as far as possible, with the unaided intellect, without taking account of any sense of sight in his thinking, or dragging any other sense into his reckoning–the man who pursues the truth by applying his pure and unadulterated thought to the pure and unadulterated object, cutting himself off as much as possible from his eyes and ears and virtually all the rest of his body, as an impediment which...prevents the soul from attaining to truth and clear thinking? Is not this the person, Simmias, who will reach the goal of reality, if anybody can?

Simmias replies that what Socrates has just said about "truth" and "reality" is indeed the case. Then, at 78d-e, Socrates describes the Forms thus:

> **Passage 2F** ...each of these uniform and independent entities remain[s] always constant and invariable, never admitting any alteration in any respect or in any sense....

25 Unless otherwise noted, all passages from Plato are from the Hamilton and Cairns *Collected Dialogues of Plato*. For additional commentary on this passage from the *Phaedo*, see White 1989, 46-8, 107-8.

And a few lines later, at79a, Socrates contrasts the mode of existence of Forms with the mode of existence of particular entities as instances of Forms:

> **Passage 2G** And these concrete objects [particular entities] you can touch and see and perceive by your other senses, but those constant entities you cannot possibly apprehend except by thinking; they are invisible to our sight. [Cebes] That is perfectly true.

We take Socrates at his word in these passages and will assume that he refers to what reality really is as such (the concerted redundancy at the end of this part of the sentence is by intent).

2.5 Formality: Appearance and Visibility

For Derrida, a form is "indissociable" from "appearance, sense, evidence, or essence" (**Passage 2C**). We suggested that this set should be taken disjunctively and the *Phaedo* has indicated why.

In the *Phaedo*, evidence is the broader, more inclusive concept; thus the appearance of something (more accurately, a certain kind of something) is a type of evidence for the existence and character of that something, but not all evidence is available through appearance. Consider then that in a related vein, Derrida asserted (**Passage 1E**) that formality "is what is presented, visible, and conceivable of the thing in general." Assume that this locution, "what is presented, visible and conceivable," is a set of properties and that this set should be taken as a unity but with a potentially intricate part-whole relation. Then assume that "visible and conceivable" represents (a) a logical conjunction and also that (b) Derrida is amplifying, by means of this conjunction, how presence should be understood in the context of formality. In other words, formality–formality as such, regardless which philosopher incorporates the concept of formality within a given metaphysical system–comes to presence by being "visible and conceivable," i.e., *both* characteristics are necessary, in conjunction with one another, in order to equal presence.

However, if we take "visible and conceivable" as a logical conjunction applicable to *all* types of form, then the passages cited from the *Phaedo* illustrate that Derrida is simply wrong about the Platonic approach to form. For Plato, a Form is *not* visible (i.e., capable of being seen); it is only conceivable. The *Phaedo* explicitly distinguishes between these two types of knowing (**Passages 2D, 2E**) and just as explicitly identifies which of the two possibilities appearing in Derrida's text–"visible and conceivable"– is correct (**Passage 2E**) if what is known is a Form in the canonic Platonic sense stated in the *Phaedo*. Whether or not the dialogue's sharp distinction between visual and cognitive evidence is sustainable epistemologically is not the issue; the question is *what the dialogue explicitly and literally says as a Platonic text* relative to the nature of the Forms as an instance of formality and, ultimately, with respect to the relation between formality and being as presence.

2.6 Formality: Presence and "Sight as *eidos*"

In **Passages 1B, 1C** and **2A,** Derrida appeals to *eidos,* intending this term to refer to one way the dialogues typically identify Forms. In **Passage 2B,** Derrida more narrowly describes Plato's contribution to the history of metaphysics as "presence of the thing to sight as *eidos.*" Here Derrida recalls for the reader the fact that *eidos* is derived from the verb *eidein,* one meaning of which is "to see"–hence the etymological connection between Form and sight.

When Derrida specifies (**Passage 2B**) that a Platonic Form is to be understood as "presence of the thing to sight as *eidos,*" the *Phaedo* again establishes that Forms are precisely *not* present to sight or to any mode of sense awareness. The *Phaedo* irrefutably distinguishes between evidence based on sensory experience, of which the visible is a cardinal instance, and evidence based strictly on the ability of the mind to come into cognitive contact with a Form. The texts cited above clearly and unambiguously show that the *Phaedo* has distinguished between the linguistic roots employed to name the Form and, from a metaphysical perspective, what the Form is as such.

The reason for this distinction derives from the metaphysical character of Forms–Forms are uniform, immutable, invisible and, as a result, accessible only to the mind in its capacity to reason about such reality (**Passage 2F**). The Form is never seen in the purely visual sense–indeed, Forms are not available to any sensory avenue. We see a statue, a material object configured in a certain way, which we appreciate and identify as beautiful; we do not see the Form beauty the participation in which makes that statue, that material object, beautiful. The Form beauty, as all Forms, is known by the mind, not perceived by the senses. Socrates sharply distinguishes between using the senses to perceive and using the mind to think, the latter as a prerequisite capable of being in position to "track down reality pure and by itself" (**Passage 2E**). No element or aspect of any human sense, including sight, can be included if a seeker wishes to know reality–thought alone provides the necessary medium for accessing reality. Thus *eidos* is not accessible to sight (presumably it would be accessible to sight working in conjunction with the mind but only the mind as such can discern what the being of the *eidos* truly *is*). The contrast between the two types of cognition, thinking and seeing, distinguished by their very different objects, is neatly glossed during the interchange between Socrates and Alcibiades toward the conclusion of the *Symposium*: "The sight of the mind begins to see sharply when that of the eyes starts to grow dull..." (*Symposium*, 219a).[26]

26 The translation from Plato's *Symposium* is by R. E. Allen 1991, 165. See Bibliography.

Despite this sharp split in Plato's text, however, it will be instructive at this point, especially by way of anticipation of the argument in Chapter 3, to note that the designation of an *eidos* as "presence of the thing to sight" (**Passage 2B**) encapsulates what Derrida apparently intends presence to accomplish metaphysically. Derrida has clearly added a human component into the content of being as presence when the being in question is a Platonic Form. This passage from "Signature Event Context" (an essay originally published in 1972) supports the point, i.e., when Derrida refers to (**Passage 2H**):

> ...the entire field of what philosophy would call experience, that is, the experience of being: so-called "presence." (MP, 317; cf. alternate translation at LI, 9)

Thus presence here may be taken as "the experience of being"–*not* being as such. It would follow then that Derrida's "form is presence itself" (**Passage 1E**) is equivalent to "the experience of form is presence itself." This implication, if allowed as a legitimate reading, has crucial consequences with respect to Platonic metaphysics and its reliance on Forms. For if Derrida can categorize a Platonic Form as "presence of the thing to sight as *eidos*," then he can readily subsume Platonic metaphysics–insofar as that metaphysics derives from the Forms–under the descriptor "being as presence." The question becomes whether this extension effectively destroys the nature and integrity of the Forms as they are depicted in Plato's texts.

2.7 Presence and the Autonomy of Platonic Forms

Human beings may or may not be disposed to apprehend the being of the Forms. But this mode of apprehension, if elected and pursued, must be distinguished from the *content* of particular apprehended perspectives on being insofar as those perspectives animate human consciousness. The Platonic position with respect to the being of a Form–the *eidos*–provides an apposite and dramatic example of the need to establish and preserve this distinction. Forms exist outside of space and time and thus are invisible and immutable. By contrast, human beings exist in spatio-temporal settings. Can it then be maintained that human beings also exist–in a sense–in the "presence" of these Forms?

According to the *Phaedo*, Forms and particular entities differ fundamentally, as do the modes of cognition by which humans access and experience each type of being. The autonomous existence of the Forms is clearly presupposed by this epistemic contrast and by the separation Socrates introduces in the *Phaedo* between Forms and instances of Forms. But if we consult the section of the myth in the *Phaedrus* which R. Hackforth identifies as "the soul's vision of true Being," this independence is explicitly and dramatically drawn. We enter the myth at the point when soul has ascended by means of winged chariots to the rotating rim of the heavens and, in company with divine soul, has reached a position of access, ready to behold "true Being." The following three passages (Hackforth's translation) give the most pertinent details of this vision:

a. It is there that true Being dwells, without color or shape, that cannot be touched; reason alone, the soul's pilot, can behold it, and all true knowledge is knowledge thereof (247c).
b. ...[soul] is carried around with the gods in the revolution, but being confounded by her steeds she has much ado to discern the things that are; another now rises, and now sinks, and by reason of her unruly steeds sees in part, but in part sees not (248a).
c. ...many [souls] have their wings all broken; and for all their toiling they are baulked, every one, of the full vision of Being, and departing therefrom, they feed upon the food of semblance (248b).[27]

In the region beyond the heavens, "Being dwells," the "things that are"–that is, the Forms. According to the myth, certain souls behold some Forms while other souls witness different Forms and thus have a different experience of Being. But some souls, due to their lack of control over soul's mythically depicted good and evil steeds, are deprived altogether of "the full vision of Being" and must rest content with, in Hackforth's stirring locution, "the food of semblance."

The language is mythical but the metaphysical dimensions of this section of the dialogue are straightforward. The Forms exist on their own and, pertinent to Derrida on being as presence, they continue to exist independently whether or not they are, or ever will be, experienced by human souls. Thus if Derrida wants to maintain both that "only a form presents itself" and, immediately afterwards, that this "is a point of certainty that no interpretation of the Platonic or Aristotelian conceptual system can dislodge" (SP, 108) and if he also wants these two assertions to apply to *all* Platonic Forms, then the text of the *Phaedrus* forces the advocate of Derrida to face a difficult question. How can a Form "present itself" and yet also exist completely apart from all types of human experience? Whatever the "presenting" may be (in order to preserve Derrida's point) will require analysis in modes of explanation containing no reference to human experience (in order to preserve the sense of the myth in the *Phaedrus*).

There is another alternative: that Derrida's sweeping assertion concerning the necessary connection between Platonic Forms (as one type of formality) and being as presence is in fact a mistake. Whether or not coming to presence transpires, the myth teaches that each and every unexperienced Form still remains what it is *qua* Form. Derrida's position seems to maintain that the Form must come into presence in order for the Form to exist as a Form at all. As we have shown here, the text of the

27 For commentary on the metaphysical aspects of these passages from the *Phaedrus*, see White 1993, 107-39.

Phaedrus apodictically precludes this possibility, as did the text of the *Phaedo* by way of presupposition. The point is worth repeating: according to the *Phaedrus*, Forms can and do exist completely apart from any apprehension of them, in any sense of apprehension (whether human or divine!). Thus if being as presence applies to these Forms, then presence must mean something that "comes to presence" completely apart from any mode of human awareness, however broadly that awareness may be construed.[28]

2.8 Plato, Presence, Particularity

Derrida's treatment of Platonic metaphysics is also problematic in other ways. At the end of **Passage 1B**, Derrida claimed that logocentrism "would thus support the determination of the being of the entity as presence." But if we take "the being of the entity" to refer to the mode of existence of a particular thing (where "entity" and "particular thing" are identical), then Derrida's notion of being as presence also fails to preserve a fundamental difference within the *Phaedo*'s metaphysics. The *Phaedo* draws a crucial distinction between properties of a Form and properties of an instance of a Form. Thus the latter have their being only in relation to the being of something other than themselves and it is possible for their being to be destroyed or, as the text states it, to "withdraw (102e)." But Derrida's blanket claim about being as presence seems to dissolve this difference; if a particular being exists, then it would apparently follow that the mode of existence of a particular coming into presence is indistinguishable from the mode of existence of a Form coming into presence.

From the perspective of Derrida's vision of the history of metaphysics, all being–whether of a Form or of an instance of a Form–is presence. In order to maintain the principle that being is presence and that this principle encompasses every distinguishable moment throughout the history of western metaphysics, Derrida must assume that the way a Form comes to presence is indistinguishable from the way an

28 In the Preface to *Speech and Phenomena*, xxxii, Newton Garver asserts that "for Husserl and *the tradition* [my italics], the sense of being has always been interpreted as *presence*" [italics in text] and in two respects (cf. **Passage 2H**), one of which, the "interpretation of being as objectivity," is that "something *is* insofar as it presents itself or is capable of presenting itself to a subject –as the present object...of a sensible intuition or as an objectivity presented to thought." But according to the passages introduced from the *Phaedo* and *Phaedrus*–preeminent moments in "the tradition" (and the "interpretation of being as objectivity")–the "sense of being" proper to the Forms is in no way accessible to any type of "sensible intuition." Furthermore, even if the capacity of "presenting itself to a subject" were allowed to be a potential characteristic of a Form, the being of that Form taken as an "objectivity" does not consist in the realization of that capacity. Finally, if being as presence constitutes an invariance (**Passage 1C**), then there must be a commonality identical to all manifestations of this "sense of being." Garver's formulation usefully enunciates the polar parameters of the problem but does not resolve it.

instance of a Form comes to presence. Any qualitative distinction regarding types of being a metaphysician, Plato or anyone else, might introduce becomes invisible and otiose. (I emphasize at this point that this inference appears legitimate given the texts examined so far; the argument and its conclusion with respect to being as presence will be qualified below.)

The critical stance just established depends for its cogency in part on the distinction between sense and reference–an exemplary instance of the kind of fundamental opposition which Derrida intends deconstruction to upend and dismantle. The distinction between sense and reference is preserved here, however, in order to indicate that if a Platonic dialogue posits claims about Forms, then the Forms, as referents within the linguistic universe of that dialogue, display a character in some sense apart from the penumbra of language which surrounds them. The texts in the dialogue *point to* the Forms, and they do so in such a way as to acknowledge that these realities exist independently of language, whether the language in question be discursive or mythical (or a complex blend of both). This referential function of language, if accepted as legitimate, allows the interested student to reflect on the Forms, to note their nature, to determine their structural relations to one another and, crucially, to derive implications from this mode of existence for other, distinct types of reality (e.g., particulars constituted in part according to spatio-temporal considerations). Although these reflections will be necessarily couched in language, the assumption animating the sense/reference distinction is that such reflection does not merely catalog characteristics of the dimension of language (insofar as language can be directed at some object of interest); rather, the mode of existence of the referent is sufficiently supple and receptive to enable language to determine characteristics of that object itself existing separately and apart from language, although an apartness which can be described linguistically.[29]

2.9 Formality and Essence

We have already noted that "essence" often operates differently from other avenues Derrida lists along the gamut of modes that characterize the exhibition of being as

29 Derrida found Plato's dialogues to be "an abyss, a bottomless pit in which I could lose myself" generating possible meanings (Kearney, 145). Yet this claim becomes hagiographic haze since however these texts are or will be construed, all such readings will seemingly reduce the interpretive products to variations on being as presence. The putative inexhaustibility will always and necessarily dissolve any apparent interpretive novelties, i.e., differences, derived from these texts and locate them on the flat interpretive plain–"being as presence." Derrida's notion of "irruption" might seem to offer an outlet to this entailment (see Chapter 9 for discussion) but then the question becomes whether even a text construed as an irruption must for the sake of consistency also fall under the uniform scope of an omnipresent being as presence.

presence. "Essence" in its typical formulations refers to what something ultimately *is* and in this respect is metaphysically dominant in ways that other modes of commensurate priority are not. The following discussion should clarify some of the problematic aspects of this area.

In **Passage 2C,** Derrida is careful to assert that "only a form has or is an essence." This distinction between "having" an essence and "being" an essence (a clarification Derrida does not develop in this context) is important. I suggest that the most accurate deployment of the distinction is as follows: form for Aristotle "is" an essence whereas the Platonic Form "has" an essence. But this identification must be immediately qualified. Students of *Metaphysics VII* are intimately aware of the considerable difficulties in determining as precisely as Aristotle's texts allow the connection between form and essence. So although Derrida would have secondary authorities as support, the claim as asserted in this simplistic way is problematic at least from the standpoint of the level of dispute of existing scholarship on this matter. Furthermore, "essence" as a metaphysical name is typically inapposite in Platonic contexts (perhaps because "essence," in Latin *essentia*, is an attempt to compress and translate Aristotle's multi-word Greek original).[30] But even if "essence" could be introduced appropriately in Platonic contexts, the relevant metaphysical consideration is that for neither philosopher is it the case that essence, or the properties that cohere together to constitute essence, depend on anything other than the inherent structure of this reality. Essence as a property of or synonym for forms always remains independent of any human experience of forms.

Thus if essence refers to (a) properties of, e.g., either Platonic or Aristotelian forms, then these properties will be restricted to, in traditional terminology, predominantly realist attributes, e.g., the immutability of Forms.[31] But appearance as a parallel mode of exhibition of formality can refer (b) only to (i) what is capable of generating an appearance insofar as that datum can be experienced by (ii) a perceiving agent. The following question then becomes unavoidable: how can "being as presence" encompass characteristics deriving from both (a) and (b) above in such a way as to justify the predication of a common and singular element, i.e., "being as presence"?

As we shall see, this is just one aspect of a much more convoluted problem facing the advocate of Derrida's notion of being as presence. The problem emerges from the hypothetical application of essence within a Platonic universe and it parallels the difficulty outlined above in the context of particularity. In this case, the problem holds for Forms as such, that is, apart from particulars existing as dependent on Forms. The mode in which Form A comes to presence is apparently indistinguishable, *qua* presence, from the mode in which Form B comes to presence. If so, then the

---

30 Aristotle formulates "essence" as τὸ τί ἦν εἶναι at, e.g., *Physics* 195^{a}20-1 and *Metaphysics* 1028 ^{b}34.

31 See *Republic* 479a for the immutability of the Forms; a comparable claim about the Aristotelian approach to form appears at *Metaphysics* 1033^{a}27-^{b}18.

uniformity of being as presence dissolves all differences with respect to the content of however many Forms inhabit the metaphysical universe of the *Phaedo* (and the Platonic metaphysics in general). It follows that if Derrida's ascription of essence to Platonic metaphysics pertains to the Forms, then Forms become indistinguishable from one another at the level of essence if all the Forms instantiate being as presence. To appeal to the essence of one Form in order to differentiate that Form from another Form (which will, putatively, differ essentially from every other Form) will be unavailing, since the omnipresence of presence, as it were, will conceal differences at the level of essence. In general, the attempt to discern the sameness justifying the predication of a uniform being as presence to the totality of western metaphysics threatens to dissolve differences which are both textually prominent and necessary to the internal structure of each system as a distinct member within the complete history of metaphysics.

2.10 Plato and the Possibility of Presence

It might be objected at this point that Derrida is not doing fine-grained, scholarly exegetical work on a classical text; he is, in fact, philosophizing about the supposed import of that text. But it is crucial to appreciate this stance as far as evaluating the cogency of what Derrrida wrote. Recall Derrida's coy comment cited above, i.e., asking forgiveness from the reader for "demonstrating so little and being so elliptical." Being "elliptical" when writing about an important philosopher is one thing; being manifestly wrong about that philosopher's texts is something else again. What we need here is considerably more "demonstrating" that Platonic and Aristotelian texts can be read the way Derrida claims they should be read, a claim delivered with the highest degree of certitude. We have seen Derrida contend that "no interpretation of the Platonic or Aristotelian conceptual system can dislodge" the necessary connection between formality and presence (SP, 108). But no textual support was given to justify this strong claim, either from primary or secondary sources. And as this section of the essay has shown, all it takes to "dislodge" this theoretical position as far as Plato is concerned is a moderately close reading of the *Phaedo*.[32]

---

32 In a lengthy footnote, 170-1n1, Russell Daylight wonders "whether or not to expand upon and interrogate Derrida's engagement with Heidegger and the ontology of temporality." However, since such a project would involve a consideration of the "canon of temporality" from, according to Derrida, "Parmenides to Husserl," Daylight decides against its feasibility on the grounds that it "...would be subject to the powerful compression of Derrida's engagement with metaphysics." Instead, Daylight contends, "we could assume the result in advance: that whatever differentiation is found within the history of temporality–between subject and objective temporality, for example–does not disturb the continuity of its metaphysical conceptuality."This is a remarkable inference. In his book, Daylight has shown through careful and informed reading of Saussure that Derrida omitted and overemphasized much, and got a fair amount of Saussure wrong. But "Derrida's engagement with metaphys-

Form is precisely *not* "presence of the thing to sight," unless Derrida intends to claim that in a Platonic context, an observer experiencing a particular instance of a Form (e.g., a beautiful statue) is in the presence of the Form because the observer is seeing an instance of that Form. Furthermore, a Form as such can come to presence but it need not do so. The myth in the *Phaedrus* shows that a Form could very well exist and not be known or experienced in any way whatsoever, or could exist and not be known fully–in the former case, it becomes difficult to assign a meaning to "form is presence itself;" in the latter case, if Form is presence itself, then presence must admit of degrees, i.e., more or less presence depending on whether the Form is more or less known. It seems then that Derrida is open to the criticism that his notion of being as

---

ics," is with *the entire history of metaphysics* from Plato and Aristotle to Bergson and Heidegger. This is a much more voluminous and considerably more complex body of work than anything Saussure produced. Why then should we presume that the conclusions of Derrida's "powerful compression" of texts comprising an entire tradition are more trustworthy than those derived from his indifferent reading of Saussure? If we assume that "whatever differentiation" within the history of metaphysics' reflections on time will "not disturb the continuity of its metaphysical conceptuality," why not make precisely the same assumption–which Daylight obviously did not do–about Saussure's *Course* and all the "differentiation" it contains on matters and concepts pertaining to a rigorous study of language?It is surely minimal scholarly courtesy (i) to *read* the texts comprising the history of metaphysics, (ii) to *study* those texts in light of Derrida's conclusions about that history *en bloc*, and then (iii) to *evaluate* those conclusions critically and fairly. Indeed, Daylight might consider authoring a sequel to his book on Saussure entitled *What if Derrida was wrong about Plato*? This would be the first in a lengthy series of volumes with the same questioning title traversing all the requisite identities defining the history of western philosophy. In fact, Daylight has presentiments along these lines; he criticizes Derrida's lack of evidence in subsuming Aristotle as an exponent of the "metaphysics of presence," 178. But reservations concerning Aristotle in this regard must–or should–be directed at every seminal figure in the history of metaphysics.
Various commentators (Christopher Norris on more than one occasion) have assumed that Derrida's reading of Plato is justified and rigorous. See Norris 1989, 189, 193; Norris 2012, 75, 97, 125, 134; also Stocker, 18. In a related vein, see Ch4n46 for sources critically evaluating Derrida on Husserl. As noted, one may contend that such scholarship should be aimed at every major thinker Derrida discussed in his extensive canon.In general, the concern for and apparent display of rigor in Derrida has been marked by many: Culler 1983, 85; Daylight, 21, 29, 30, 42, 173; Geoffrey Bennington 2000, 11; Hill, 116, 117-8; Niall Lucy 1995, 36; Marian Hobson 1998, 193–i.e., a special sense of rigor; 234; Norris 1983, 3, 6-7; Norris 1989, 193-4; Naas 2008, 125. Christopher Norris has recently repeated his belief that Derrida writes with the "utmost degree of formal, conceptual and logical rigour" (Norris 2012, 66); Norris then restates this evaluation on 69 (twice), 70, 71, 77, 78, 90, and 97–all within the confines of a single chapter. (In Chapter 5 of the same work, Norris also attests to Derridean rigor at 123, 127, 137, 139, 144, 145, and 149.) However, the unsupported claim that a thinker's work displays undiminished and undeniable rigor does not make it so; indeed, Daylight's book and this chapter show otherwise, and in a variety of interpretive contexts. Thus Norris' global adulation in this regard should command no allegiance by itself. To assert that Derrida's thinking has rigor is one thing; to demonstrate this rigor by (i) detailed analysis of Derrida's readings of texts from the history of philosophy compared with the letter and contexts of these sources as well as (ii) critical evaluation of the intelligibility and coherence of Derrida's positions in response to those readings is a different matter altogether.

presence fails to distinguish between (a) the character of a determinate kind of being (in this case, a Platonic Form) and (b) the mode of apprehension a receptive agent directs toward that determination. When it does become present to an observer or someone intending to know the Form, the Form as such remains irreversibly distinct from the experience of the Form.

For a Platonist, if we assume that "being is presence" and then construe this assumption as equivalent to "the experience of being is presence," then what appears as directly observable to sight or any other medium of sensation is a blend of being and not-being, as the *Republic* famously puts it (478d); the reality of the thing perceived is and remains the underlying Form in which that particular thing participates. Therefore, Derrida's introduction of presence in a strictly Platonic context is misguided–unless he means something by "presence" which can allow the Form to be (a) inaccessible to sight or any other sense but (b) still available to experience in a way which justifies the introduction of presence taken as a crucial technical term. The exponent of Derrida's position must detail what is common or the same in two very different experiences: (a) purely mental awareness of a Form canonically characterized as immutable and invisible and (b) apprehension of an instance of a Form through some everyday avenue of perception. What is "present" which is the same in both the concept of a Form as well as a percept of an instance of that Form? If "present" means what is present to a concept when it is experienced cognitively, then Derrida's position requires that this same element must also be present to a percept when it is in the process of being perceived.

2.11 Formality: Plato, Aristotle, Presence

An additional problem should also be made evident. To review: first, in **Passage 1A**, Derrida insists that the history of metaphysics may be considered as a unity with respect to logos as the origin of truth "despite all differences" marking that history; and second, in **Passage 1B**, "all the subdeterminations" of the history of metaphysics "depend on this general form," i.e., presence as the source which organizes that history into a sequence and a system. Therefore, Derrida is consistent in **Passage 2C** when he asserts that "...only a form has or is an essence, only a form presents itself as such." Derrida concludes this line of thought with the "point of certainty" about the necessity of form as presence in any and every interpretation of the Platonic or Aristotelian conceptual system.

This claim, juxtaposing Plato and Aristotle not only with respect to forms as a single component in their thinking but also with respect to their entire "conceptual" systems, attests to Derrida's confidence in his position concerning the fundamental and undeniable character of presence as the controlling factor in the history of western metaphysics. The final two sentences of **Passage 2C** imply that what comes to presence for Plato is the same as what comes to presence for Aristotle insofar as the position of each philosopher, whether that position deals with metaphysics or any other topic of philosophical concern, falls under "being as presence." Any differences between Plato and Aristotle, not only with respect to forms but any other elements in their respective conceptual systems, are merely "subdeterminations" of being as presence and do not affect the fundamental character of both systems insofar as each belongs to western metaphysics, from which it follows that each system represents a legitimate exemplification of being as presence.

However, Derrida's account of being as presence is at least ostensibly not only incompatible with the text of the *Phaedo* but it is also incompatible with the texts of Plato and Aristotle on form since for Aristotle, forms do not exist or are not real the way Forms are real for Plato. For Aristotle, the form as essence does not exist separately, apart from particular instances of essences, the way Forms exist separately for Plato.[33] This is an especially crucial difference between the Platonic and Aristotelian metaphysics. But, again, this difference does not affect Derrida's notion of being as presence since, as he explicitly asserts, "only a form presents itself as such" and again, no interpretation of either Plato or Aristotle can dislodge this certainty. However, Derrida can maintain this claim with respect to that phase of the Platonic or Aristotelian conceptual system which encompasses formal reality only if both Plato and Aristotle connected their respective approaches to forms as such with experience of forms as such. But neither philosopher in fact made such connections.

2.12 Formality and Presence–Question and Quest

Derrida establishes a tight intimacy between presence and form, especially form as it was developed at the beginnings of western metaphysics. He also explicitly added the factor of evidence as an integral feature to the structure of being as presence. Indeed, the fact that Derrida lists various ways to realize being as presence strongly suggests that he thought of being as presence necessarily in relation to the evidence offered to justify the particular ultimate realities of a given metaphysical system. But evidence is, as such, an epistemological notion since evidence must be experienced by an observer in order to qualify as an instance of this concept. It follows then that being as presence, viewed as a technical term, has become both complicated and enriched. It now consists in (a) a characteristic (or characteristics) of being *in relation to* (b) the evidence offered to justify and clarify that element or elements within a systematic and discursively articulated network.

---

33 For the inseparability of the forms for Aristotle, see, e.g., *Metaphysics* $1039^{b}22$-27.

In order to pursue the clarification and refinement of being as presence, this interpretive conclusion should be linked with the provisional statements of presence maintained in *Speech and Phenomena*. These passages were discussed in Chapter 1 (**Passage 1E, 1G**) and one of the two passages is relevant at this point in Chapter 2.

> **Passage 2H** The factor of *presence*...is itself modified, without being lost, each time there is a question of the presence (in the two related senses, of the proximity of what is set forth as an object of an intuition, and the proximity of the temporal present which gives the clear and present intuition of the object its form) of any object whatever to consciousness, in the clear evidence of a fulfilled intuition (SP, 9–italics in text).

Prior to disclosing "two related senses" of presence, Derrida introduces a feature of the "factor of presence" which has not been emphasized, arguably not even hinted at, in any of the passages on being as presence examined so far. Derrida says that the factor of presence "is itself modified, without being lost, each time there is a question of the presence...of any object to consciousness, in the clear evidence of a fulfilled intuition." As just noted, the immediate context of this passage is a commentary on Husserl, but I submit that it is legitimate to take this passage as a continuation of the account of presence in general, regardless who the philosopher might be–especially when the philosophers in question depend heavily on the notion of form.

By "any object," I suggest that Derrida is referring to objects across the complete spectrum of metaphysical thought. Although the reference to "the clear evidence of a fulfilled intuition" is Husserlian in tone if not in content, it could also be Platonic if the "object" in question is a Form. Thus if a cognitive agent knows a Form, then that agent has a "fulfilled intuition" of that Form. The aspect of presence which is new is the fact that it can be "modified, without being lost" each time a question arises with respect to the intuition of an object. The claim that presence can be subject to modification, "without being lost," is important. In fact, this characteristic of presence can be applied to several problematic areas indicated earlier in this chapter. First, the approach to presence as a uniformly undifferentiated, blanket phenomenon can now be recognized as an oversimplification. Presence will in fact be "modified" depending on whether the frames of reference include (i) two distinct Forms; (ii) a Form and a particular instance of that Form; (iii) forms divergently defined by two (or more) philosophers. These contexts all incorporate significant differences within their respective defined limits; this interpretive application of "modified" presence allows Derrida's position to accommodate these differences as they animate given metaphysical systems while concurrently preserving the presence of presence, so to speak. As a result, despite its diversified applications in a variety of metaphysical circumstances, presence never gets "lost." For example, we may now infer that the presence of a Form is not identical to the presence of an instance of a Form. And the same contexts of presence so differentiated apply to any and all other variations within and between metaphysical systems.

Although Derrida has anticipated the general thrust of these criticisms with the notion of the modificational status of presence, serious questions nonetheless remain. The "two related senses" of presence detailed in Derrida's text on Husserl indicate that (a) both senses emphasize temporality and (b) the first sense appears to claim that it is the "temporal present" which gives the "object" of cognition its "form." This sense of form differs dramatically from the perspective on form applied in Derrida's analyses of Platonic and Aristotelian metaphysics. Indeed, the case could be made that this usage of form is approximately Kantian in that the object of experience with respect to being as presence is shaped, as it were, by the fact of its appearance in and through temporality. However, it is also crucial to note that nothing is asserted about the substantive nature of *what* appears in and through this temporality. Being as presence must include some phase or aspect of being insofar as it emerges "in" this "present" temporality.

Consider this possible determination of substantive content in a Platonic context: Envision the "leading edge" of a Form as it makes contact with or becomes tangent to the corresponding leading edge of an observer of that Form. We know (from the *Phaedo* and other dialogues) that a Form fundamentally differs in its reality from the reality of an observer of that Form. But Derrida's position on being as presence requires that as the pair of leading edges "touch" or become tangent to one another, the context of tangency (to give the area of inquiry a name of sorts)–what he calls "presence"–results. But if it is impossible to penetrate the context of tangency in order to discriminate a more determinate feature proper to that distinctive context, then presence reduces to the sheer juxtaposition of such divergent elements. The constitution of element A is merely being held in proximate suspension to the constitution of element B but without any feature of A or B yielding a distinctive character in such a way as to merge into a commonality to produce the required "invariance." This commonality, especially with respect to the sameness of being which must be identified and preserved, appears necessary in order to justify predicating the single term–presence–as appropriate nomenclature for this invariance.

I suggest therefore that if "being as presence" is intended to identify at least some elements of *being* rather than have its entire center of gravity defined by *presence* as the ensemble of conditions necessary for the *experience of* being, then the following considerations emerge as pertinent to the specification of being as presence.

The related senses in **Passage 2H** emphasize the subject-side of a subject-object relation (cf. Derrida's phrase "object of an intuition"). The sameness that pervades the entirety of the history of metaphysics and has been in question now includes both (a) the content of explicitly stated elements constituting a given metaphysical system as well as (b) considerations pertaining to the evidence introduced to justify the relevance of those elements within a systematic setting. But if this mode of affectivity has both a subject and an object pole, then the sameness becomes necessarily doubled in structure–that is, it must be possible to discern shared elements which are the same (a) on the side of the subject as well as (b) on the side of the object.

Furthermore, this shared unity must hold not only for Plato and for Aristotle, but also must obtain in such a way that the relation between these shared elements remains the same throughout the entire history of western metaphysics.

In the metaphysics advanced in the *Phaedo*, for example, sensation and thought as distinct cognitive experiences establish a kind of unity which must be preserved so that each of its elements remains distinct from one another (in order to reflect accurately the text of the *Phaedo*). However, each side of the relation must possess the same characteristics as the other side of the relation. Thus whatever constitutes the subject-factor in a given system of metaphysics (whether sensation or thought or a combination of both) must have the same characteristics as whatever constitutes the object-factor of that system of metaphysics (whatever form "object" may take in that system, e.g., Platonic or Aristotelian forms).

The conditions controlling the sameness necessary to fund being as presence are therefore extremely stringent. But the specification of these conditions is not yet complete. If formality is granted the paramount status Derrida claims it enjoys throughout the history of metaphysics, we offer a conclusion worth careful scrutiny as a statement of parameters which must be met in order to articulate an adequate critical response to Derrida's basic position: being as presence must be understood as what is common throughout all instances of (i) appearance, (ii) sense, (iii) evidence and (iv) essence insofar as each of these concepts names a mode of evidentiary content relative to the existence of form. The problem then is to identify an element in *each* of the members of the disjunction or indeed any comparable disjunctive description of the manifestations of formality and then to show that each of these elements is *the same as* all the other elements identified in the subject-object correlation just described above. This commonality must be both discernable and at least theoretically capable of articulation.

Finally, the sameness prescribed by Derrida's notion of *différance* must pervade the *entire history* of metaphysics read as one unified narrative. No single system of thought can be an instance of metaphysics and not also display the sameness which, according to Derrida, *différance* has bestowed on the history of metaphysics as a sequential unity. Derrida's insistence that being as presence maintains itself throughout the history of metaphysics despite all constitutive "subdeterminations" of that uniform phenomenon makes it all the more imperative to isolate and identify those features which are and remain the same throughout that history in order to give content to and eventually to justify the very existence of this single descriptor.

Chapter 3 develops what might seem initially to be an unorthodox approach toward describing this sameness.

3 Being as Presence: Transcendental Dimensions

The preceding analysis of texts from the *Phaedo* (and related passages from other Platonic dialogues)–texts bearing a decided metaphysical dimension–might suggest that Derrida's categorizing of the western metaphysical tradition as nothing but variations on one given theme sharply and fundamentally distorts that tradition. This conclusion appears warranted especially as it emerges from a textual event toward the origin of that tradition which, from the perspective of historical significance, is surely seminal in the fullest sense. But does it matter to the principal point of Derrida's position that it ignores the blunt distinction in the text of the *Phaedo* between sensory evidence and mental or cognitive evidence? To what extent, if any, does the distinction between (a) consciousness when it is functioning in the presence of a Form and (b) the specific content of the Form affect Derrida's notion of being as presence? Is it possible that if certain conditions obtain, Derrida can justifiably assert that presence will always continue to be *form*, understood as that through which all metaphysical content is both experienced and realized?

3.1 A Transcendental Strategy

Derrida's generalized accounts of being as presence are incompatible textually with at least one moment within the history of metaphysics, as Chapter 2 has demonstrated. The following extrapolation from this fact also seems warranted: the more that distinct elements comprising each and every metaphysical system are juxtaposed with one another, the greater and more dramatic are the textual incompatibilities which will arise. However, sustained caution toward the feasibility of Derrida's project suggests that this rampant proliferation of textual and logical tensions along with the coordinate disassociation within as well as between different metaphysical systems does not, by itself, invalidate the potential cogency of that approach toward being as the metaphysician's ultimate concern. For it is possible that the intended sense of presence should be understood at a deeper level of specification. We have tentatively prepared groundwork for some possibilities in this regard toward the conclusion of Chapter 2, possibilities emerging from reflection on different ways to approach being as presence when the contexts were circumscribed by texts drawn from Plato and Aristotle. These possibilities are now extended, systematized and explored.

We will analyze Derrida's descriptions of being as presence at what might be called a transcendental level and articulate a series of necessary conditions intended to establish the possibility of presence standing as the animating principle of western metaphysics. Consider the following as a terse summarization of Derrida's position on being as presence:

Passage 3A (cf. **Passage 1E**) …all the concepts by which *eidos* or *morphē* could be translated and determined refer back to the theme of presence in general (SP, 108).

But since Plato and Aristotle used *eidos* and *morphē* to refer to forms of very different metaphysical definition, it becomes crucial to determine what conditions must obtain in order that "the theme of presence in general" might indeed encompass *both* Plato *and* Aristotle, much less the subsequent totality of western metaphysics. Although Derrida built such differences into the structure of being as presence, this via the notion of *différance*, the problem is, again, to identify what is common or the same and, as Derrida put it, "invariable" as derived from the texts of Plato and Aristotle in order to justify the ascription of the single heading–"being as presence." If these conditions can be stated, or indeed even approximated, then this more distant and abstractly stated perspective might justify Derrida's broad generalization about being as presence with respect to its defining function within that totality.

The account in this chapter originates with Derrida's position, incorporates the lines of criticisms advanced in Chapters 1 and 2, then restates that position in terminology representing the direction of Derrida's formulation of being as presence but in a way which at least by intent deflects, if not rebuts, the force of those criticisms. To reduce the possibility of terminological confusion, it should be noted that "transcendental" in this chapter's title instances a meta-terminological gambit, a term introduced only as a marker to highlight the higher-order abstraction pursued during this phase of the essay. In fact, however, the structure of the transcendental possibility introduced in this chapter is derived from hints in Derrida's own texts so it should not be looked at as a purely *ad hoc* maneuver.[34] Moreover, the strategy in this chapter is not to attempt to outstrip Derrida in speculative range and novelty regarding the structure of being as presence. Rather it is merely to make more explicit conditions which reinforce what appears to be the scope of Derrida's project regarding the history of metaphysics. This theoretical extension will have two significant effects with respect to the goals of this essay–first, it broadens the contexts of concern which surround being as presence in order to provide the student of Derrida with a greater range of concepts and lines of inquiry for assessing the basic intelligibility, cogency and relevance of that notion; second, it grounds an interpretive and evaluative response to the apparent ends of deconstruction, as will become evident in Part III of this essay.

---

34 Derrida frequently notes connections between transcendental dimensions of argumentation and the statement of his own positions: P, 13; 111-2n44; MP, 219; SP 130, LI, 59. For commentary on Derrida as a "transcendental" philosopher, see Norris 1992, 32, 50; Norris 2012, 127; Fynsk, 169n26; Hill, 30; Wood, 65; Caputo, 102; Rudolf Gasché 1986, 273-6; Stocker, 176; Neal Oxenhandler 1989, 267-8; DFL, 219n11. Derrida also introduced the term "quasi-transcendental" with respect to his own thought: LI, 127, 152; Jacques Derrida 1992, 71; but cf. Jacques Derrida 2001, 103, where Derrida admits that by itself, the qualifier "quasi-transcendental" does not provide much in terms of clarification. For discussion, see Caputo 1993; see also DFL, 41-2, 219n12. The term "transcendental" will be employed in a specialized sense when we evaluate Derrida's positions on language (Part II).

In the Summary concluding this chapter, I shall outline the substantive conclusions established in Part I, both as review and as anticipation of Part II. It will then become evident that the implications derived from the questions raised in Part I within contexts almost exclusively shaped by metaphysical factors become more penetrating and more problematic when Derrida situates his positions on being as presence in relation to the more limited phenomena associated with language, whether spoken or written. The two main purposes of the transcendental survey in this chapter are, in sum, retrospective and anticipatory: to clarify the importance of precision in determining the meaning of being as presence in order to justify the significance of this notion with respect to the totality of the history of metaphysics; to give point to the pressure that being as presence exerts (a) on the structure of concepts relevant to traditional approaches to language and (b) as a prelude to the asserted necessity of reading texts deconstructively.

Note by way of preamble that Derrida voiced warnings against the primary goals of this chapter, and by extension of this essay as a whole. In the Hegel article, commenting on the "theory of the sign," Derrida asserts that a "continuum" of reflection exists on this topic and that if taken as a whole, its

> **Passage 3B** ...concepts are certainly no longer the same; and it would be more than foolish to erase the differences...to produce a smooth, homogenous, ahistorical, all-of-a-piece cloth, an ensemble of invariant and allegedly "original" characteristics (MP, 72).

Furthermore, in order to appreciate what Derrida calls the "chain" constituting the elements of any concept, extensive historical labor must be done:

> **Passage 3C** For as long as the great amplitude of this chain is not displayed, one can neither define rigorously the secondary mutations or order of transformations, nor account for the recourse to the *same word* in order to designate a concept both transformed and extirpated–within certain limits–from a previous terrain (MP, 72–italics in text).[35]

The movement in these passages is away from uniformity of meaning and toward ever-present variations in sense. Derrida warns against erasing differences and producing "an ensemble of invariant and allegedly "original" characteristics" in dealing with a matter as recondite and as historically pervasive as philosophizing on the article's immediate topic at this point, the theory of a sign.

But in spite of these qualifications, it must be kept in mind that Derrida has been careful to emphasize that "being as presence" generates an "invariable" characteristic throughout the history of metaphysics (**Passage 1C**). This affirmation and denial of

---

35 Derrida's use of "chain" is crucial for his position on the existence and status of concepts; the term will be critically analyzed later in Pt. II, especially in Chapter 5. For a representative treatment of the function of a chain, see Derrida 2001, 117-8n15. See also Ch5n58.

invariance may appear to be a contradiction, and in an especially fundamental context. We can, however, remove Derrida's texts from this indictment of logical difficulty if we insist that the passages on the invariability of "being as presence" are unique since they involve the entire metaphysical tradition whereas in the Hegel article, Derrida is concentrating on determining "original" characteristics adhering to the sign as evidenced by those philosophers who have concerned themselves with this matter. Similarly, the recourse to the "same word," a theoretical temptation which Derrida also warns us against, need not be taken as a rejection of using the same phrase–"being as presence"–in assessing the import of western metaphysics. And even here, Derrida admits that the same word–he is apparently referring to "sign"–does evoke a concept which, although it has been "both transformed and extirpated," nonetheless remains the same as it was from "previous terrain," albeit within "certain limits." The ascription of limits to the elementary sense of flux and change in this context mirrors a parallel sense of limits insofar as such boundaries apply to the uniform character of being as presence–hence the concentrated concern in this essay to isolate and identify what lies within the limits of the sameness which animates *différance*.

The question then becomes, again, whether it is possible to stipulate conditions which would allow being as presence to function at the uniquely fundamental level Derrida ascribes to it.

3.2 Form and Consciousness

For Derrida, the "meaning of being in general" (**Passage 1B**) is presence. This meaning pervades the entire history of western metaphysics. It is clear then that Derrida subsumes being in the intended sense *within* the experience of those individuals capable of and interested in addressing this reality. If the phrase "the meaning of" being is emphasized, then presence must include a subjective or human component, since meanings are derivative considerations depending on the reaction of a being capable of such response. Thus the intended sense of the "meaning of being in general" includes the factor which is capable of assigning meaning in the first place, i.e., human consciousness. Consider in this regard the following passage:

> **Passage 3D** Presence has always been and will always, forever, be the form in which, we can say apodictically, the infinite diversity of contents is produced. The opposition between form and matter–which inaugurates metaphysics–finds in the concrete ideality of the living present its ultimate and radical justification (SP, 6).

Derrida can assert apodictically that being will appear as–or simply "is"–presence in part because human beings, given their primordial interest in reality (in which of course they themselves participate), must necessarily be present in order to interact with reality in the very process of thinking about reality and voicing–whether to others or themselves (or both)–the thoughts so produced. It is asserted then that this human construal occurs within a generic notion of consciousness. We use this common term in a transcendental context without intending anything more by it than to stand as a place holder for the ultimate grounding of being as presence. This living source, conscious and cognitively aware, is capable of appreciating that something exists which is other than the thinker and also capable of approaching that otherness in a way which allows the possibility of asserting a distinction expressed in thought and language between, to use highly artificial and abstract terminology, the form and matter of that otherness.

At this point, being as presence means (a) something existing in a certain specifiable sense *necessarily related to* (b) something capable of experiencing whatever mode of existence has been identified as proper to (a). We must read "being as presence" so that it contains the phrase "in the presence of, " which in turn denotes a relation stated abstractly as "X is in the presence of Y." "Being as presence" thus names a relation as such–aRb–but with distinguishable poles; it does not refer to a reality of limited and determinate cast taken in isolation from the experience of an observer or discussant existing apart from that reality. In sum, when Derrida refers to "being as presence," what he means is "consciousness aware of being as presence."

A pair of caveats: first, this approach to consciousness should not be taken reductively as merely a doublet of Hegel, e.g., the strategy pursued in the *Phenomenology of Spirit*. We shall see that this is only the initial phase of three concentric dimensions of transcendental refinement aimed at clarifying the notion of being as presence. No grand speculative "payoff" awaits in the guise of an Absolute of uniquely Derridean cast. We posit only a bare relationship between an undifferentiated consciousness and what may be a wide diversification of objects fulfilling that relationship. These objects will include the penumbra of types of evidence Derrida has introduced as structural features of being as presence. Second, the reader will recall that one of Derrida's lists of exemplars drawn from western metaphysics includes "co-presence of the other and of the self" (**Passage 1B**). Derrida must have a particular philosopher (or philosophers) in mind with this descriptor, although strictly speaking the extreme generality of "co-presence of the other and of the self" could potentially apply to many philosophers. But again, if we take "co-presence of the other and of the self" to refer to one philosopher, as the other designations on this list are clearly intended to evoke, then the first transcendental dimension ranges much more broadly in its intended scope. We are positing a dimension which includes all metaphysicians who would want to separate theoretically the activity of thinking (which is, of course, executed by the "self") from the object of that thinking (which is "other" than the self, and in a variety of ways–e.g., the Platonic Forms and the notion of form in Aristotle, neither of which are mind- or self-dependent).

The defender of Derrida on the fundamental character of being as presence can now assert with regard to the *Phaedo* that when a Form is known (to the extent such awareness is possible), this knowledge must be present to the mind in attendance. If therefore a Form is known in the fullest Platonic sense, its nature becomes "present" to the mind of the inquirer into that Form in such a way that its nature can enter the consciousness of the knower and then be articulated as a definition. The adequacy of this definition will then presumably be measured by the extent to which it can withstand legitimate critical evaluation which other thoughtful individuals can direct toward the content of this definition. However, it still seems open to maintain, as argued in the discussion of the *Phaedo* in Chapter 2, that the "becoming present" of the Form to an attentive mind must be distinguished from the content of what has become so present. No distinction is asserted in the text of the *Phaedo* between (a) the instant at which the knower is in contact with a Form and (b) the specific content of the Form, whatever that content may be. As a result, we are apparently compelled to conclude that what is present in the case of being as presence when the being in question is a Platonic Form cannot be reduced to sheer temporality–a slice or cross-section of consciousness, as it were–as a dimension of reality divorced from the existence of something "in" that temporality (a conclusion reinforced by the observation about presence as time in the comment on **Passage 1D** in Chapter 1). It is vital then to appreciate that in this case an essential element in presence is presence *of a Form*, that is, presence with a specific content.

The connection between consciousness and its participation in the specific type of reality exemplified by a Platonic Form must be generalized, however, since being as presence encompasses *the entire history of metaphysics*. Therefore, the initial phase of the transcendental approach to being as presence may be stated thus: being as presence involves consciousness and consciousness is necessarily related to something other than itself–in short, being as presence cannot be reduced to a kind of an internally constituted or purely subjective idealism.

3.3 Relationality: Presence and Contrast

Being as presence encompasses human beings insofar as they have capacity for language and thus, via this medium, insofar as they can interact with what is other than themselves. But we recall that in **Passage 2C**, Derrida maintained that the concept of form "...is, and always has been, indissociable from the concepts of appearance, sense, evidence, or essence." Furthermore, the history of metaphysics with respect to the concept of form includes many positions, starting with Aristotle, in which the stated content of being differs from that attributed to the Platonic Forms.

Consider then the second phase of a transcendental approach to Derrida's notion of being as presence. This phase can be exemplified minimally by the pair of relations between (a) consciousness and a Platonic Form and (b) consciousness and form in one of its Aristotelian incarnations. The first phase of our transcendental survey of Derrida's approach to being as presence has shown that Derrida's position is not idealistic if being as presence is, necessarily, a relation between consciousness and something other than consciousness. But now we can appreciate that it is *the relation itself* which allows Derrida to read the entire history of western metaphysics as a unity, as presence. If this relation is taken as a constant, one which permeates the entire history of metaphysics, then a measure of cogency surrounds the assertion that being is presence even in the face of this remarkably variegated history. Let us assume then that this reading of "being as presence" has merit. Thus the differences between and among all the distinct texts characterizing this history–differences embodying considerable contrasts with respect to one another concerning the metaphysical content of forms–may fall, without loss of intelligibility and distinctiveness of that content, under the single notion "being as presence."

We then state the second phase of this transcendental account: being as presence is a relation encompassing consciousness and *a series of contrasts* involving various metaphysical objects other than consciousness.

3.4 Intersectionality: Presence, Sameness, Difference

When Derrida asserts that "all the names related to fundamentals, to principles, or to the center have always designated an invariable presence" (**Passage 1C**), the modifier "invariable" justifies taking "being as presence" as a theoretically determinate entity preserving itself as a unity throughout the history of metaphysics. Furthermore, although the explicit content referred to by "being as presence" will vary from metaphysical position to metaphysical position (**Passage 1B**), whatever constitutes presence remains the same throughout this sequence. This inference is justified on the grounds that if such sameness did not obtain, then "being as presence" taken as a unifying descriptor would in fact lack any import spanning that history and thus be clearly inadequate as a name for that sequence of thought. Furthermore, this sameness must be discernible in some way, it must be open to articulation and specification for otherwise Derrida's claim remains arbitrary and the phrase "being as presence" loses significance. Evidence of some sort must therefore be available to support this comprehensive and exhaustive predication.

Being as presence as a unity radiates in two directions: toward the human subject experiencing it and toward something, some "other," being experienced. Furthermore, both poles of this relation are constants, that is, both poles must always be present, as it were, in order to justify the existence of the full relation–being as presence. Now if all that being as presence signifies is that consciousness intersects with something other than consciousness so that the person becomes aware of this fact, i.e., that something *exists* other than consciousness, then "being as presence" hardly has the significance Derrida attributes to it. Derrida cannot be referring merely to the presence of a particular position in the history of philosophy insofar as this position becomes present in the mind of the person reading it or hearing about it. This is a purely private and personal presence. Although it is in a sense "being as presence," the being in question resides solely within the consciousness of the individual paying attention to a given metaphysical account. In fact, all that has happened by virtue of the inclusiveness of this ascription is that some determinate element of different metaphysical systems "comes to presence" whenever each system begins to be comprehended by anyone who studies and initiates an understanding of that system. As a point pertinent to philosophical education this eventuality is true, of course, but it is hardly informative about the history of metaphysics *per se*, nor about the relevance of the claim that this history must be subsumed under the uniform designation "being as presence."

The locus of being as presence–the sought-for sameness–may therefore be named intersectionality, i.e., the intersection of consciousness and a series of contrasting metaphysical transcriptions of reality. The unity of this sameness cannot be limited to either a subjective side–e.g., consciousness as the receiver of as well as (*pace* Kant) a contributor to the experience of presence, or an objective side–the specific character of what, existing other than consciousness, consciousness as such is experiencing. The combinatory function of the two poles necessitates inquiry into what remains denotationally identical *within* this sameness in order to establish and preserve the sense of semantic unity which the word "sameness" entails.

The unity of this intersection may for the sake of critical investigation be depicted as a continuum rather than as a point, since in this way the unity has room, as it were, to allow the admission of *patterns of differentiation* within the unity. We will interpret being as presence as coordinated around the two poles already identified–very broadly stated, a relation between a subject and an object pole, that is, between a human source of experience and something constituting what is other than this source and, presumably, existing independently of the human source.[36] If "being as presence" denotes *something being present*, then it is essential to the intelligibility of the phrase "being as presence" to attempt to specify characteristics of this "something" relative to consciousness so that the phrase has more significance than merely as a neutral "to be in the presence of" something. Derrida must be referring to presence existing in some sense *in* each metaphysically substantive position throughout the history of metaphysics. If so, then written texts refer to something–something other than the

36 "Consciousness" traditionally names a subject pole (in the subject/object relationship), although Derrida insisted on the need to confront what the tradition has called consciousness with stringent deconstructive analysis (e.g., SP, 147). See also MP, 283; SP, 62-3, 147; LI, 76, 105; Derrida 2002, 23; Derrida 2004, 156. But cf. Derrida 1987b, 254 where Derrida says that "the problem of consciousness is a very local and marginal problem for deconstruction." Contemporary analytic philosophy has devoted considerable attention to the concept of consciousness. For an accessible and vivid account of difficulties and opacities involved in such analysis, see the Chapter "Explaining Consciousness," 21-42, in Daniel Dennett's *Consciousness Explained* (see Bibliography). Dennett's ultimately Wittgensteinian approach to the "nature" of consciousness–i.e., anti-essential and multivariant in its articulation–may resonate with advocates of deconstructionism. For more recent analytic treatments of consciousness, see the third edition of Paul M. Churchland's *Matter and Consciousness* (see Bibliography).

writing itself–displaying a characteristic or characteristics which can actually and fully substantiate "being as presence."

As the content of what is other than consciousness varies throughout the history of metaphysics and as the contributing factors in consciousness to this relation also vary (although presumably within a limited range of possibilities), the relation between these two elements remains constant, a unity in itself. However, when the intersection is considered from the perspective of what is other than consciousness, that is, from the perspective of the various positions advanced throughout the history of western metaphysics, a series of concerns emerges with respect to the funding or substantiation of the relation in question. (The statement of these concerns should be taken in conjunction with parallel problems raised in Chapter 1, in the discussion of *diffèrance.* The concerns elaborated now emphasize critical approaches taken toward what is, for Derrida, the fundamental status of being as presence.)

First, when non-temporal elements from a given position enter the relationship, does presence in "being as presence" in some way affect the stated character of these elements? For if presence refers to temporality in any sense of the word, then "being as presence" becomes incompatible with a given metaphysical position if that position asserts that being, as such, is non-temporal. The question then becomes whether it is possible to identify an element (or elements) from any metaphysical position advancing a theory of being in which being exists in non-temporal circumstances and to insert this element into the relation of "being as presence." For it appears plausible to contend, against the tenor of Derrida's conviction concerning the comprehensiveness of being as presence, that if a metaphysical position suffers substantive alteration if it were to become a component of this relation, then the relation negates any effectiveness it might have by disassociating the integrity of precisely what it is intended to relate to consciousness at the point of intersection. Thus would the Platonic Forms become inherently in motion–regardless what kind of motion is putatively applicable–in order to satisfy the conditions of being as presence? If so, then being as presence can be accepted as a commonality throughout the history of metaphysics but only by ignoring the fact that it destroys a central and explicitly asserted tenet in Platonic metaphysics, one of the cornerstones in the edifice of western metaphysical reflection (a position it enjoys quite apart from its inherent tenability as a viable theory).

The second concern revolves around the fact that when the various positions in the history of metaphysics are viewed serially, as a connected sequence running from, say, Plato through Heidegger, each position differs in some way or ways from all the others.[37] As the history of metaphysics plays itself out, significant differences obtain at both end points of the poles which "in the presence of" collectively and uniformly

37 Recall Derrida's appeal to the "unfolding of presence" (**Passage 1D**). Does this mean presence unfolding as (i) an unbroken continuum *throughout* the history of western metaphysics, (ii) presence unfolding discretely, i.e., *within* the positions of each figure, figure by historical figure, or (iii) an unfolding in both senses?

represents. According to Derrida, to the extent that being as presence names a constant, an unvaryingly unified name, none of these differences are significant with respect to that constancy and invariability, although none of them is denied and all are preserved as stated characteristics of each system. Therefore, even if it were possible to render some element of a given position amenable to presence so that this process of accession would not affect the character of that element as explicitly stated within the system, all elements so processed would threaten to become identical to one another. This exercise in radical reduction would follow insofar as the entire set of positions constituting the history of metaphysics must share the same characteristic in order to exemplify Derrida's claim that invariable sameness obtains (by virtue of *différance*), thereby justifying the singular term "being as presence."

Consider, for example, Plato and Aristotle at the dawn of the metaphysical epoch. Imagine the philosophical awareness of being when it is construed as (a) a Platonic Form dialectically analyzed by consciousness and as (b) an Aristotelian form intersecting with consciousness. For Derrida, being as presence results from (a) and (b)–minimally, of course, since the rest of the history of metaphysics is temporarily discounted. As we have argued, presence means presence *of* something and presence *to* someone, both taken together as an indivisible unity. But for Aristotle, form, at least as this concept is developed in his *Metaphysics*, does not exist independently of substances determined by form. The question then is whether the Aristotelian form "presents itself as such" in the same way that the Platonic Form "presents itself as such." For only if this uniform presencing is in fact the case would it be possible to specify a quality or factor common to both positions.

A third concern now appears, related to but distinct from the second. Add to the history of metaphysics a position which construes being as primarily if not exclusively matter (Democritus, Hume, etc.). It then becomes evident that the claim "form" means "a center of evidence" implies that form refers only to evidence taken at the level of direct immediacy, as in the immediate experience of materially existing entities. Consider two instances of this relation: (a) consciousness–Form and (b) consciousness–sense impression (of something material). Is it possible to identify what is common to both instances so that the factor of presence can be recognized as, in fact, the same?

The following line of argument seems open to the critic.The materialist is immediately aware (through perceptual experience) of the presence of individual things, the Platonist is immediately aware (through cognitive activity) of the presence of immutable Forms insofar as they constitute the ultimate reality of individual things. But such presence, although encompassing the thoughtful environments of advocates of both metaphysical positions, fails to bring out the fundamental differences between what is present to the materialist as opposed to what is present to the Platonist. Therefore, it seems fair to conclude that Derrida does not distinguish between (a) the processing activity of consciousness in which something is perceived, etc., an activity which is undeniably present to that consciousness since consciousness exists *as* that activity while the activity is ongoing and (b) the content of what is experienced within consciousness. In short, presence as the *medium* of experiential accession to objects of thought has been equated with–and thus effectively reduced to–the *content* of those objects.

The argument continues. Distinctions embedded in this comparison must be brought into the open. For a materialist, being is presence in that reality consists in what is directly observable through sensory appearance. Furthermore, class terms–such as, for example, Hume's notion of an abstract idea–are typically mere names for the empiricist and, as such, do not possess the reality of the things which those terms denote. It may be concluded then that the manner in which a Platonic Form exists as an element in the being-presence relation is not identical to the manner in which, say, a purely sensory impression, *à la* Hume, exists as an element in this same relation.

This comparison reveals that the "presence" in "being as presence" refers to very different phenomena. The center of evidence for the materialist is sensory appearance; the center of evidence for the Platonist is the mind's ability to cognize an underlying Form. For both positions, something must be "present" in order to determine how reality is configured–in this respect, Derrida's claim that being is presence can be justifiably asserted. But what is the same when consciousness is in the presence of a Platonic Form in contrast to consciousness when it is in the presence of a given sense impression? Are both experiences in contact with some feature or element in their respective objects which can be discerned and then identified as "the same"? The relation connecting sensation with the entity sensed, this connection taken as a unity, must share at least one characteristic (there may be more than one) with the relation connecting thought with the entity thought. In other words, these relations must be isomorphic with one another in terms of some specifiable shared content in order to justify the claim that "being as presence" will encompass *both* epistemic/metaphysical states of affairs. But can a sameness be indicated in order to account for presence when the requisite sameness includes both sensory and mental evidence?

Note the divergent positions of Plato, Aristotle, and Hume. If the language each philosopher uses is allowed traditional denotational scope, then the reductive force apparently endemic to being as presence is eliminated. Plato's word "Form" refers to an independent being existing immutably and outside of space and time; Aristotle's word "form" refers to a dependent being with a more complex mode of existence with respect to space and time; Hume's word "idea" refers to a mental copy of a given percept and his notion of an abstract idea, if taken as a counterpart to form in the Greek metaphysical sense, refers to an existent solely in the mind of the perceiver of impressions.

There are many problems of interpretation, consistency and cogency in each of these three concepts (two of which employ the same naming word–form), but these issues are not relevant here. In terms of clarifying being as presence, the salient point is that what is present to consciousness differs considerably with respect to evidentiary content; furthermore, and even more crucial, what that evidence refers to throughout the totality of the history of metaphysics is wholly distinct in terms of its metaphysical character. This problem is compounded exponentially if we gather (a) all the various characteristics (present to) experiencing a Platonic Form and juxtapose this set of characteristics with (b) the set of all the characteristics (present to) experiencing a form in some other non-Platonic respect as well as, for all empiricist metaphysics, the modes of experiencing a physical object. Derrida's description of form as a center of evidence and implications directly derivable from this description glosses over fundamental differences obtaining *between and among types of evidence*. As a result, the specific content of being indicated by various types of evidence functioning within given metaphysical systems becomes submerged, effectively vanishing within a continual onrush dubbed, then accepted as dogma by Derrida and his epigones: "being as presence."

3.5 Directionality: Presence on the Move

The advocate of deconstruction might object that the analysis argued above, especially with its stark if not dire conclusion, is too beholden to the conceptual and argumentative rhythms of traditional metaphysics; more salutary possibilities remain open pending a more congruent and conceptually sympathetic approach to the principle at issue. Here is one such possibility. Consider the relation "X in the presence of Y" as a process, not as a static or linear linkage of two distinct entities or accounts of entities. Then introduce the notion of *directionality*, that is, movement along this process from the core or center area of the relation to either of the two end points. If this movement is toward X, or consciousness, then "being as presence" tends toward a species of idealism, since what is emphasized is the being of consciousness insofar as it exists in the presence of something other than itself. If the movement is toward Y, being as such existing apart from consciousness, then "being as presence" tends toward a kind of realism in the sense that what is being emphasized is the content of what consciousness is conscious *of*. Of course, this directional realism is only a threshold variety since there would be many distinct examples or instances of being construed as real, from Platonic realism envisioned as Forms outside of space and time to Kantian realism, the categories as necessarily real elements instantiated within a given Newtonian space-time. What then would be common to these two instances of realism (there are, of course, numerous other possibilities) which would justify the sameness necessary for the unity of "being as presence"?

Consider directionality from another approach. If the process is held in suspension so that it lingers, and necessarily lingers, *somewhere between* these moving tendencies as they are in motion toward a specific end point, then "being as presence" becomes essentially a surd, where the emerging reality is the relation as such, constantly oscillating between its two defining poles. If the best that can be done is to situate consciousness *tangent to* the accounts of being which constitute the history of metaphysics but without either pole of the relation "in the presence of" effectively intersecting with its counterpart, then being as presence is reduced to a name, a locus of possibility which, as argued here, cannot be actualized in terms of a discursive account since its intended referent does not exhibit existence with sufficient stability to be evoked in language.

To conclude, at least tentatively: directionality as developed above and taken by itself does not stand as a defensible ploy in the attempt to deliver a more generalized, transcendental account of being as presence.

***

These three approaches toward formulating transcendental conditions for appreciating the complex structure of being as presence result in a series of problematic concerns which can, in turn, be employed to evaluate the general intelligibility of the notion. These problems arise from implications which follow attempts to interpret and deepen the contexts for determining the range and efficacy of being as presence as Derrida formulated that notion in a number of early writings. Attempts at evaluating implications derived from these problems appear in each of the three discussions. However, the reader need not necessarily embrace the arguments and conclusions in the form in which they have been stated and developed above (although I contend that at some point in interpreting Derrida, the critical concerns arrived at by these arguments must be faced and discussed). At this juncture, these concerns may be taken as tentative rather than as fully established. But to deal adequately with these concerns would require the defender of Derrida to demonstrate that the texts examined so far in this essay as well as the reasoning interpreting those texts are jointly not relevant to a more accurately detailed account of what Derrida meant by "being as presence."

This caution is both procedural and substantive. However, what the essay has demonstrated is sufficiently persuasive to justify concluding Chapter 3 with a summary outlining material relevant to appreciating the scope of being as presence and to pose a series of questions and quests which should assist in directing additional inquiry. The interested student of Derrida can apply pertinent elements of this summary to the set of problems engendered earlier in this chapter, although the import of this summary is sufficiently expansive to fund interpretive and evaluative possibilities ranging far beyond the scope of the particular issues made explicit in the commentary so far.

This supplementary material concentrates on being as presence in two respects: first, insofar as being as presence represents a putative contribution to the history of metaphysical thought (even if part of its significance is to announce the closure of that history); second, as a schema suitable for development when approaching the phenomenon of language so that the deep structure of language becomes fully analyzable only when the traditionally identified elements of language are encountered through the strategies of deconstruction. The following restatements will integrate and reinforce the immediate relevance of the points made above within the more abstract contexts of the concepts and realities now introduced.

3.6 Being as Presence: Structure, Questions, Quests

The following discussions consolidate our inquiry.

3.6.1 Foundational Elements

The texts examined so far, including the interpretive extension comprising the set of transcendental conditions argued above, yield the following foundational elements constitutive of being as presence:

i. **A totality field enveloping the various systems of the history of metaphysics.** If being as presence encompasses the entirety of western metaphysics, then it may be inferred that every element in each metaphysical position within that history falls under this descriptor. But most if not all the positions defining western metaphysics have advanced a theoretical concern for coping with and accounting for totality, that is, their respective treatments of being included all being–individually, collectively and in terms of formal categories (if principles of formality were part of a given system). Therefore, being as presence must encompass totality, where totality refers to the sum of everything explicitly construed as existing and every inference entailed by this summation insofar as these summations pertain to any and every given metaphysical position constituting the history of metaphysics.
ii. **A sameness grounding differences which characterizes instances (or "subdeterminations") occurring throughout the history of that field.** Many differences inflect being as presence but these differences in no way disturb its status and its name as a single and uniform component, hence Derrida's insistence on the invariable sameness driving *différance*–being as presence's fundamental principle and source of unification–a sameness Derrida asserts to be distinct from identity.
iii. **A necessary connection between evidence, broadly characterized, and form as a primary manifestation of being as presence.** From this perspective, being as presence becomes a relational vehicle which includes the evidence–presupposing an agent experiencing this evidence–for a particular character of being insofar as that character is introduced and developed by a given systematic metaphysics.

3.6.2 Structural Factors

If we approach Derrida's position on being as presence from a suitably detached perspective, it becomes evident that its possibility as a relevant reading of the history of metaphysics rests on three classic metaphysical concepts: unity, sameness, and difference. Only one of these three elements has been explicitly subjected to scrutiny in Derrida's texts–difference as a component of what, for Derrida, is the more fundamental factor of *différance*. However, unity and sameness are equally crucial and the extent to which they function without explicit or, at most, minimal examination is the extent to which their integration within Derrida's texts will affect the ultimate intelligibility of the notion of being as presence.

Unity, sameness and difference must be taken as fundamental factors in the organization and development of positions stated in texts, including Derrida's own. But the more that these realities as such, as independent factors, remain unanalyzed or examined only marginally, the more that conclusions drawn by applying these realities as they concern being as presence become affected by their residual metaphysical auras.The danger then is the intersection of mutually incompatible lines of thought producing an extended series of logical tangles which affect, perhaps to the point of rendering incomprehensible, the cogency students of Derrida assign to being as presence and, by extension, to its ultimate effects on the philosophical enterprise.

Concerns regarding such incompatibility become apparent through detailed critical scrutiny developed in the following sections. The goal at this juncture is not only to continue attempting to elicit the content of the sameness which Derrida, by his appeal to the notion of being as presence, has ascribed to the history of metaphysics–it is also to identify what he has assumed procedurally, that is formally with regard to the structure of discourse and interpretive thought, which allows him to make such assertions about the totality of that history in the first place, much less to invest them with an air of apparent decisiveness.These assumptions derive in a variety of ways from the structural assemblages proper to each of the three realities listed in the previous paragraph. The following accounts will encompass (a) conceptual and argumentative content insofar as such content can be determined from Derrida's texts and (b) assumptions essential for the possibility of discursively developing that content.

A useful reminder drawn from the history of metaphysics. In the Platonic dialogue bearing his name, Parmenides has scrupulously interrogated Socrates concerning the intelligibility of the relation between Forms and particular entities participating in Forms. After the young Socrates does not acquit himself well in defending that classic moment in metaphysics, Parmenides reacts with a blend of sympathy and sternness:

> Believe me, there is something noble and inspired in your passion for argument, but you must make an effort and submit yourself, while you are still young, to a severe training in what the world calls idle talk and condemns as useless. Otherwise, the truth will escape you (*Parmenides*, 135d–Cornford's translation).

Parmenides' warning about "the truth" escaping the youthful philosopher Socrates may be rejected as just another naive instance of Platonic foundationalism. But if the advocate of Derrida retains at least a vestige of concern for the preservation of that venerable notion, then Parmenides' advice concerning the pursuit of truth may be important to note in this context.[38] Parmenides is about to embark on an extended series of paradoxical implications revolving around the notion of the one or unity (in a sense or senses open to a variety of readings). He tells Socrates that this "severe training" is crucial for Socrates' development as a philosopher but he also mentions that what they will discuss occupies a level of abstraction which "the world" would call "idle talk and condemns as useless," assuming the world would consent to enter this intellectual arena in the first place. The following observations are lightweight to be sure, but they remain sufficiently removed from the general run of much contemporary philosophical discourse to be dismissed by many as "idle" and "useless." The reader with an affinity for Derrida and deconstruction is asked to be as open as possible to the relevance and implications of what follows.

Unity as Substantive Consider the feasibility of Derrida's project regarding the history of metaphysics. Any thinker confronting for analytic and deductive purposes the entire history of an activity as broadly diverse as philosophy in general and metaphysics in particular always risks being defeated in the quest to isolate a single common feature of this history simply by sheer textual magnitude of the existing positions, not to mention the equally evident divergences in content and style marking all the participants of this totality. This empirical consideration and its consequences hold regardless of the purpose driving the philosopher's inquiry. Thus Hegel's quest for "the end" of history, a question of avowedly larger scope than Derrida's determination of presence as the nerve of western metaphysics, is also open to parallel concerns based on the myriad events which constitute history writ large.[39]

It seems appropriate then to wonder why there is one and only one way to interpret the history of metaphysics, as Derrida affirms so often with the designation that being as presence is the sole singular description of what is common throughout this history. That the 2500-year history of metaphysics can be gathered into a *unity*, and that this unity–if, again, its magnitude is interpretively manageable–can then yield a *single property* open to identification and description as its defining characteristic must be considered allied assumptions of epic scope.

From this perspective, two distinct but recognizable senses of unity are presupposed–the first gathers and unifies the history of metaphysics, the second

38 At MP, 39 (cf. OG, 10), Derrida puts the concept of truth into question but elsewhere he insists on the preservation of truth–e.g., P, 105n32; LI, 44, 150. For discussion, see Hill, 38-9, and Norris 2012. At LI, 44, Derrida asserts his ongoing "concern with the question of truth." What does this mean? Is the advocate of Derrida to develop a modification of the correspondence or coherence theory of truth? Or is the goal a completely new approach to truth? If the latter, is it possible to discern any hints of what Derrida might have had in mind? I suggest that until the new inscription pertaining to truth takes tangible form and its difference is clarified with respect to traditional theories of truth, Derrida's claim is only a broadly speculative promise.

39 Recall **Passage 2A** and its ready conjunction of the history of metaphysics with "the history of the West." For similar assertions of global scope, see MP, 134-5; WD, 282.

elicits one characteristic as proper to that history. Derrida's entire project with respect to philosophy as such depends on an implicit and foundational trust in the unwavering status of unity, albeit a unity of organic dimension.[40] The possibility that this singularly monolithic reading of the history of metaphysics taken as an organic unity can, if not must, be broadened is addressed, with implications, in Chapter 9.

The same reservation holds when Derrida announces that the history of metaphysics has been "closed." For if there were many possible cogent and coherent histories of metaphysics or, alternately, many viable versions of only one such history (its content defined by a series of accepted texts), then it would not be possible to identify a single characteristic common to and defining the history of metaphysics simply because the distinct histories or the single variant history lack a doctrinal dimension of unity and are themselves a highly diverse multiplicity. Thus to canvas the entire history of metaphysics, as Derrida clearly intends to do, would require separate investigations of each viable version of that history with the coordinate possibility that the characterizations resulting from these investigations–if juxtaposed in such a way as to constitute a set–would not be logically compatible with one another while concurrently, each taken as an individual account, enjoying status as legitimate readings of that history. But even without this logical possibility based as it is on the generation of contradictions, to claim that the epoch has been "closed" would presuppose that every single perspective derived from feasible readings based on historical data had been rigorously examined and shown to have "ended" in ways which would justify the ascription of such completion.

Finally, the notion of being as presence depends for its plausibility on self-identity and difference merely to be capable of being formulated as *one logically coherent reading* of the history of metaphysics. Being as presence exists as a unity with its own self-identity; indeed, it has preserved this unity throughout the entire history of metaphysics to such an extent that it has, for Derrida, defined that history. Derrida can indicate the possibility of a realm of reality "outside" that history which in some sense both exists and is available for entrance into language as contoured or styled in some way (or ways) only on condition that the entire history of metaphysics can be situated as one continuous movement, formed with its own beginning and circumscribed by its own domain of closure.

In sum, Derrida assumes a number of dimensions of unity in his approach to the history of metaphysics as the source for the generation of "being as presence" and implications he draws on the basis of so understanding the history of metaphysics.

Unity as Heuristic The sense of unity embodied in these assumptions along with their applications in Derrida's texts is necessarily substantive and, as we have just seen, deployed in several dimensions. But consider the Platonic dictum, voiced in the

40 Recall the sources cited at Ch1n7, primary and secondary, which appeal to unity and its importance in philosophy proper and also in the more restricted context of interpreting Derrida's thought.

Republic (525a): "...the study of unity will be one of the studies that guide and convert the soul to the contemplation of true being." One need not be a Platonist to respect the importance of unity and correlative general terms (an importance acknowledged in different ways in the *Republic* as well as the *Parmenides*) both in metaphysics proper as well as in the analytic treatment of the most discriminating and subtle points relative to philosophical dimensions of language (as we shall see in Part II of this essay). Thus a relevant sense of unity, running in a different direction from the substantive unity doing foundational work for Derrida, will be unity employed as a heuristic parameter. "True being"–a phrase ringing with foundationalist overtones–is not equal to being as presence, but being as presence must be taken into account as a constant throughout the history of metaphysics as an essential adjunct to "true being," assuming that that phrase has a modicum of significance.

The opening sections of this chapter have demonstrated that this possibility includes specifying transcendental conditions for establishing the kind of missing or distorted metaphysical properties Derrida has attempted to excite via the notion of being as presence. Another important heuristic application of unity derives from formulating stable concepts and principles as investigative prerequisites to study being and language. For if contemporary philosophers continue to function within the sphere and influence of the history of metaphysics, then it seems prudent if not necessary to incorporate at least some phases of the influence of such stability into any future thought on metaphysical subjects. This influence will hold regardless whether the working agents of philosophical activity are or are not aware of Derrida's overall position concerning what they are doing or what they believe themselves to be doing. Chapter 9 will analyze the necessity of such stability; Chapter 10 will then illustrate it.

Sameness In his account of *différance*, Derrida distinguishes between sameness and identity in order to emphasize the demands he places on *différance* as a neologism of unique speculative dimension. Sameness and identity, if advanced in a theoretical context without additional clarification, will typically be taken as synonyms. But for Derrida their distinction is essential since sameness–not identity–drives the complex functionality of *différance*. Thus Derrida's sameness names a unity performing a unique derivation affecting multiple metaphysical systems falling under that unity. The unity applicable to sameness must therefore be distinguishable in some sense from the unity in Derrida's counterpart use of identity; without some distinction of this sort, the contrast between the two terms becomes difficult if not impossible to preserve. Clarification of the distinction between sameness and identity–established at an equivalently high order of abstraction–is therefore vital in order to appreciate how sameness functions in the context of being as presence.

In general, to claim that sameness pertains to a given state of affairs is to commit to a plurality of entities as well as a respect (or respects) in which two (or more) entities share an identical property. Thus Derrida and Descartes are the same in various respects (speaking French, writing philosophy, etc.). Derrida's technical sense of sameness employed in *différance* differs from this standard usage in that sameness generates a specific substantive content–presence–and thus is not indeterminate or open-ended regarding the possibility that the relevant perspectives justifying the predication of sameness could be virtually anything. However, Derrida's technical usage of sameness is similar to the traditional usage of the term in that the relevant modes of discrimination *within* sameness include all the extant differences which mark the relation between any two (or more) systems in the history of metaphysics. In fact, since the differences in question encompass not merely any two systems or any number of multiple systems, fidelity to Derrida's position requires that the differences in question encompass *all* systems taken individually with respect to one another and collectively, given that the totality of systems constitutes the history of metaphysics.

But the sheer juxtapositions of different systems is not sufficient by itself to identify the respect governed by the predication of sameness. Derrida stipulates that "presence" is *the* specific respect which justifies the predication of sameness. But so predicating the single term "presence" presupposes, at least traditionally, that analysis can elicit an element of sameness pervading the entire sequence of the history of metaphysics.

Furthermore, this unity, following Derrida's own terminology, is invariable. I suggest the following rationale for this striking and perhaps unexpected appeal to stability, assuming that stability is a minimal condition for the predication of invariability to sameness: without this invariance in place, nothing will remain when the constitutive elements defining the extent of realities identified throughout the history of metaphysics are all gathered under the aegis of sameness. The resulting collection becomes a sequence of conjoined elements each sufficiently different from all others in such a way that, if these elements are viewed collectively, it would be impossible to identify a single characteristic or property (much less a set of such properties) belonging to and shared by every element.

It is necessary then to identify or at least to approximate *what presence means* when it establishes and substantiates the sameness which, Derrida insists at the outset of his essay on *différance*, animates the sequence and content of western metaphysics. If this putative sameness were not itself existent or if it was existent but not determinable or open to linguistic expression, then the generalization Derrida intends to apply to the history of metaphysics would lack theoretical force and relevance to contemporary thinkers. Advocates of Derrida's position must therefore account for at least one feature in sameness which is (a) invariable throughout all instances of metaphysical systems–with respect to both subject- and object-poles–given that these systems all differ with respect to each other and (b) other than any correlative features which may be found in the notion of identity if that notion were applied to the same sets of conditions.

Sameness must be given a determinate content for two reasons: (a) to distinguish sameness from identity. Without the specification of at least one property belonging only to sameness, no distinction obtains between Derrida's uses of sameness and identity, from which it follows that the contrast between the two terms becomes difficult if not impossible to preserve. And with respect to notional function (b), sameness as the core of *différance* controls the predication of "being as presence" over the entirety of western metaphysics. If being as presence itself has content–i.e., the substantiation of "presence," whatever substantiation that may be–then sameness must have at least sufficient content to generate the possibility of presence emerging as the omnipresent property, as it were, belonging to the history of metaphysics.

The appeal to one name, "presence," does not necessarily preclude the possibility that presence, if it has a structure and the structure is open to elucidation, admits of several, perhaps even many, constitutive components. But even if upon investigation a plurality of components does emerge, it seems legitimate to expect that uniformity of name presupposes uniformity of structure. Without identifying the element (or elements) underlying sameness, Derrida's concept of sameness as the invariant factor in being as presence becomes vacuous because of its indeterminacy. If so, then all that remains substantively of the concerted appeals to being as presence is the comprehensive, universalizing scope of the appeal.

It may be noted that, as if to anticipate these questions, Derrida presented several hints concerning the content of the sameness constituting being as presence; the following comments situate the potential relevance of these possibilities. The analyses are provisional and represent only an initial foray into complex issues of interpretation and theoretical content.

Sameness and Temporality Textual evidence from passages cited in Part I indicates that presence is not limited to temporality as the now or the present as mid-section between past and future. In this regard, the description of the Aristotelian moment, the now, as "temporal presence as point" (**Passage 1B**) suggests that presence is more inclusive than the temporality attributed to the now. But does being as presence necessarily occupy temporality in some sense? The identification of the Aristotelian now as *an instance of presence* implies that presence in its broad-based technical sense includes the temporality embedded in the now as a moment "in" time. The issue becomes that of detailing the character of temporality irrespective of the now or the moment insofar as temporality necessarily pertains to presence. There is also the question of identifying those factors which (a) belong to being as presence while they are concurrently (b) not reducible to temporal qualities.Thus, the tentative conclusion is that whatever presence may be, it encompasses temporality but cannot be reduced to sheer temporality. It is worth noting, however, that as argued in Chapter 2, if it necessarily includes a factor of temporality, then being as presence fundamentally distorts any metaphysical position (e.g., Plato's) in which being is described as essentially non-temporal.

Sameness and Relationality Textual evidence suggests that the relation between the observer and the mode of existence of something other than the observer underlies Derrida's notion of being as presence. Furthermore, this relation is a constant permeating the entire history of metaphysics and identified by Derrida as a type of sameness.

If being as presence is relational, then one localizing property would be identification of the relata. If, for example, being as presence included experience of a human observer, then this factor would be necessary to being as presence but, again, presumably not exhaustive of being as presence. Thus being as presence would always include, whatever its distinctive content may be with respect to *différance* and the differentiation exemplified by individual metaphysical systems, the factor of being experienced by an observer. In addition, if being as presence possesses at least one recognizable factor on the side of being, that is, a factor which would always be determinable yet also not exhaustive of being as presence, then this condition would (a) justify that being as presence had at least some sort of discernible structure and also (b) reinforce the relevance and necessity of preserving the factor of self-identity insofar as it is necessarily predicable of being as presence. It may be observed as a corollary that even without (a) obtaining, the necessity of (b) still stands. This logical requirement will be developed separately later in Chapter 3.

Sameness and Consciousness We have interpreted being as presence to be a kind of tangency between (a) consciousness–understood in a sense distinct from that identified by Derrida as an instance of being as presence–and (b) an object "other than" consciousness, a tangency holding throughout the history of metaphysics. It follows therefore that the relation in question will have variable instances for both relata. And the degree of variation will in some cases be extreme. Thus if consciousness is related to being–that is, "being" understood as what exists in modes other than consciousness itself–in different ways, e.g., through a predominantly conceptual medium (as in Plato) and through a predominantly perceptual medium (as in Hume), then it is necessary to identify what is the same when the relation involved includes both (a) accounts of otherness exemplified by distinct metaphysical positions (e.g., Plato and Hume) as well as (b) the avenues of reception within consciousness toward these diverse instances of otherness insofar as they differ from one another.

Consider also the case of Hegel. In the essay on Hegel discussed above, Derrida identified two instances of being as presence. If these two designations are significantly different from one another and if the second designation indicates that this mode is limited to the sphere of consciousness, then as we argued in Chapter 1, the parallel defining characteristic of "presence in the form of the object" must be something *other than* consciousness. If this otherness were not the case, then both evocations of being as presence would include some necessary dimension of consciousness and no difference would characterize the two designations of being as presence.

We develop this line of argument: If consciousness is self-generating with respect to presence and if we allow consciousness, as self-generation, to exist necessarily as *in motion* in some basic, fundamental and broadly characterized sense, then the self-presence of consciousness has, minimally but necessarily described, the characteristic of motion. However, the problem of property-identification for sameness as *the* single underlying reality constituting the history of metaphysics remains fully at hand.

Self-presence does not automatically provide the relevant property just from the ongoing presence of consciousness itself; that property must be identified and we have done so here, provisionally, by the appeal to motion (in, again, a very broad sense of the term). But now the same problematic incompatibility derived from the factor of relationship reappears. Here is an example: if the characteristic funding the sameness of being as presence for Hegelian metaphysics is motion, which we have assumed is endemic to consciousness, and if being as presence also pertains to, e.g., the Forms in Platonic metaphysics, which are never in any sense in motion, then how can sameness include both motion (for Hegel and, according to Derrida, the self-presence of consciousness) and the complete absence of motion (for Plato and the Forms as they also are subsumed under "being as presence")?

Further analysis is required in order to justify Derrida's "being as presence" as the singular notion controlling the entire history of western metaphysics when implications from consciousness enter the domain of critical inquiry.

Sameness and Uniformity It will be instructive to generalize the problem. For Derrida, the uniformity of being as presence as a trans-epochal phenomenon is represented by the element of sameness which animates *différance*. Let us approach the articulation of sameness with respect to the desired element or elements of uniformity by construing the relation between any two different metaphysical systems (*a*) and (*b*) as a process which we name "presencing." Let A with respect to B and C with respect to D represent parallel relations of explicitly designated elements in each of the two different metaphysical systems and, following Derrida on the function of *différance*, also posit that these relations respectively manifest being as presence in these two systems. Thus if "presencing," the name for the relation A–coming to presence–B (for system *a*) is the same as the presencing of the relation C–coming to presence–D (for system *b*), then this sameness as a unity negates any distinction between (i) A with respect to B and (ii) C with respect to D. In fact, the precise character of the relation embodied by each of these two instances cannot be preserved. A parallel instance illustrates this seriously destructive conclusion. If A is the father of B and C is the mother of D but the presencing of A to B is the same as the presencing of C to D (insofar as presencing names the juxtaposition of variables constituting "being as presence"), then the difference between "being" a father and "being" a mother disappears since both relations are merely–and solely–instances of being as presence.

To attempt as a clarifying device an appeal to "being" a parent–the more inclusive concept with respect to paternity or maternity–is also unsuccessful since being a parent conceals whether the parenthood is that of a father or mother, and this difference is precisely the analogue to the differences between (and, by extension, among) metaphysical systems which, according to Derrida, being as presence preserves. Therefore, according to this analysis of being as presence, it remains unclear what presencing consists in other than some sort of proximal correlation between two distinct entities. The precise character of this correlation must be identified and described in order to endow being as presence, when construed relationally, with determinate content.

If it is impossible in principle to specify what dimensions of consciousness animate the relation and it is commensurately impossible to identify the character or quality of being of which consciousness is *of*, then it is not evident that "being as presence" conveys any significance which should actively or even procedurally concern the working philosopher. Consider extreme examples in order to appreciate this point: a nihilist or a skeptic both illustrate this very broad relational construal of being as presence–the nihilist says in relation to what is other than the nihilist, "Nothing of any substance is really there" whereas the skeptic says in relation to what is other than the skeptic, "Things exist but they cannot be known." If both these positions legitimately illustrate the import of "being as presence," there is strong reason to believe that "being as presence," taken as a theoretical element intended to accomplish important philosophical results, becomes vacuous and uninformative.

Sameness and Self-identity We must also recognize the unstated but crucial dependence of sameness on unity as an instance of self-identity, that is, that the sameness in question is and will always remain one in that it will always be identical to itself throughout all the variegations marking the history of metaphysics. Even if the sameness animating being as presence should turn out to be a surd, that is, incapable of yielding any further specification in language as far as defining properties or characteristics, "being as presence" as a single name representing, presumably, a single sameness at this level of speculative generality must remain identical to itself in preserving its surd nature while it also continues to function so pervasively throughout the textually based history of metaphysics.

This inference, as stated, rests on a logical formality–the principle of self-identity. Derrida's thought axiomatically generalizes over the entire history of western philosophy. His analysis of this history as a subset of philosophy elicits a single property spanning the entirety of that history; the appeal to the multiplicity of names subsumed under this unity does not shatter or even fragment the unity. Furthermore, Derrida assigns grave consequences to the intended mission of metaphysics regarding the elucidation of being if this single property is not duly recognized by those in philosophy who take seriously metaphysics in all its diversified guises.

As a result, the formal property of self-identity must be in place in order to guarantee that (i) the *entirety* of western metaphysics will remain the same as itself; (ii) the *sequence* of individual metaphysical systems will remain in the same order; and (iii) all *individual systems* of metaphysics remain intact and the same as themselves throughout any and all thoughtful transactions with any given system within that history or with any considerations derived from or attributed to the history of metaphysics taken as one narrative whole. For if self-identity did not obtain in each one of these formal respects, then it is logically possible that the history of metaphysics could become something other than itself, whether in whole or in part. If such fragmentation from the original unity were to occur, it would follow that nothing could be inferred with any measure of certainty about that history taken as a unified whole, or any moment within the boundaries of that narrative whole. Derrida would never have been in a position where it was discursively possible to identify and state *one name* as a single, uniform designator, much less to insist that this one and the same name prevails over the entirety of western metaphysics.

Even if Derrida invests sameness with a technical sense which accomplishes considerable theoretical work for his overall position, this special sense–however it may be characterized or even hinted at through the wispiest traces–does not affect the logical necessity that such sameness must remain the same as itself. Thus self-identity as a concept of formal scope must be recognized as both determinate with a fixed structure and also as necessary; indeed this formal sameness as self-identity is in its own way more fundamental than the purely theoretical development of sameness since formal sameness establishes self-identity for every constituent element in the formulation of Derrida's extensively variegated discourse–*différance*, difference, trace, supplement, etc.[41] As such, this dimension of sameness contributes to establishing the possibility of intelligible discourse for any speaker as an individual agent of language as well as for a plurality of speakers to interact meaningfully with one another in any linguistic context.

In order to accommodate all the differences which inflect the history of metaphysics, this sameness must be unique. As such, it exhibits its own unity and also its own self-identity. Furthermore, this necessity presupposes the reliability of the principle of identity, a point worth emphasizing here since Derrida has not only ignored recognizing any special status this principle might have, he has, as we have seen, contended that identity is the paramount logical principle engendering being as presence (**Passage 1L**). The question of priority is crucial in this context. Without the underlying and stabilizing force of self-identity, the name "being as presence" would not attach to anything which remained sufficiently stable to be named at all, much less to be given a name which "sticks" throughout all the permutations and peregrinations of an exquisitely variegated history, a name coined by an observer of that history from a vantage point occurring at a juncture marking the putative closure of that history. Derrida can allege that identity is the "founding" principle of being as presence but it is precisely because of that principle in its formal guise as self-identity that Derrida is allowed to articulate the phrase "being as presence" in such a way that it generates its own autonomy as a stable notion of theoretical significance.

Difference Difference has already been introduced as an exemplar of a logical notion exhibiting unity in a formal sense–a sense sanctioned by millennia of philosophical approval as well as by apparent intrinsic rightness. However, at this point we examine

41 Especially useful in these contexts is Simon Morgan Wortham's *The Derrida Dictionary*, a lexicon of technical terms ranging over the entire corpus of Derrida's work. See Bibliography.

difference both (a) as such and (b) in relation to Derrida's formulation of *différance*. For Derrida, there is a fundamental priority of the latter over the former–difference derives from *différance* and therefore the latter is the more fundamental element. He bluntly states the point:

> **Passage 3E** To reduce différance to difference is to stay far behind in this debate (P, 101n13).

But even if this priority were the case as far as metaphysical projections are concerned, such a priority would not affect the status of difference represented as (in Platonic terminology) a "greatest kind" existing in such a way that it would necessarily have to be the same as itself, i.e., retain the formal property of being self-identical.[42] Furthermore, additional concerns arise if we approach difference with specific textual data in hand and consider implications which follow when we examine in detail what differs from what in metaphysical contexts.

Normative Dimensions X is different from Y–the question is how the difference-factor functions when X and Y represent distinct kinds of realities. Here are five contexts for analysis, all involving Plato:

i. Two Forms (e.g., Beauty and Justice);
ii. A Form and a particular instance of that Form (Beauty and a beautiful entity);
iii. An intermediate and a Form (Eros as described in the *Symposium*, i.e., intermediate between a human being and the Form of, e.g., Beauty);
iv. Forms and "new tools" (the early theory of Forms in the *Phaedo* vs. the quaternary account of reality in a later dialogue, the *Philebus*–limit, unlimited, measure, mind).
v. Forms developed and described in distinct metaphysical systems (e.g., Plato and Aristotle).

This set illustrates the following: intra-system differences (i, ii, iii); alternative or distinct approaches within a given system (iv); and differences between two (or more) distinct metaphysical systems (v). On Derrida's own principles, all these types of difference must be preserved in order to reflect the spectrum of givenness that characterizes western metaphysics. Thus the greater the number of distinct kinds of elements in a given metaphysics, the greater the number of distinctions which must be conveyed by the notion of "different."

How then to understand this array of differences at an appropriate level of abstraction? The following discussion develops two approaches to difference; first, the description of contexts which depend for their reliability on a formal sense

42 For detailed discussion showing, against Derrida's claims, the priority of difference in the traditional sense, see DFL, Chapter 4, "*Différance* and Difference," 77-99. See also the Chapter "Derrida's *Différance* and Plato's Different" in Samuel Wheeler 2000, 231-48.

of difference; second, the statement of problematic implications from Derrida's theoretical preference of *différance* over difference.

Each element differs from the other element in proposition i. The same factor of difference holds for the other four propositions in this set. It could be argued then that there are different kinds of difference for each of the five propositions because the realities determined as different from one another in any one of the five propositions (e.g., Beauty and Justice–in i) are distinct from the realities determined as different from one another in the other four propositions (e.g., Beauty and a particular instance of Beauty–in ii). This approach to the content of the propositions suggests that difference itself exhibits a *qualitative dimension* depending on the metaphysical character of what is being differentiated from what. Briefly put, there are different kinds of difference. However, an alternative rendering is to follow the more generic Platonic approach and to take difference to mean simply "other than." Thus qualitative distinctions between X and Y in each of the five examples listed above depend entirely on the determinate character of the kind of beings instantiated as X and Y. From this perspective, the "is different from" relation would then be construed as purely formal and as functioning identically in each of the five propositions.

The possibility that difference should be construed from a normative perspective indicates an important feature of difference. Derrida emphasizes the fact that *différance* includes "subdeterminations" and "differences." And as we have seen, he insists that *différance* is more fundamental than difference. But *différance* as an integral component in Derrida's systemic thought includes both sameness and difference inextricably united with one another. Furthermore, if sameness yields the specific content of being as presence and if sameness governs all differences within any one given metaphysical system, then all the differences which animate the structure of a given system are necessarily related to sameness *within the limits of that system.* But it then follows that every individually determinate difference within that system is also related to and in fact constituted by being as presence. This inclusion must be the case since this system–indeed, any system–as an approach to the determination and resolution of questions about being is and will remain a single, unified expression of being as presence.

But if each determinate difference is also an element in the expression of being as presence, how can *différance* support and account for the normative distinctions represented by the list of examples given above?

Consider Beauty and Justice as different Forms. Derrida and the followers of deconstructionism would surely not deny this difference. But both Beauty and Justice, as Forms, also jointly exemplify being as presence. What then is the status of Beauty and Justice with respect to their instantiation of being as presence if Beauty and Justice fundamentally differ from one another? Thus if Beauty as such is indeed affected by its existence as an instance of being as presence and if Justice as such is also commensurately affected by its existence as an instance of Being as presence, then Beauty and Justice are indistinguishable from one another insofar as both Forms (distinct from one another as such, or so it was assumed) are both instances of being as presence. If the deconstructionist wants to maintain that (a) Beauty and Justice are different from one another as well as (b) they are equally affected by their instantiation of being as presence, then with respect to being as presence no distinction exists between Beauty and Justice. If this argument can be generalized, the deconstructionist loses *either* the uniqueness of diverse metaphysical entities within a given interpretive system *or* the uniform sameness putatively resident in being as presence.

Derrida could preserve his reading of the history of western metaphysics with respect to being as presence if he would allow being as presence to differ *qua presence* when it affects different elements in a given metaphysical system. Thus if the way Beauty came to presence differed with respect to presence from the way that Justice came to presence, then (a) the difference between the basic metaphysical elements of Beauty and Justice is preserved and (b) being as presence exhibits a commonality with respect to this system but does so in a way which respects the resident differences of the elements in that system. Now we have in fact seen Derrida appear to posit exactly this feature of being as presence (**Passage 1G**). According to the factor of accommodation described in this text, the way in which the Form Beauty *qua* an instance of being as presence exhibits presence is different from the way in which the Form Justice *qua* an instance of being as presence exhibits being as presence.

If it is assumed that being as presence can be analyzed in the abstract and that such analysis will not distort its uniqueness, then it follows from **Passage 1G** that being as presence *as such* admits internal differentiation. However, it is crucial to realize that this type of differentiation does not result from the directive agency of the sameness animating *différance*; rather, it originates from the unique character of a given Form in relation to another, distinct Form given that both Forms are instances of being as presence. Thus *the Forms themselves* generate the unique ways in which being as presence becomes differentiated within a given metaphysical system. The alternative explanation requires showing how being as presence has, by itself, the suppleness, range and guiding agency to incorporate a given Form (i.e., all the Forms!) in such a way as to preserve the character of being as presence as well as concurrently reflect the uniqueness of all specified elements within a given metaphysical system.

It is not evident that Derrida analyzed the implications of being as presence with respect to the preservation of differences to a point where this consideration emerged as relevant.

Particularity The status of particulars was raised in Chapter 1 in context of determining the appropriate way to account for particulars both as individual entities and as members of a plurality. Does being as presence permit an account derived from deconstructive principles and whatever gestures of particularity may be available if the latter represents a standard metaphysical issue and problematic area when characterizing the history of metaphysics?

If Derrida's thinking remained at least to some extent within the scope of being as presence when viewed from its totality-perspective, then an important conclusion may be drawn regarding this question. On Derrida's principles, it is not possible to *compare* different metaphysical systems from the perspective of normative adequacy, that is, whether of two given accounts of particularity, one of the two develops a more compelling description of particular entities than the other. All moments in the history of metaphysics are apparently of equal value with respect to their intended end as linguistic evocations of reality. Derrida gives no indication that any one metaphysical system reveals being as presence more or less adequately than any other system. In this respect, being as presence functions within Derrida's own project only as a principle of grouping; it gathers all the various systematic metaphysics under one descriptor, then interprets the results with respect to sameness and difference. Thus no one system is in any way privileged with respect to the stated ends of metaphysical thinking. As a result, an advocate of Derrida's texts on being as presence cannot select some one candidate from the history of metaphysics and then contend that its perspective on being as presence is "more penetrating" or "more informed" than any other.

Observe the following implication: If it is a common goal of metaphysics throughout its history to develop an articulated account concerning the existence and structure of particular entities and if any account of particular entities depends on the underlying vision of being which animates a given metaphysical system, then Derrida's emphasis on being as presence as the overriding principle that drives the history of metaphysics blocks endorsing as somehow privileged in terms of explanatory power *any* single account of particular entities insofar as they exist in relation to a given account of being. The advocate of Derrida cannot advance a position on particularity, at least, again, from within his or her target framework of scrutiny–metaphysics as a medley of accounts all of which are variations on being as presence. Particular entities exist, but *how* they exist relative to metaphysics can never be asserted since to do so would presuppose that one position in the history of metaphysics is preferable to all other such positions. Since all metaphysical systems lack any determinable sense of value with respect to explanatory succinctness and success, any system will do as far as accounting for the distinctive character of particular entities.

We have argued that analytical awareness cannot penetrate and identify what is common to being from within the "being as presence" framework as it extends throughout the history of metaphysics. But now we can appreciate that an equivalent opacity pertains to the character of individual beings, particular entities, insofar as their existence will always and necessarily be a function of the grounding vision of being which animates a given metaphysical system. Bluntly put, Derrida loses particularity within the historical sweep of being as presence. If particularity is taken as an unarguable given, then it follows that Derrida cannot offer his own metaphysics, or perhaps more appropriately, perspectival metaphysical vision, since presumably any configuration which Derrida would have advanced as an attempt to cover the same considerations as those addressed throughout the history of metaphysics will itself be conditioned–that is to say, tainted by–the lingering residues of that very history.

In fact, it is difficult to see how an advocate of Derrida's position can establish particularity in the path of being as presence since any reliance on language or concepts that were excessively affected by previous usage in the history of metaphysics would presumably infect the attempt to evoke particulars according to a fresh, liberated vision, one not subject to the repressive strictures derived from the history of metaphysical thought. The only effective alternative: a completely new vocabulary coined and developed to address metaphysical ends or, if that were not feasible because the advocate of Derrida would remain to some degree, perhaps to a considerable degree, entangled in the history of metaphysics, a vocabulary as distant in language and deployment as possible from the distorting effects of traditional concepts and terminology. We recall Derrida's warning early in *Of Grammatology* that philosophy, after deconstruction has had its say, may be "monstrous" in both form and content.[43]

At this point then, the relevant contexts of difference must include not only the senses in which difference obtains between particular entities but also how difference characterizes distinct accounts of particularity, as well as the difference between one systematic and comprehensive account of the structure of particular entities and another such account. In Part II, we will demonstrate how the theoretical loss of particularity as a consequence of implications derivable from his position on difference will indirectly ground one of Derrida's cardinal conclusions in his critique of standard concepts in the philosophical treatment of language.

3.7 Difference, *Différance*, Deconstruction

What is the difference between (a) traditional accounts of philosophical concepts and distinctions pertaining to language and (b) the descriptions of language following the agenda of deconstruction?

If these differences are epochal and significant, is it necessary to have conditions in place in order to establish the possibility of stating these differences? *Différance* has controlled the history of metaphysics. But does it follow that *différance* also affects, even if it does not control, deconstruction itself in its critical function when placed in juxtaposition with the history of metaphysics? I submit that it is necessary to determine how *différance* and all aspects and dimensions of the account of being as presence, which as such is a venture into pure metaphysics or at least metaphysics

43 Ch. 10 discusses the possibility of a new vocabulary to remedy the distortions of western metaphysics. Derrida comments often on the need for and possibility of producing new language and terminology for philosophical purposes: WD, 19-20, 28, 88, 240, 272; MP, 60-1, 134-5, 219, 259, 292; SP, 158; P, 111-2n44; Derrida 1987, 255; PS, 13, 22-3; Derrida 1992, 73-4. Rorty 1982 develops incisive comments in this regard, especially at 177.

when its assertions are accompanied necessarily with epistemological riders, affect the conclusions Derrida intends to draw about the elements of language.

The question can be framed as a set of alternatives: if *différance* continues its effects beyond the "closure" of metaphysics and penetrates even if only to an extent the articulation and application of deconstructive gestures, then both the formulation of these gestures apart from their application as well as the actual applications themselves will be determined by precisely the elements of western philosophical thought which deconstruction intends to reveal and reorient. But if we were to assume that difference as a greatest kind necessarily remains intact throughout, its residual governing force will sanction the possibility that Derrida can advance deconstruction as an ensemble of interpretive techniques without concern that *différance* will extend beyond the "closure" of metaphysics and affect whatever principles of organization and coherence characterize the formulation of deconstructive strategies. I state this issue in the summary statement of the dimensions of being as presence which have emerged so far in this essay; the issue proper–an interpretive crux–will be addressed in Chapter 8.

3.8 Being as Presence and Logic

We have detailed crucial implications regarding the structural necessity of logical concepts within Derrida's notion of sameness in the context of *différance*. But of course these concepts and related principles exist in their own right and with extensive application to professional activity of philosophers as well as to practical demands of the lifeworld.

What logical concepts are presupposed (and presumably necessary) in Derrida's formulation of being as presence as well as the theoretical gestures introduced through deconstructive avenues in order to characterize being as presence? If identity and contradiction depend on "formal" concepts, then what connection exists between this formality and being as presence? Consider this dilemma: if being as presence necessarily encompasses and perhaps distorts all concepts falling under its purview, then all logical principles either affect or skew whatever they pertain to; but if logical principles traditionally defined are necessary for intelligible discourse, then being as presence must apparently allow exceptions when it ostensibly pertains to elements underlying the concepts and principles of logic.

As so stated, this dilemma has considerable import however it may be resolved. Consider in this regard the principles of contradiction and identity.

3.8.1 *Contradiction* We must question the status of the principle of contradiction with respect to its functionality within the "organic" unity of the history of metaphysics. The history of metaphysics is not merely a series of discretely different systematic treatments of being. There is the obvious fact of historical influence, e.g., Aristotle's concepts of being were influenced considerably by Plato, Aristotle's immediate predecessor, and Plato's concepts of being helped determine the views of A. N. Whitehead over 2,000 years later. But the relevant feature regarding differences between and among metaphysical systems is not the fact or degree of influence characterizing this history; rather, it is that juxtaposition of certain pairs of systems produces contradictions. Therefore, if the history of metaphysics is an organic unity as Derrida avowed, then given the fundamental status of being as presence, it is ostensibly a unity countenancing all contradictions which arise when various incompatible metaphysical systems are juxtaposed with one another. Thus if system A and system B assert logically incompatible positions, then how should the application of the principle of contradiction emergent from this juxtaposition be interpreted with respect to the primordial character of being as presence? For example, would Derrida's appeal to a kind of pluri-dimensionality (as described in Chapter 1), which apparently transcends the temporality of simultaneity, obviate the generation of contradiction as a consequence of juxtaposing two (or more) logically incompatible metaphysical systems?

A formal mode statement of contradiction–"p.~p"–incorporates and depends on difference in the deployment of the principle's components. We may then readily pose the question that arises from the introduction of contradiction as a principle requiring the active agency of difference: Can the formal unity of logical concepts and principles be deconstructively analyzed and in some sense "shaken" or is the best we can do simply follow Aristotle and itemize the consequences which ensue if these principles are denied or even asserted to be capable of denial?

The problematic character of contradiction arises assuming two conditions: first, that self-identity obtains with regard to individual metaphysical systems and also, second and importantly, that difference is granted privileged status over *différance* in Derrida's technical sense. As noted, the principle of contradiction presupposes a purely formal configuration of difference. The juxtaposition of individual metaphysical systems configured in certain determinate ways (e.g., p.~p) will produce contradictions. However, the organic character of this phase of history is apparently unaffected by the fundamental cognitive dislocation produced, at least as traditionally understood since Aristotle's trenchant analysis in Chapters 3 and 4 of *Metaphysics* IV.[44] For if difference is theoretically dependent upon *différance*, i.e., if *différance* displaces difference, then it is open to the advocate of deconstruction to maintain that, at least with respect to the juxtaposition of distinct metaphysical systems, contradiction does not substantively apply. The reason: *différance* has been defined in such a way as to *accommodate* all differences. As a result, the negating gap that emerges so destructively between the "p" and the "~p" does not arise in such juxtapositions. By contrast, if the advocate of Derrida wishes to preserve contradiction

---

44 For discussion of divergent interpretations of Aristotle on contradiction, see Graham Priest 2006, 7-42.

as a purely regulatory principle even within the epochal gamut of western metaphysics controlled by being as presence, it will be necessary to juxtapose the formal realities constitutive of the principle of contradiction with the gestures of deconstruction and then safeguard the privileged status of the former in the face of the "deconstructing" purposes and applications of the latter.

As we have seen, Derrida emphasizes form as well as the intimate connections between form and being as presence. Therefore, insofar as difference as traditionally interpreted (following the Platonic designation of difference as a greatest kind) marks the structure of purely formal logical principles, e.g., contradiction, the question is whether deconstruction can reinterpret the notion of form so that the fundamental character of contradiction is obviated when applied to any segmented sequence within the history of metaphysics.[45]

Self-identity Derrida has stated axiomatically that identity is the "founding" principle of being as presence (**Passage 1L**). But as we will discuss in later chapters, the principle of identity (or self-identity, its more common formal name) and the identity conditions for a given entity are quite distinct logically. It is necessary therefore to distinguish between the predication of self-identity–e.g., the unity of the history of metaphysics insofar as it remains necessarily identical to itself as a certain narrative sweep–and self-identity as such, as a concept with necessarily binding logical implications for the structure of discourse and intelligibility.

If this distinction is allowed, then *self-identity* can be isolated as a second-order property, what logical theory often refers to as "formal" in character. When understood at this level of abstraction, this property is impervious to variance with respect to its pure formality (although self-identity can be predicated of many things and many different kinds of things–e.g., each system within the history of metaphysics) and as such must remain identical to itself. In other words, the logical conditions which underlie the possibility that *any one* metaphysical system can (and, for Derrida, must) be construed as an instance of being as presence will remain identical throughout the history of metaphysics in order to guarantee that *all* metaphysical systems can legitimately be so classified as instances of being as presence. But how should this

45 Derrida has a guarded respect for the application of contradiction in some areas–cf. LI, 127, 152, although he appeals to contradiction when it serves his interpretive ends, e.g., in the critical assessment of Foucault at WD, 40. Later, Derrida also writes: "The expression "to say madness itself" is self-contradictory" (WD, 43). In general for Derrida, "norms of minimal intelligibility are not absolute and ahistorical, but merely more stable than others" (LI, 147). Thus Derrida refers to "apparently contradictory gestures" in his own works; see Derrida 1992, 51. See also Kearney, 144-5; cf. P, 74-6 for observations on connections between the principle of contradiction and the history of philosophy. Contradiction is discussed below in Chapter 8, i.e., the section "Logic, Self-reference and the Rise of the Rhetorical." For additional discussion of the principle of contradiction from a more formal aspect, see DFL, 129-49, 161-76, 241-2n9. For discussion on Aristotle and Derrida concerning the limits of reason, see White 1992.

formal necessity be construed in light of being as presence? The reader will observe that this question mirrors in structure the one posed at the end of the discussion of the principle of contradiction.

Difference A final thought on difference. The principle of contradiction *qua* principle is a whole of parts and as suggested above, the ineluctable part of that principle is difference since the presentation of elements of that principle as distinct from one another presupposes effective formality of difference as an underlying structural and regulating concept. We therefore also assume that *difference* can be isolated as another second-order property, also impervious to variance when difference functions predicationally at a formal level. If self-identity and difference are granted this status, other realities of equivalent formality may also exist and function with equivalent necessity regarding the intelligibility of discourse as well as anything discourse referentially indicates. Again, the advocate of deconstruction must consider the effect or lack of effect of being as presence on difference as such in addition to the same consequences with regard to whatever other formal realities come into play in ordering significant discourse.

3.9 Summary and Provisional Conclusion

In this chapter, we have shown that Derrida's position on being as presence when articulated (to the extent that articulation appears open) and with due attention paid to consequent implications can be–perhaps must be–read as a relation of tangency between consciousness and a given account of being. Furthermore, if the texts introduced throughout Part I reflect the core of Derrida's thinking and the accompanying reasoning in the interpretation of these texts is sufficiently fair and rigorous, then it may be maintained with some force that "being as presence" becomes more of a slogan than an accurate, precise and informative watchword. When juxtaposed with the explicitly stated metaphysical systems which, for Derrida, constitute the source of its characterization, being as presence remains only a lingering invitation for additional analysis, a trope without ground and without cogent determination.

The argument in this essay has been that some aspect, phase or element of being in "being as presence" must be specified, whether it be temporal or non-temporal in character (or, perhaps, both intertwined in some way), in order that what has come to presence can be recognized as "the same" throughout the history of metaphysics. Without such specification, "being as presence" becomes, in a quasi-nominalist way, a name for a singularly abstract and undifferentiated formulation of reality. Although for Derrida "being as presence" is the only accurate summarization of the history of metaphysics, the phrase names, at least if construed as developed in Part I of this essay, at most a heuristic adventure akin to Aristotle's maxim that being can be said in many ways (cf. *Metaphysics* 1028^{a}10).

We conclude Part I: Derrida's axiomatic assertion that being as presence permeates the history of metaphysics–a history marked by wide divergence in doctrinal position–is insufficient by itself to justify that a student of philosophy in general and metaphysics in particular should approach every subsequent issue as if it were characterized at its base according to these theoretical surmises. The qualification "by itself" is crucial in this evaluative formulation; Part III will outline an approach to being as presence which in fact yields substantive results.

The array of problems and criticisms developed so far presupposes the continuing relevance of traditional concepts and lines of argument. If these problematic features are legitimate, they solidify and enhance the character and implications of being as presence as an intended contribution to the richly diverse tradition of western metaphysics. As such, these problems may or may not be resolvable, either by adroit application of preferred or sanctioned stances drawn from within that tradition or by resolutions employing strategies which are innovative and thus not reducible to any lingering vision from that history.

A second, related conclusion may also be noted as a point of departure for the investigation pursued below. The drive toward abstraction and a high level of generality codified by "being as presence" has in its discursive applications a direct bearing on the cogency and significance of Derrida's approach to language. In Part II, questions raised in Part I concerning the ground and justification for Derrida's notion of being as presence become crucial for assessing the implications of Derrida's deconstructive account of traditional concepts and distinctions attached to philosophical reflection aimed at revealing the structure of language. Thus if parallel and commensurately forceful criticisms obtain for his approaches to language–in addition to the set of concerns registered in the sections devoted to questions and quests in the present chapter–this cohesively structured and reasoned response to Derrida's fundamental thinking on being as presence will help direct the arguments developed in Parts II and III.

The ultimate conclusion will be that deconstruction itself to the extent that it depends on considerations pertaining to being as presence is weakened to the point of becoming untenable as a methodological option for pursuing philosophy. This is clearly a serious contention. However, whether students of Derrida will embrace or even just entertain this criticism remains to be seen since its actualization requires explicit arguments to establish its internal coherence and persuasive power. But now, at the conclusion of Part I, we can appreciate the centrality of being as presence not merely as a notion conceived in response to reading the history of philosophy but also as a theoretical filter of systemic dimension for the deconstructive description of the elements of language. The arguments developed in each of the four chapters of Part II extend and support this foundational dependency.

Part II: Being as Presence and the Aggregations of Language

4 Signification: Meaning and Referentiality

Being as presence represents an interpretive perspective on totality. Indeed for Derrida being as presence is constitutive of totality when totality is taken at the most fundamental level–"being as such"–at least with respect to the aspiration of metaphysics to encompass totality within its written accounts. But although being as presence is apparently intended to have precise denotative force, that is, to elicit a common factor running throughout the history of metaphysics, the argument so far in this essay has been that close reading of a series of thematically linked passages in which being as presence plays a crucial role indicates that this notion lacks specificity. We have concluded therefore that until being as presence representing both a straightforward instance of unifying nomenclature and a seminal element in Derrida's overall project can be supplied with sufficient content to justify ascribing this descriptor to the entire history of metaphysics, "being as presence" remains only a name without substantive import as well as appropriate evidentiary support to justify its uniform applicability to that history.

4.1 The Question of Language: Introduction

The analysis in Part I concentrated on examining being as presence primarily in terms of its status as a contribution to the history of metaphysical reflection. We justified this seemingly inapposite approach from various perspectives even if Derrida's intention in introducing being as presence was to set that entire history in relief and then to interrogate its hidden or suppressed moments. Furthermore, as noted at the conclusion of Chapter 3, this evaluative reading of being as presence has ramifications for Derrida's thinking extending far beyond the realm of metaphysics as traditionally characterized.

For Derrida, being as presence functions not simply as a foundation on which to pursue a corrective to what metaphysicians have overemphasized and overlooked throughout history but also, in a profound and incisive way, as a pivotal component in his philosophical approach to language. When we examine his texts synoptically, it becomes evident that Derrida moves freely from contexts in which being as presence is treated according to metaphysical concerns to contexts in which being as presence drives Derrida's thinking about language. The latter discussions describe how the apparent structural and theoretical components of language when viewed according to standard philosophical templates have been consistently distorted and thrown out of balance by the history of metaphysics. For Derrida, this branch of intellectual history is an extended exercise in protracted and undue emphasis on certain types of privileged concepts and their deployment. One of the most important such concepts is that of *opposition* and, more specifically, what for Derrida has been the traditionally sanctioned enshrinement of one opposite to the detriment of the other member of that pair. As we shall see, this oppositional imbalance has continued in traditional areas of thought devoted to the philosophy of language.

In this essay, we have separated being as presence into a predominantly metaphysical construct (Part I) and an essential component in the critical analysis of the theoretical aggregations of language (Part II). As noted in Chapter 1, this approach enables the student of being as presence to establish a more precise conceptual framework for informed appreciation and critical review. Furthermore, conclusions reached in Part I concerning basic metaphysical considerations pertaining to being as presence will stand as necessary parameters for guiding responses to Derrida's treatment of language as a series of phenomena all occurring within a field of presence.

The commentary in Part II combines critical observations on the cogency of Derrida's reasoning as well as questions inferred from this framework of criticism. The purpose of this diversified approach to Derrida on language is to elicit strains of thought embedded with sufficient if presently inchoate critical frameworks for more developed lines of argument to be pursued in future study. The questions and problems introduced in Pt. II, concentrating on Derrida's deconstruction of traditional concepts pertaining to language, will thus extend the ramifications of the metaphysical considerations regarding being as presence developed in Pt. I. The interlocking transcendental factors detailed in Chapter 3–relationality, intersectionality, directionality–contribute additional principles of organization for the critical discussions of Pt. II. This transcendental prism helps to stabilize and ground being as presence insofar as its core elements establish Derridean perspectives on language.

In Part I, we saw that the formulation and development of "being as presence" is problematic if the notion is critically assessed from suggestions derived from basic metaphysical positions.The fact that philosophy–including Derrida's–and the world generally remain within the epoch of metaphysics and its attendant implications, both theoretical and practical, justifies this interpretive approach. It would then be reasonable to expect that the extent to which being as presence informs Derrida's position on language will parallel, if not drive and animate, the extent to which that position also becomes problematic. Indeed, the ramifications from the range of difficulties described in Part I so permeate the substance of Derrida's approach to and conclusions about language that the entire project of deconstruction, at least as formulated on the basis of Derrida's texts, becomes an inherently dubious exercise. This strong conclusion is tenable if we interpret deconstruction as a position–or, less statically represented, as a set of gestures aimed at achieving related ends–characterized by theoretical clarity and overall coherence. Part II of this essay deploys a set of interpretations of Derrida's texts in conjunction with a reticulated series of arguments intended to justify this assessment.

The criticisms and commentary in the sequence of chapters constituting Pt. II gradually reveal a shift in Derrida's texts *from* a vigorous skepticism regarding the cogency of a number of pivotal concepts in the traditional delimitation of the structure of language *to* a palpable but subtly present array of generalized concepts and principles aimed at positive, substantive content. A primary goal of the four chapters of Part II is determining the contours of this shift, and discussing in a preliminary way how a student with multiple interests–Derrida as a thinker, the activity of language itself, the venerable sector of philosophy known as metaphysics–might react to these contours. To assist in achieving that goal, each chapter in Part II concludes with a summary recapitulating dominant features in the arguments of that chapter; these summaries also point to anticipations of arguments introduced and pursued at various junctures in Part III.

A final prefatory observation. The phases dividing investigation of language with respect to being as presence exhibit planes of increasing abstraction and generality, starting with the present chapter and proceeding to the final chapter of Part II. We develop two types of criticism: those based on concepts derived from systems which remain viable in the history of metaphysics and those generated by implicit acceptance of positions which, as stated in Derrida's texts, seem inconsistent with the development of being as presence. These criticisms are multiple and serious. The challenge to the advocate of Derrida's position is to show that texts on being as presence can be interpreted to address these criticisms and then adapted through argument so as to deflect their force.

4.2 Setting the (Linguistic) Scene....

The questions arising from the totality-aspect of Derrida's reading of the history of metaphysics are, as noted earlier, directly relevant to his theoretical position concerning the formative character of language. "Formative character" refers to the ensemble or aggregation (one of Derrida's preferred terms of referential unification) of terminology introduced to approach language in ways which, at least by intent, minimize the distorting effect of traditional concepts of linguistic analysis and open up the reality of language to more discriminating articulation.

The notion of *signification* serves as a locus for inquiry. A useful strategy for appreciating its meaning with regard to Derrida's overall agenda is to expand the term into a relation between *signifier* and *signified*. This pair of relata controls much of the discussion in this chapter as well as throughout Part II. We shall see that the subtle connection between signification and linguistic meaning gradually becomes evident as inquiry advances.

To initiate development of the scope of these concepts, it is useful to think of signifier as a word and signified as a concept. Thus for Derrida, one pivotal investigation is the relation between word and concept. Can concepts exist without being linked to words? A related concern is whether a concept, as signified, can exist by itself as transportable, i.e., from one speaker and cognitive agent to another. A third context of inquiry involves the various relations linking words, concepts, and extra-linguistic entities. It is evident that many crucial aspects pertaining to an informed philosophical grasp of language are in play within these inquiries.

In order to coordinate and react sympathetically yet critically to the disparate data discussed below, we introduce a setting, apparently straightforward in content and description. For Derrida, however, what underlies this surface simplicity is more complex than what might initially meet the philosophical eye. By design then, the contextual structure of this setting is intuitively obvious and paradigmatic of many linguistic events; in addition, the terminology used to describe its elements is traditional and readily applicable. This setting (or something similar) may be kept in view as a template to facilitate a critical response to Derrida's positions.

We name this miniature narrative "Forest Scene." Two language-users, A and B, are walking through a forest; A perceives one tree and says to an attentive B, "That tree is tall." In this scene, there is perception, a tree as a natural object, words, a meaningful proposition, concepts (existing in the mind of A and then apparently in reciprocal fashion in B), and communication (between A and B); we also have (a) referentiality–the word "tree" referring to the tree perceived by A and B, as well as (b) an instance of truth if the tree is indeed tall. For simplicity's sake, assume the correspondence theory of truth, i.e., a proposition, "That tree is tall," is true if it corresponds to a state of affairs, the extra-mental tree of a certain height standing in the forest. Signification as a technical term thus involves, minimally, words and concepts. However, the extent to which this relation between words and concepts is also theoretically connected to cognitive awareness of speakers of those words and objects referred to by propositions which speakers utter involves much of Derrida's deconstructive attention.

4.3 Language: Signification and the Possibility of Meaning

We begin with a brief statement of purpose followed by a lengthy, complex passage, both excerpts from the Introduction to Derrida's early work, *Speech and Phenomena*. The general context revolves around Husserl's phenomenology with particular emphasis on laying a foundation in order to

> **Passage 4A** ...see the phenomenological critique of metaphysics betray itself as a moment within the history of metaphysical assurance (SP, 5).

Despite Husserl's insistence to the contrary, his methodology is, Derrida insists, shot through with precisely the kind of metaphysical content–in particular, reliance on being as presence–which Husserl was at pains to circumvent. Whether Derrida's criticisms are legitimate is not the point–Husserl scholars will address this question if they deem it worthy[46]; relevant here is the theoretical basis Derrida posits to support those criticisms.

The more extended passage to be analyzed occurs while Derrida critically evaluates foundational assumptions in Husserl's *Logical Investigations*, in particular conditions necessary to establish the possibility of meaning in language:

> **Passage 4B** This nonworldliness is not another worldliness, this ideality is not an existent that has fallen from the sky; its origin will always be the possible repetition of a productive act. In order that the possibility of this repetition may be open, *ideally* to infinity, one ideal form must assure this unity of the *indefinite* and the *ideal*: this is the present, or rather the presence of the *living present*. The ultimate form of ideality, the ideality of ideality, that in which in the last instance one may anticipate or recall all repetition, is the *living presence*, the self-presence of transcendental life. Presence has always been and will always, forever, be the form in which, we can say apodictically, the infinite diversity of contents is produced. The opposition between form and matter–which inaugurates metaphysics–finds in the concrete ideality of the living present its ultimate and radical justification (SP, 6–italics in text).

This passage functions on two levels. First, Derrida initiates his critique of Husserl showing him to be a philosopher remaining squarely within the western metaphysical tradition; second, Derrida lays a foundation for his own substantive responses to issues Husserl poses and, at least by intent, resolves. This subtle expositional duality does not become evident until later in this chapter of *Speech and Phenomena*. We shall see, for example, that the pair of references to possibility at the outset of **Passage 4B** identifies an element which contributes essential features to Derrida's deconstruction of language as traditionally conceived (a revamped vision of language not reaching its complete realization until Derrida's explicit account of "iterability," which we discuss in Chapter 7). The commentary at this point addresses both levels of analysis, albeit in an introductory way.

4.4 The Transcendental Factor

For Derrida, the possibility of signification establishing meaning in language depends on "one ideal form" which assures the unity of the indefinite and the ideal. This "ideal form" is "the present, or rather the presence of the living present." The "presence of the living present" is referred to as "one ideal form" because presence "has always been and will always, forever, be the form in which, we can say apodictically, the infinite diversity of contents is produced." Derrida reads Husserl as sharing in a version of this "form" even after all the eidetic reductions Husserl proposes methodologically are fully in place; as a result, phenomenological analysis will eventually produce its

46 See Leonard Lawlor 2002 for a general review of Derrida's reading of Husserl. For a critical assessment of Derrida on Husserl, see J. Claude Evans 1991.

own versions of being as presence. Derrida's axiomatic claim about the omnipresence of presence indicates that he is not averse to making transcendental assertions of his own (even if the purport of his analysis at this juncture is a critique of the reliance on "transcendental" considerations in Husserl's approach to philosophical inquiry). We note also that Derrida's mention of the "origin" of ideality as involving "endless repetition" is a brief anticipation of one of his cardinal themes in the context of iterability (about which more later in this chapter). Any given instance of discourse must be capable of endless "possible repetition" (in, presumably, a wide variety of both present and future contexts) in order to be meaningful in any one present context.

Derrida is analyzing a complex situation–the origin of signification and, ultimately, the existence of meaning in language–by explicitly taking a step back from the phenomenon as such in order to state how the phenomenon is possible. This approach is decidedly transcendental in format, that is, Derrida is stating conditions which must be met in order for meaning in language to exist at all.

Subsequent elements in the passage reinforce the rightness of this reading: the description of the "ideal form" as "ideal" suggests conditions which govern the relevant sense of "form" according to strictures of completeness and uniformity. Derrida insists that "all" instances of repetition insofar as repetition applies to language are governed by what has been asserted here. Furthermore, Derrida contends in a claim of remarkable universality that "living presence" has "always been and will always, forever, be" the form in which meaning will occur. The temporal properties must be noted–the pivotal factor Derrida identifies, living presence, has always been–the present perfect indicating movement from the past, as far back as meaning has ever existed, into the present–and will always, forever, be the requisite formal channel for meaning. This projection into a future without even the possibility of an exclusion and without a potential temporal limit (i.e., "forever") of that futural temporality supports with the aid of a basic metaphysical component (time) Derrida's subsequent qualifier, "apodictically," with regard to the logical force of this claim.

We pose this question at the outset: what foundational perspective can assure Derrida that what he has just asserted is (a) possible to assert at all, given the limiting conditions which he is in the process of applying to language via the agendas of deconstruction and (b) justified as a transcendental account conferring apodictic necessity on the logical status of this claim? Even if Derrida's strategy in this work on Husserl is ultimately to dislodge as theoretically unreliable the reliance on such transcendental formality as espoused by *any* philosophical position (not just Husserl), it is a question whether Derrida's formulations of the necessary conditions for meaning and signification in language are accurately deployed and, if so, whether criticisms launched against these formulations are indeed justified. We will return to these questions later in the chapter (and evaluate the implicit structure of deconstruction as a unified approach to determining the nature of language, both as such and as interpreted in texts, in Chapter 8). But the immediate insertion of "presence" as the key notion in the opening statement of Derrida's critique of Husserl adds to the importance of clarifying just what exactly presence means, both when applied to Husserl as one example and to western metaphysics in general.[47]

4.5 Presence: Evidence and "the other"

The transcendental modality of this passage sets in relief the interpretive necessity to discuss and analyze key elements in this passage. Derrida's appeal to "possible repetition" is crucial to his position and this notion, especially with respect to possibility, is sufficiently intricate to be treated by itself (we shall do so in Ch. 7, in association with Derrida's concept of iterability). For now, we develop the usage of "presence."

First, the appeal to "living presence" is not supported with argument or evidence of any sort–Derrida merely asserts the pivotal connection between the cardinal notion of presence, the submerged hallmark of western metaphysics, and establishing the possibility of signification and meaning in language. Philosophers are always free to make such speculative claims; the point is only that the reader of this philosophizing should appreciate the central function of presence in contexts characterized by language. We see here blunt evidence of the fact that Derrida's interpretation of the history of metaphysics interplays with his thoughts on language to such an extent that this text in *Speech and Phenomena* stands as the Derridean equivalent of a "self-evident" truth.

Second, it seems evident that presence in this sense presupposes that *something* must be present (i.e., "something" understood in a very broad sense). This relation would link (a) language as a phenomenon with its own self-containing limits whether as deployed in spoken or written discourse and (b) something other than language, whether that something exists in the arena of cognition or "outside" the events of language as such, whenever these events occur or, indeed, do not occur. Depending on how this "otherness" is conceived and then integrated into traditional categories and modes of thought, it is possible that the neat, sharply separated elements laid out in the Forest Scene above may all require rethinking and redeployment.[48]

---

47 See sources cited in Ch3n34 on the embedded transcendental dimension in Derrida's approach to the concepts in question. Additional strains of a transcendental character will emerge as Derrida develops his account of language.

48 The notion of "otherness" and appeals to "the other" appear variously in Derrida's earlier writings: Derrida 1992, 66; Kearney, 140, 149; P, 45; PS 15, 39. For commentary on the other in Derrida, see J. Hillis Miller 2002, 327 and Simon Critchley 1992, 29.

4.6 Presence and Opposition

It may initially be inferred that "living present" cannot be reduced to either form or matter taken individually and separately, since the claim in **Passage 4B** is that "living present" justifies, radically and ultimately, the *opposition* between form and matter. Is it possible then to describe any of the properties of "living presence" by itself, apart from whatever its object may be?

Analysis will indicate that living presence assumes definable shape. For Derrida, repetition which grounds the possibility of meaningful discourse generates instances of meaning in which "one may anticipate or recall all repetition." Now if anticipation and recollection are understood in their usual senses, then living presence is a *certain kind* of living presence, i.e., that found in human beings, as long as we assume that only human beings can fully experience anticipation and recollection with respect to language. If the claim that the "infinite diversity of contents is revealed in presence" referred to the fact that whatever can be said or written is infinitely diverse when such discourse emerges on a continuum of instances of possible language, then despite this unlimited range, we may safely assume that presence will be the necessary condition for the existence and subsequent revelation of this content, whatever it might be, as well as its actualized elevation into the realms of signification and meaning.

Derrida contends that the "opposition between form and matter" marks the starting point of metaphysics and also that this inauguration receives its "ultimate and radical justification" in the "concrete ideality of the living present." The use of "metaphysics" must refer to that branch of philosophy bearing this name, since "form and matter" are decidedly abstract concepts and as such constitute one of metaphysics' foundational components. But it is crucial to recognize that it is the "opposition" between them, i.e., the fact that form and matter *differ* from one another and also that this difference is qualitatively such that they become *opposed* to one another, which for Derrida marks the onset of metaphysics. In sum, it is not merely the introduction of the notions of "matter" and "form," whether taken singly or as a simple conjunction (as, say, in the specific metaphysical positions of a philosopher such as Aristotle), but rather the opposition obtaining between these notions that generates the epoch of metaphysics. Although "matter" and "form" are recognizably Husserlian concepts, I submit that at this juncture in **Passage 4B**, Derrida clearly has in view the prevalence of this dyad throughout the history of metaphysics.[49]

If the "opposition" between form and matter receives its "ultimate and radical justification" in the dimension of reality which Derrida names "the living present," then

49 It would be both necessary and instructive to juxtapose Husserl's matter/form interplay with matter and form in Aristotle, then attempt to discern the "sameness" which both share as exemplars of being as presence. The same kind of juxtaposition would apply to any thinker who employed the same pair of fundamental metaphysical notions.

two additional inferences may be drawn: first, if the "living present" is equivalent to "being as presence," then the only relevant sense of "living" is human life. For if "form and matter" are taken as concepts, then living presence must include or be equivalent to consciousness in which such concepts can arise in the present and be studied in the present. Even if "form and matter" are read as neutral referential counters, the fact that living presence is the source for the justification of metaphysics must mean that "living" refers to human beings capable of acknowledging this difference, even if they are unaware of the formal terminology philosophers often employ for expressing it. Furthermore, the claim that repetition emerges either in recollection or expectation (or necessarily in both forms of awareness) must also refer to human consciousness as the intended referent of the appeal to "living" as the qualifier of presence.

Finally, although presence in the epochally defining phrase "being as presence" remains ambivalent as far as precise specification is concerned, the necessary addition of a human factor clarifies the ultimate metaphysical context in which being as presence plays itself out, so to speak. This interpretation also reinforces the place of consciousness as an essential factor in being as presence. In fact, this reading, all but conclusively asserted in the progression of interpretation in Chapters 2 and 3 above, may now be taken as fully established–being as presence is not merely a claim about being as such, i.e., apart from consciousness of being. Rather, we will assume throughout the rest of this essay that "being as presence" means "*the relation between* (a) consciousness and (b) what is other than consciousness." But even with this clarification, identifying and describing the sameness which generates the definitive epochal reality of being as presence remains as much a question as it did throughout the investigations pursued in Part I.

4.7 Presence and the Origin of Metaphysics

Derrida's primary concern in this context is twofold: identifying the fact of opposition as the underlying support for the distinction between matter and form as well as asserting the claim that this opposition contains the origin of metaphysics, presumably *all* metaphysics. Derrida is not concerned with the evidentiary consequences if matter is kept distinct from form and questions are directed at matter as an object of philosophical concern and then form as another, separate object of such concern. Metaphysics for Derrida is characterized at the most fundamental level genetically, as ordered inquiry revolving around opposition. One could then argue from this passage that Derrida's position implies that without the possibility of deploying distinct phenomena according to the logical schematic of opposition, metaphysics as practiced for nearly three millennia would never have existed at all.

4.8 Opposition, Language, Signification

If "living presence" refers to being as presence inclusive of consciousness, in particular human consciousness, then the matter/form distinction is the primordial unity of the subject-object dichotomy–that is, speaker and content of what the speaker has uttered all taken as an organic whole, a unity. How then is the opposition between form and matter–which, for Derrida, "inaugurates metaphysics"–justified in this capacity insofar as its "concrete ideality" is found in the "living present"?

The answer, I submit, is that "matter and form" refer immediately to a purely *linguistic* context–i.e., matter is the concrete instance of discourse which can take "indefinite" guises and modes of expression, form is the "ideality" extending to infinity within which sectors of possibility this particular instance of discourse–and by extension any and every instance of discourse–must be capable of being uttered in order to be meaningful on this or any particular, concrete occasion. Thus the initial event of meaningful discourse is paradigmatic for the possibility and onset of metaphysical thought. Such thought, regardless of its specific content within a given system, must be construed as basically an ongoing interplay of matter and form, where "matter" refers to the subject of a particular linguistic venture, whether its content be metaphysical or immediately and bluntly pragmatic, and "form" refers to the implicit appeal that this matter will–or could (we detect here a presentiment of Derrida's notion of iterability)–pertain to and occupy the full extent of future temporality. The elements in the Forest Scene, especially those with manifest metaphysical provenance, must be filtered through "living presence" in order to appreciate how, after appropriate reexamination, they might exemplify a less encrusted level of bias imposed by the weight of traditional metaphysical reflection.

4.9 The Unity of Language

The previous sections indicated that many disparate elements constitute Derrida's approach to the possibility of signification as it pertains to the realization of meaning in language. This multiplicity of content makes it all the more essential to appreciate the range of applicability Derrida ascribes to the description of signification:

> **Passage 4C** The unity of ordinary language (or the language of traditional metaphysics) and the language of phenomenology is never broken in spite of the precautions, the "brackets," the renovations or innovations (SP, 8).

It is evident that at this point, Derrida is writing *about* Husserl's position although he is doing so in a mode of general acceptance of at least a segment of the claim. Once this allied approach is established, Derrida's appeal to unity in this passage becomes dramatic and rich with import. Consider the claim's logical structure. Derrida asserts a disjunction between "ordinary language" and "the language of traditional metaphysics." He then says that *all* instances covered by this disjunction–i.e., any particular occurrence of ordinary language regardless of extenuating circumstances pertaining to context as well as any instance of traditional metaphysics, again regardless of major or peripheral differences characterizing individual systems–can be considered as a single phenomenon, as a "unity." The objection might be raised, following the strictures of the later Wittgenstein considering monolithic theoretical claims about language, that even if ordinary language is deemed sufficiently "family-like" in its manifest diversity to be considered as one unique kind of linguistic phenomenon, the language of traditional metaphysics is commonly taken to be a very different type of linguistic activity.[50] Furthermore, a strict Husserlian would insist that to follow phenomenological practice according to Husserl's own specifications is to achieve a level of discourse far different from either ordinary language or the utterances of any of the previous figures in the history of metaphysics. But for Derrida in this passage, all such type-distinctions–and presumably any other which a discriminating philosopher of language may wish to draw–vanish into an unbroken "unity."

This is the second appeal to the feasibility and structural necessity of unity in fundamental contexts. In **Passage 4B**, Derrida refers to the "one ideal form" which "must assure this unity of the *indefinite* and the *ideal....*" Derrida remains within predominantly Husserlian contexts at this point but the use of unity is crucial since it serves to combine the two basic components which, as a unity, ground the possibility of signification in language. Derrida has severe reservations about indiscriminate applications of the notion of unity when it is a matter of deploying the elements of his own systematic approach to the unveiling of being in relation to the history of metaphysics (see Chapter 6 for discussion). And as we have seen (**Passage 3B**), it is a fundamental error to level out the history of metaphysics in order to seek and then articulate a common, unified transcription evocative of being as presence. Apparently no "one" term can capture the multifold character of being as presence because

50 For parallel passages combining the import of ordinary language with technical philosophical language, e.g., WD, 292; MP, 121n15, 292; LI, 89-90; PS, 11-2; Derrida 2001, 18n19. At WD, 284, Derrida says: "...language bears within itself the necessity of its own critique"–an axiom presumably applying to all language, whether "ordinary" or specialized for philosophical purposes. Thus Derrida is unimpressed with predominantly analytic approaches, e.g., the mention/use distinction: PS, 11-2, LI, 32, 79, 82, 98; for a similar reference to the type/token distinction, see LI, 52. For commentary on the status of ordinary language from Derrida's perspective, see Harvey, 110; Hans-Georg Gadamer 1989, 95, 99; Habermas 197; Norris 2012, 219n15. Also useful in this context is Lynne Rudder Baker's *Metaphysics of Everyday Life: An Essay in Practical Realism*; as its name suggests, this work analyzes the immediacy linking abstract theoretical concerns and the nuts-and-bolts of the world everyone–even the most speculatively resplendent philosophers–resides in most of the time (see Bibliography). And for discussion of potential application of deconstruction to ordinary language fixed within legal contexts, see Drusilla Cornell 1997 and William M. Sokoloff 2005.

unity as an intended second-order concept will necessarily falsify its intended object at this theoretical level.

But Derrida shows no hesitation whatsoever in construing language, both in its ordinary settings and as articulated according to the wide-ranging technical vocabulary proper to the history of metaphysics, as embodying a "unity" when interpretive strategy requires confronting and reducing Husserlian projects so they can relevantly (and conveniently) fit a deconstructive agenda. Nor is the introduction of unity theoretically inappropriate when it is a matter of articulating the "unity" of necessary elements in a transcendental account of signification. We recall that since the historical Parmenides' extant poem and Plato's own *Parmenides*, the notion of unity, or "the one," has enjoyed classic status at the outset and throughout the history of metaphysics (e.g., as a category for Kant's analysis of experience). We must wonder then whether this joint rejection and endorsement of unity as a procedural operator functioning as an integral element in fundamental contexts can be allowed, or whether its occurrence in Derrida's seminal texts should be questioned as an instance of radical inconsistency (more on the status of unity as an abstract element in Derrida's position in Chapters 5 and 8). Simply put, to what extent does Derrida presuppose and exploit dimensions of unity in his own writings?

That question aside for now, the drive toward unification and the resultant collapse of distinctions based on differences within the domain of linguistic phenomena mirror the parallel dissolution of differences within the epochal manifestations of being as presence via the subtle agency of *différance* (as discussed in Part I). And what ultimately confers that unity is almost immediately identified. Although the context is Derrida-on-Husserl, we shall see that Derrida's claim continues to range far beyond the doctrinal limits of just that one thinker.

4.10 Language, Logic, Presence

The next phase in the analysis of language from deconstructive perspectives extends and concretizes language in the broad sense which characterizes Derrida's approach:

> **Passage 4D** [This is due to the fact that]...being interested in language only within the compass of rationality, determining the logos from logic, Husserl had, in a most traditional manner, determined the essence of language by taking the logical as its telos or norm. That this telos is that of being as presence is what we here wish to suggest (SP, 8).

Derrida's claim that Husserl determined the "essence of language" in "a most traditional manner" suggests, by its patronizing tone, that Husserl was ineluctably caught in the wake of being as presence even if he intended his thought to "bracket" all metaphysical considerations in the development of a method for eliciting eidetic natures. For Derrida, Husserl's "taking the logical" as "telos or norm" of language is nothing more than relocating this telos under the aegis of "being as presence," effectively reducing Husserl to just another moment in the history of western metaphysics defined in its totality by being as presence. The fact that Husserl added a considerable amount of speculative thinking and content to the purely "formal" character of "traditional" logic is, for Derrida, apparently irrelevant. Since all this content is subsumed under the standard rubrics of logical concepts and principles, all of it is as metaphysical as anything which preceded Husserl's work and its intended strategy.

It should be noted that "logical" is ambivalent: Is Derrida referring to the formal ordering of sequences of propositions–e.g., *modus ponens* or the principle of contradiction–or does he intend a more inclusive sense incorporating the dimensions of logic but also additional characteristics covering modes of behavior in achieving ends (where "logical" could refer to the word used in propositions of the sort, "if you want to be happy, it is logical to act this way")? If we recall Derrida's dictum that the principle of identity is the fundamental origin of being as presence (**Passage 1L**), then this supports taking the reference to "logical" in **Passage 4D** more, perhaps exclusively, in the formal rather than the rational, i.e., content-driven sense; however, the ambivalence stands.[51]

This inference has crucial consequences not just for Husserl as one philosopher among many but for all philosophers, at least to the extent that they have couched their thoughts according to the controlling concepts and principles of formal logic. For it may be inferred that anything logical as Derrida views logic is equally determined by being as presence. This is the broader and more telling consideration advanced in this passage, since what phenomenology in Husserl's extensive work could accomplish as far as revealing eidetic essences is concerned remains within the scope of Husserl's *ouevre.* In this respect, Husserl is a philosopher and, if Derrida is correct, as such he remains ultimately just another metaphysician working within the tradition regardless what he, Husserl, maintained he was doing in the face of that tradition.

Furthermore, if logic (in the sense described above) is just another version of being as presence, then it follows that *all* philosophers who coordinate their thinking according to the rules of rationality exhibited by the principles of formal logic–regardless what philosophical area this thinking occupies–are beholden to the limiting pressures of being as presence simply by virtue of the fact that their thought has been couched and advanced according to the formally generated strictures of logic. This conclusion encompasses virtually the entire history of philosophy as well as the evolution of all its various sub-disciplines and specialities, not just the efforts of philosophers to account for the nature and structure of reality through metaphysical

---

51 For a vigorous example of a contemporary version of the distinction between being logical and being rational, see Dennett, 76-7, 263-70.

modes of thought. Furthermore, although nothing obviously philosophical is transpiring in the Forest Scene, we must question whether this implication concerning philosophy and logic will also extend to the lifeworld and its resident instances of language. For if A had said to B "That tree is tall and that tree is not tall," how would B have reacted?

If we accept this reading and implications from Derrida's claims concerning Husserl and logic, then the relevant question given the context at hand is straightforward: what is the connection between logic, especially when expressed in purely formal terminology, and being as presence? If philosophical discourse is "logical" in satisfying formal requirements for intelligibility, does its content necessarily becomes falsified and distorted precisely because of the procedural concern to establish such intelligibility on behalf of its intended audience? And if the "logical presentation" of a philosopher's thoughts is retained (for whatever reason), then is there a way to coordinate the restrictive shells of logical concepts and principles with novelty and freshness of insight so that the results speak to the concerns of contemporary philosophers? These are serious questions regarding the implications derivable from Derrida's position. More will be offered in this regard below (Chapters 9 and 10).

4.11 Reasoning to the Roots of Language

The limits of logic in relation to language as discourse raise the question of the limits of language itself. In this regard, one of Derrida's most provocative dicta is often rendered: "There is nothing outside the text."[52] Whatever "text" means here

52 This formulation appears in OG, 158. Cf. P, 59; LI, 136-7, 148, 152-3. The typical translation of Derrida's "*il n'y a pas de hors-texte*" as "nothing exists outside the text" has been challenged. Alternate translations and commentary appear in Hill, 46 ("nothing is extra-textual" or "text knows no bounds"); see also Culler, 105; and Stocker, 29. See also Norris 2012, 145; for Norris 2012, 87, the trope means that "...contexts must be thought of as *jointly and inseparably* textual-linguistic-discursive on the one hand and material-concrete-experiential on the other" (italics in text). We shall see that this reading has considerable merit, although what "material" and "concrete" signify here must be discussed. Also, Norris' appeal to the apparently self-evident relevance of "contexts" in this regard should be noted (see Chapter 6). In a related vein, consider Stocker's interpretation of the relation involving writing, language and consciousness, 74: "...what must apply to the structural necessities of writing must apply to all language, and all consciousness. There is no consciousness before writing for Derrida, in the sense that the structural necessities of writing (repeatability, repetition of an absent origin, contextuality and indeterminacy of meaning) are all in consciousness. It is a necessary strategy to define consciousness in terms of writing rather than the other way round". Daniel Dennett, working in a different tradition and presumably operating with a different set of assumptions, also insists on the need for language, whether spoken or written, as a necessary condition for consciousness. See Dennett's discussion in *Consciousness Explained*, i.e., the section "Words, Pictures and Thoughts," 297-303, esp. 300-2; also relevant are 416-8.

(and the meaning is subtle), it surely must include the ordinary connotation of "text," i.e., words, language, and probably written marks on a page, assuming that "text" would not typically apply to discourse if it was only spoken. And if "text" meant a book or at least something written, the Forest Scene as described above exists without any text whatsoever! Thus the meanings of "text" as well as the spatial (taken in a broad sense) indicator "outside [of]" control the cogency of this claim; and its seemingly paradoxical excess receives clarification and support from a passage early in *On Grammatology*.

> **Passage 4E** For some time now, as a matter of fact, here and there, by a gesture and for motives that are profoundly necessary, whose degradation is easier to denounce than it is to disclose their origin, one says "language" for action, movement, thought, reflection, consciousness, unconsciousness, experience, affectivity, etc. Now we tend to say "writing" for all that and more: to designate not only the physical gestures of literal pictographic or ideographic inscription, but also the totality of what makes it possible; and also, beyond the signifying face, the signified face itself (OG, 9).

This passage is crucial for Derrida's overall views regarding language as well as the implications derived from the supposed privileging of spoken over written discourse, a state of affairs which, since spoken language always occurs in the present, goes far in establishing the "existential" origin for the predominance of "being as presence" throughout the history of metaphysics. We may therefore justifiably examine the passage with care, especially regarding the shape of its reasoning. Consider first the section of the passage concerning language:

4.11.1 Language–Questions and Observations

Derrida claims that "for some time now" people are saying "language" to cover a multitude of apparently non-linguistic phenomena. First, what evidence is cited to support this empirical claim? Which people, where and in what contexts do they say these things? Second, why should we accept that the motivation for this extended use of the word is "profoundly necessary"? Why not instead merely view this broadened usage–if indeed it is a fact–as one way popular discourse is affected with faddish turns of phrase inscribed with a finite life span? Third, even if it were true that the motivation was in fact profoundly necessary, would it not be possible for this motivation to change, losing its necessity and, along with this modality, the empirical fact that the word "language" was being used in this extended way? Derrida has presumably addressed the point concerning the potentially transitory character of this mode of speaking with the qualifier "profoundly," but no evidence is offered to explain why the necessity in question exhibits this status.

Finally, assume everything Derrida asserts is true–all he refers to is how people are talking, i.e., in particular the way in which they are using a word. What inferences of any sort, much less inferences of cardinal significance, can be drawn from the way in which people talk? Even if "language" was being used to describe, for example, perception as a function of consciousness, would ascription of this word affect either the psychological or philosophical analysis of perception as modes of consciousness distinct from, say, imagination? If so, the connection must be established by argument and evidence, not merely imputed on the basis of a linguistic change of habit. In sum, Derrida has shifted *from* the way people talk, presumably in popular or non-philosophical secular contexts, *to* conclusions applicable and tenable within highly theoretical and crucial philosophical settings.

4.11.2 Writing–Questions and Observations

Derrida on writing parallels Derrida on language, but with even more extended scope. Thus he claims that "we tend to say "writing" for all that and more." Again, who are the "we" in question and where does Derrida derive his evidence for this broadly empirical assertion? Be that as it may, "writing" now replaces "language" for all extended references cited in Section A above, with the addition–and this new point is crucial–"the totality of what makes it [language] possible," including not only the "signifying face" but the "signified face itself." When Derrida refers to "the totality" of what makes it possible, including within that totality both "signifying face" as well as "signified face itself," I take the referents of "totality" to be not only all the mechanisms and conventions of language but also all extra-linguistic objects (taking "object" here in the widest possible sense, such that even a concept is an object). Only if such objects exist "outside of" or at least "in relation to" language would there by any motivation or provocation on the part of a human speaker to vocalize a reaction to the "faces" such entities display to their observers. Furthermore, the modal qualification "what makes it all possible" testifies to Derrida's continued interest in pursuing an account of signification at a transcendental level of analysis. The question then becomes whether we can identify everything transcendentally necessary which should be included under the heading "what makes it all possible" (on which more presently).

If this reading is sound, then the implications for this phase of Derrida's position are crucial. To state only the most obvious question: at least according to traditional formulations, language about consciousness differs from consciousness and writing about consciousness also differs from consciousness. If Derrida wants to claim that "writing" is the word now used to refer to consciousness, then the advocate of Derrida's thinking must explain what is common to writing in the narrower sense of "pictographic or ideographic inscription" and writing in the more inclusive sense Derrida intends. After all, a written account of memory differs in many respects from the object of that account when memory actually functions as a psychological process.

To assert that the objects of writing, that is, what the writing is about or what it refers to, "are" writing must therefore be clarified. Derrida has in fact introduced what appears to be a stipulative definition for writing in order to facilitate (a) a theory about language which will include (b) an assertion of normative priority between two expressions of language, speech and writing. He is within his rights to make this move in the sense that nothing logically precludes it, but he also faces a host of obvious questions in order to substantiate the relevance and potential incisiveness of such a definitional gambit.

This point is especially important when Derrida insists that the totality in question includes both the signifying face and the signified face. Note that both end points of the contrast introduced within signification are said to belong to writing; also, if "signified face" refers to a concept or some sort of mentalistic carrier, then this element is also subsumed under the prescriptive use of the word "writing." Furthermore, it becomes vital to be as clear as possible on whether "signified face" in this context also encompasses a referent. This extension of the scope of the phrase in question is warranted on the grounds that both words and concepts are not self-referential, i.e., they are in fact *about* something other than themselves. Thus any extra-linguistic object–or, rather, any putatively extra-linguistic object–has now also been situated under the scope of writing. At this juncture then, we recognize a foundational move for asserting that "nothing exists outside the text." For if all objects referred to become instances of writing (in this new, stipulative sense of the word), and if texts (used in the usual sense) are composed of writing (in both the usual sense as well as the new sense), then it follows that nothing exists outside the text since nothing (i.e., supposedly extra-linguistic objects) can exist apart from writing, at least in the sense that extra-linguistic objects necessarily exist in relation to the capacity to experience those objects and to transcribe that experience in language. Although nothing "textual" appears in the Forest Scene as described at the outset of this chapter, *everything* in that narrative falls under "text" in Derrida's stipulative sense of the term.

It is now possible to discern a strategy emerging from (a) the close juxtaposition of speaking and writing and (b) the insistence that the history of metaphysics is nothing but theme and variations on being as presence. Speaking has been privileged over writing; Derrida insists on emphasizing this privilege and correcting it. The reason for this elevated status is that speaking is, by necessity, a phenomenon defined as presence since it occurs only in the present. Writing is "absent" or at least not present or capable of becoming present in the same way as the spoken word. Therefore, the written word must be investigated using new methodologies and suspending traditionally privileged dichotomies since the written word may have–and, for Derrida, will in fact have–hidden or suppressed residues of significance not readily available, indeed perhaps not available at all, through the reliance on spoken discourse. The history of metaphysics with its excessive emphasis on what is present has been complicit with the millennial bias in favor of the presence of speech and against the non-presence of writing. But now we know that writing is, as Derrida has asserted at the outset of *Of Grammatology*, in fact a synonym for "objective" or referential reality insofar as it becomes accessible within language.

As noted, one of the most dominant and controlling configurations in the depiction and schematization of reality is opposition. Human beings appear to have a natural predilection to assimilate the complexity of what is "out there" in terms of the conceptual neatness resident in opposites. Therefore, the more that oppositions such as matter/form–a relation taken as an exemplar of solidified written discourse–are inverted (or, indeed, rejected altogether) and shaken to their roots, the more that the concerned investigator will uncover facets or dimensions of reality currently concealed by a long tradition of privileging spoken discourse and ignoring what has been hidden in the "deep" structure of the written words of western philosophers, in particular the unconsidered reliance which the tradition has placed on concepts arranged according to the structural characteristics of opposition.

If Derrida's project of desconstruction in this regard is so deployed, then the premise extending the import of writing becomes crucial to the cogency of that project. For if writing and whatever writing is about are in fact at some level of analysis indistinguishable, then it follows that the more that analysis reveals about what has been written, the more that is known concerning whatever that writing has been about. In general, the more that metaphysics is "shaken," the greater the disturbance in the morphological considerations attributed to the ultimate referents of the language of metaphysics. If we shake the language of metaphysics, then we shake the natures of whatever the language of metaphysics has analyzed and described–i.e., the "signified face" of any and all entities cited and discussed by philosophers. After all, "what makes it possible" for language to assume a metaphysical stature and guise includes the fundamental assumption that extra-linguistic beings exist in such a way that language can surround them, so to speak, and evoke them in and through forms of discourse.

At this point, we have argued only that the evidence Derrida provides in *Of Grammatology* for this extension of significance is hardly dispositive. But even if the premise involving the radical extension of the import of writing is granted, the problems which arose in the examination of being as presence developed in Part I will reappear in Part II through the critical scrutiny aimed at the ways Derrida described the theoretical components of language and their interaction with being as presence.

The passage from *Of Grammatology* analyzed in this section establishes by concerted, compressed argument a remarkable correlation between the language people use for various purposes–philosophical and otherwise–and the extra-linguistic world to which this language refers. Derrida's text establishes an intimate, even essential correlation between writing and what writing is about–virtually everything that a human being can experience; note Derrida's "for all that *and more*"

(**Passage 4E**–my italics)[53] The theoretical "distance" between word and object has been foreshortened if not severely truncated; indeed for Derrida's deconstructive ends, this distance is almost on the verge of being obliterated altogether. But a word of caution: it then seems to follow that the very attempt to employ language in order to critique language might appear, given such preconditions, initially to be an endeavor which collapses upon itself at the very moment such an investigation is initiated.

4.12 Sign: Self-reference and the Signified

Derrida has broadened the working notion of language to where it becomes virtually co-extensive with (as traditionally stated) extra-linguistic reality. The Forest Scene thus comprises a text to be read and studied anew, shaking all standard concepts and distinctions animating its theoretical description. With this extension in hand, we may elicit additional implications regarding the conceptual underpinnings for deconstructive reading of texts.

As we have seen in Part I (Chapters 1 and 2), Derrida introduces being as presence to set in motion a foundational device which, inserted within the history of metaphysics, disrupts the smooth flow of that history by dislodging various fixed certainties, including an array of concepts pertaining to the apparent structure of language. However, Derrida also recognized that the very attempt to formulate such a radical revisionist approach to a firmly embedded system of texts and habits of thinking through and about those texts raises important questions originating from the phenomenon of self-reference–a thoroughly metaphysical notion as old as Plato. This passage from "Structure, Sign and Play," an essay published in 1967, sounds a number of the key themes:

> **Passage 4F** There is no sense in doing without the concepts of metaphysics in order to shake metaphysics. We have no language–no syntax and no lexicon–which is foreign to this history; we can pronounce not a single destructive proposition which has not already had to slip into the form, the logic, and the implicit postulations of precisely what it seeks to contest. To take one example from many: the metaphysics of presence is shaken with the help of the concept of *sign*. But...as soon as one seeks to demonstrate in this way that there is no transcendental or privileged signified and that the domain or play of signification henceforth has no limit, one must reject even the concept and word "sign" itself–which is precisely what cannot be done. For the signification "sign" has always been understood and determined, in its meaning, as sign-of, a signifier referring to a signified, a signifier different from its signified. If one erases the radical difference between signifier and signified, it is the word "signifier" itself which must be abandoned as a metaphysical concept (WD, 280-1–italics in text).

---

53 Derrida uses the term "arche-writing" to encapsulate the principal elements included referentially by "nothing exists outside the text." See OG, 60-1; P, 59-60 (on the related notion of a "general text"). For discussion, see Culler, 102; Gasché, 213-6; Caputo 1989, 263; Gadamer, 180; Habermas, 205.

Our present concern is an evaluative consideration of Derrida's treatment of the concept of "sign" and what that treatment implies about the relation between signifier and signified insofar as this relation is essential to the notion of signification.

Derrida asserts axiomatically that we have no language which can avoid functioning precisely in terms of what that language deconstructively seeks to accomplish–approaching the conceptual structures of metaphysics and "shaking" the language deployed to express those structures. In short, *all* language is metaphysically tinged, even language intended to disrupt the traditional solidity of metaphysics. Generally speaking, if being as presence includes language as a type of being, then everything Derrida says about being as presence also applies to that selfsame language insofar as it retains a kind of "present" being. Thus the reality of spoken or written language as such, apart from its philosophical function in evoking metaphysical considerations, instances being as presence to the same extent as the objects and entities subject to the conceptual configurations emergent in metaphysical language. The outgrowth of second-order vocabulary, that is, a series of reticulated concepts intended to describe, correlate and analyze the phenomenon of language itself, suggests that all components of this vocabulary–e.g., "form" as a paradigmatic instance–are no less "metaphysical" in exemplifying the effects of being as presence than is the array of concepts (including, of course, "form" itself) traditionally associated with metaphysics proper.

In general then, what holds for language about being also holds for theoretical terminology intended to describe and analyze language about being. An appeal to any already existent word, phrase or principle will necessarily throw the investigator initiating this appeal squarely within the ambit of the history of metaphysics and thus impress on the content of the investigation an allied residence within the strains of being as presence. The question then becomes identifying the best, most productive and incisive way to confront this necessity. How should the deconstructionist approach the import of language as signs?

The word "sign" represents a concept and thus is just as metaphysical as "form" or "unity." But Derrida insists that "sign" as an integral element in advancing a deconstructive agenda cannot be jettisoned just because its ineradicable roots remain within the domain of being as presence. Why? An argument develops this assertion.

For semiology, the study of signs, not all signs are words, but the link between signs and language outlines the context occupying us here. Words are signs, i.e., signs of concepts. What we as practicing deconstructionists want to accomplish is to "shake metaphysics." But any generalized claim intended to "shake metaphysics" must use the terminology of metaphysics in order to accomplish this end. The terminology of metaphysics is constituted by words, i.e., by the understanding that words are signs. Thus to shake metaphysics by way of meaningful language is, by implication, to rely on "the concept of sign." However, if "there is no transcendental or privileged signified," then it follows that one must also reject "the concept and word 'sign' itself." The reason: it is precisely by the appeal to the concept and word "sign" as embodied in words and their usage in metaphysical contexts that we are able, in these target contexts, to reach *from* the words themselves employed to speak or write about metaphysics *to* what the words-as-signs signify. Therefore, the rejection of "sign" cannot be accepted since to reject sign in this sense is to reject even the possibility of discourse about metaphysics in the first place, much less discourse which intends to shake metaphysics through a process of deconstructive analysis. (As an aside, this line of argument verges on justifying metaphysics by virtue of the implications from the very attempt to shake metaphysics via linguistic considerations; we return to this point in Part III.)

Derrida assumes that at its most basic, a sign always points to something, that is, "sign" = "sign *of*;" furthermore, the "of" refers to something which is other than the sign itself. The sign is the signifier, what the sign refers to is the signified, and signifier and signified differ from one another. In **Passage 4F**, Derrida is careful to distinguish "concept and word" (*jusqu'au concept et au mot*) with respect to "sign." Thus the word "sign" and the concept signified by that word differ from one another. Different words in different languages can signify the same concept (e.g., "STOP" and "ARRÊT"), and the word need not be present in order for the concept to exist, as in the conceptual awareness which drivers of vehicles have in mind, i.e., that they must stop their vehicles when approaching a "STOP" sign even if no such sign is, in fact, immediately visible. Thus we may conclude that "STOP" on a stop sign is a "signifier referring to a signified" in such a way that the sign is "different" from what the sign signifies: "STOP" signifies that the person *seeing* this sign must (or should) *think* and then *act* in a certain way. In the lifeworld defined by actual events, this sequence might be virtually instantaneous but as we have just witnessed, the processes characterizing signification are complex and distinguishable into clearly demarcated phases. We note then that "stop" as a mark, a visible set of four-letters, is different from both the concept of stopping and the subsequent act of stopping.

How then is it possible to shake metaphysics using language? Derrida's next step in this venture is daring–he proposes to erase the "radical difference between signifier and signified," a move which entails, according to Derrida, abandoning the word "signifier" as a carrier of metaphysical significance. Thus what the sign is "of," the signified, remains intact; the elimination of the linguistic vehicle intending or aiming at some object (in a broad sense) does not affect the object. For example, the concept of stopping remains, even if "stop" has been eliminated as a (linguistic) sign intended to be "of" the concept of stopping when a perceiver is in the presence of the entity containing a visual representation of that sign.

I suggest that erasing "the difference between signifier and signified" does not mean that the word has entirely lost capacity to connect to a concept formed or signified on the basis of the word–e.g., the concept of *tree* based on hearing someone utter a proposition about trees; rather, Derrida intends to deny a necessary connection between a given word (or naming phrase) and a specifically determinate and unique concept–i.e., the "transcendental signified"–existing by itself, independently of language. Derrida abandoned "signifier" rather than "signified" because the former refers to an object (or "mark") residing in the domain of language while the latter refers to *what is signified*. Furthermore, what is signified encompasses the realm of extra-linguistic referentiality insofar as anything within that realm can become the subject of a concept (although, again, not in the privileged and, for Derrida, inaccessible status of a transcendental signified).

According to this rendering, Derrida is not abandoning words as signs; he is abandoning words as signs of something definite at the conceptual level, i.e., a "transcendental signified." It is in this respect that a sign–presumably *any* sign–is a "metaphysical concept." Thus, as an additional consequence of considerable force, Derrida is not at this point denying the existence of concepts, at least as a theoretically feasible component with respect to, say, the possibility of knowledge. Rather, he is denying that there is any necessary connection between a given word functioning as the signifier of a given or necessarily determinate concept existing apart from language and thereby seemingly capable of transportability.

An interpretive caution. As he formulates the basic principles and arguments funding deconstruction, Derrida's language must be precise in referential identification, i.e., the particular elements of metaphysics which he intends to "shake"–e.g., the notion of opposition. Precision must also obtain in the identity of the transcendental signified (using Derrida's technical terminology), at least in the sense that the formulation of deconstruction's strategic elements and its intent must remain conceptually constant, one and the same, throughout its application to texts. Presumably deconstruction should not only shake metaphysics but *continually* do so, even after metaphysics has been shaken once. For if metaphysics could then settle down, as it were, these results would exhibit precisely the stability which deconstruction intends to uproot. Thus the sense in which Derrida's words function as signs, that each in its own way is a "sign of" something other than itself as a word or "mark" must be–in some difficult and arguably inconsistent sense–*kept apart* from the accusations and critical force which Derrida has directed toward traditional metaphysical language. We will examine in Chapter 8 whether on logical grounds the structure and implications of this theoretical separation are possible.

4.13 Signification and Opposition: the Sensible and the Intelligible

It is necessary to invoke the concept of sign in the endeavor to shake metaphysics. But when sign appears as an element in the context of language, a specific strategy emerges regarding the most appropriate way to deconstruct the oppositional solidity–and stolidity–of the signifier/signified relation:

Passage 4G The concept of the sign, in each of its aspects, has been determined by this opposition [between the sensible and the intelligible] throughout the totality of its history. It has lived only on this opposition and its system. But we cannot do without the concept of the sign, for we cannot give up this metaphysical complicity without also giving up the critique we are directing against this complicity, or without the risk of erasing difference in the self-identity of a signified reducing its signifier into itself or, amounting to the same thing, simply expelling its signifier outside itself. For there are two heterogenous ways of erasing the difference between the signifier and the signified: one, the classic way, consists in reducing or deriving the signifier, that is to say, ultimately in *submitting* the sign to the thought; the other, the one we are using here against the first one, consists in putting into question the system in which the preceding reduction functioned: first and foremost, the opposition between the sensible and the intelligible. For the *paradox* is that the metaphysical reduction of the sign needed the opposition it was reducing. The opposition is systematic with the reduction. And what we are saying here about the sign can be extended to all the concepts and all the sentences of metaphysics, in particular to the discourse on "structure" (WD, 281–all italics in text).

Simply and formally put, Derrida is using signs via his own philosophical and technical language in order to critique signs in language as such. Thus a feature necessary to the inner workings of signs must be in place in order to formulate a position, using language, to critique that very feature. Derrida's concession at this juncture indicates that he is well aware of the self-referential component in his intended conclusion. However, given that the strategy of the critique necessarily incorporates the intended point of the critique, the question again arises whether the prospective success of such a strategy is obviated simply by virtue of its mere formulation. Furthermore, if it is possible to identify an element or strand in the complicity *not* affected by the critique Derrida advances, then this phase of the complicity possesses a dimension of necessity which is self-justifying and impervious to attempts either to reduce it to simpler or even divergent components or, more radically and dramatically (as is Derrida's intention), to deny it altogether. We will return to this possibility both in this chapter as well as in Chapter 9. But for now, it is essential to consider some of the ramifications of what Derrida has specifically proposed as far as supporting the deconstructive denial of securing a given transcendental signified as a seemingly necessary feature in the notion of signification.

The problem: "erasing the difference between the signifier and the signified." Derrida's strategy for addressing this problem is–"first and foremost"–to question "the opposition between the sensible and the intelligible." Note that no opposition arises if "sensible" and "intelligible" are construed solely as words in a natural language. For isolated only as words, neither is more or less a word than the other. Therefore, if opposition between "sensible" and "intelligible" exists, it must originate elsewhere. Derrida's gambit becomes effective if the difference erased concerns words as referring agents, i.e., "sensible" refers to what is sensed (presupposing the existence of something capable of being sensed) and "intelligible" refers to what is intelligible (presupposing something in the mind of an agent existing in a way other than the existence of something sensed).

Derrida throws into question the entire relation between "sensible" and "intelligible" in order to deconstruct the notion of signification where that notion is understood to denote a necessary connection between a given signifier and what is determinately signified by that signifier. Thus the word as sign cannot be reduced to the word as thought if the distinction between the sensible (i.e., the word as sign, a vehicle of signifying) and the intelligible (i.e., the word as thought, a vehicle of what the word has signified) is "thrown into question."

However, Derrida claims much more for the strategy involving the sensible and the intelligible:

> **Passage 4H** To these roots adheres not only the distinction between the sensible and the intelligible–already a great deal–with all that it controls, namely, metaphysics in its totality. And this distinction is generally accepted as self-evident by the most careful linguists and semiologists (OG, 13).

In addition to its implications for signification, the distinction between "the sensible and the intelligible" controls "metaphysics in its totality;" therefore, Derrida reasons, if this distinction is questioned, it follows that metaphysics in its totality has been thrown into question.In **Passage 4A**, Derrida announced that "the opposition between form and matter" inaugurated metaphysics. Thus "sensible/intelligible" as well as "matter/form" are, for Derrida, both foundational to the onset of metaphysics, although the latter is evidently a particular instance of the former–thus the identification of matter distinguished from form depends on the capacity to sense the world, then formulate an intelligible encapsulation via concepts of that sensed experience. If "the sensible" indeed refers to (a) all that is and can be subject to sensory experience and "the intelligible" refers to whatever is (b) other than everything falling under (a) and which also exists "in the mind" as the result of cognitive and epistemic processes, then it becomes even more crucial to elicit the repercussions of throwing this distinction into question.

First, a small but crucial terminological point–in **Passage 4G**, Derrida writes "the *opposition* between the sensible and the intelligible" but in **Passage 4H** he has "the *distinction* between the sensible and the intelligible" (both italics mine). Distinction is a broader concept than opposition; all instances of opposition are distinctions but not all distinctions are instances of opposition. This difference becomes important because the history of philosophy has many instances when the link between sensibility and intelligibility is not cast as opposed, but as complementary. Recall Kant's famous dictum: "concepts without percepts are empty, percepts without concepts are blind" (*CPR* A51, B75). Are concepts and percepts instances of opposition? Surely not. Concepts and percepts are distinguishable elements in a theoretically unified account of knowledge; they work in harmony with, not opposed to, one another regarding their basic function. It follows that the formulation in **Passage 4G** does not apply to this Kantian position; however, the formulation in **Passage 4H** is in play given that for Kant, concepts and percepts are distinguished from even if they are not opposed to one another. The formulation in terms of opposition (**Passage 4G**) is indeed apposite with all the texts where Derrida emphasizes opposition as the pivotal linkage employed by metaphysics; as such, it embodies the conceptual mechanism which deconstruction is most concerned to shake. But in order for Derrida's strategy in this regard to be broadly effective, the formulation underlining "distinction" rather than the narrower "opposition" should control the argument.[54]

The issue concerns the meaning and implications of throwing into question the distinction between sensibility and intelligibility. For example, does this ploy mean that we also question properties which belong to *each member* of this relation as an autonomous component? The description of the forest setting at the beginning of this chapter took the difference between what is sensed (the tree) and what is thought about what is sensed (concepts of the tree) to be fair and reliable. But if, e.g., sensibility is considered by itself, apart from how it exists in relation to intelligibility, is Derrida also questioning properties ascribable to sensibility as a self-contained process? Thus is the sharp distinction between percept and concept as it underlies the Forest Scene philosophically legitimate? The same question pertains to intelligibility if considered by itself, apart from correlation with sensibility. In fact, is it a mistake to separate the two activities as far as philosophical interrogation is concerned?

Consider these questions from the perspective of an unabashed realism (e.g., the Forest Scene), that is, entities exist independently of any observation or experience of those entities. From the perspective of sensibility, is Derrida questioning whether sensibility *exists at all*, i.e., a self-contained process whereby an observer is placed in awareness of actually existing entities? But if what is sensible as an isolated phase of experience is itself cast into doubt, then what assurance do we have that we can move outside of what appears to be the sensed order of our experience? Or does this throwing into doubt assume the functional reality of sensibility but deny that any properties can be determined which are unique to that reality?

Now consider intelligibility. We have assumed that reality can be described as a stark realism. If so, then the objectivity of a concept for an extra-mental entity is not equivalent to the objectivity of that entity itself, which will maintain its own dimension of existence apart from any attempt, successful or otherwise, to conceptualize that object. In the same vein, a number of notions of high generality–"unity," "self-identity," "difference" (in quotes to emphasize that these concepts exist as products of metaphysics' inherent drive toward intelligibility)–could never have been formulated at all, given that they presuppose the possibility of being formulated in language as a function of the domain of intelligibility as such. But it seems undeniable that

54 Derrida could preserve the generalized effect of his critique of opposition if he had analyzed with sufficient attention to structural detail all relevant senses of opposition. This issue will be discussed (among other aspects of opposition) later in Chapter 4 and also in Chapter 8 below.

(a) the *unity* and *self-identity* of an extra-mental entity will maintain themselves independently of any awareness of that entity and (b) consciousness of and language about these entities will always be *different* from the entities themselves. The tall tree in the Forest Scene stands as it does whether or not either A or B (or anyone) observes its height. How then can the distinction between sensible and intelligible be questioned at this level of experience?

There is an even more radical avenue of interpretation (hinted at above). By throwing into question the difference between sensibility and intelligibility, is Derrida questioning whether a discernible difference does indeed obtain between sensibility and intelligibility?

Important implications accrue from this possibility: if the difference between sensibility and intelligibility is rejected, how does the speaker of a language "reach" an extra-linguistic object? In particular, how does this speaker intersect with an object when in the midst of a lifeworld event occurring in the present, e.g., the walker in the forest immediately perceiving the tree? Indeed, if this distinction is put into question in all respects, how can referentiality be preserved? Consciousness of a plethora of extra-mental objects–recall that our suppositional realism is still in force–presupposes sensory awareness of those objects; without such awareness, there is no evidence available to consciousness that anything exists in such a way that it could engage the agency of language. The point is not that all extra-mental reality is sensed; it is the more limited but unavoidable conclusion that significant instances of extra-mental reality are indeed sensed and that accessing this reality would not be possible if it were necessary procedurally to throw into question the distinction between sensibility and intelligibility.

Here is another approach to Derrida's strategy, less radical but with equally difficult implications. To question a distinction between sensibility and intelligibility allows for a difference between these two modes of experience but not one securely and justifiably stabilized for theoretical purposes. Under this interpretation, concepts are possible but remain hit-and-miss affairs, potentially actualizable but shadowy cognitive elements without guaranteed existence as vehicles of stable signification. Questioning the distinction between sensibility and intelligibility thus would mean acknowledging the inherent ephemerality of any seemingly established linkage between what is sensed and what is intelligibly thought and codified conceptually about what is sensed.

Let this point return to present circumstances:Derrida's stated intention is to collapse the distinction between signifier and signified. But Derrida's own discourse must clearly and unambiguously pick out, or denote, or signify language-as-signifier in contradistinction to language-as-signified in order to make evident in the mind of the reader of Derrida's own texts that this distinction sanctioned by millennia of philosophical investigation into language and the relation between language as such and extra-linguistic reality is, within the functional sway of deconstruction at work, in fact without warrant.

In sum, the status of the difference–assuming that this difference is necessary for the very possibility of philosophical inquiry–between sensible and intelligible must be determined in two respects: first in relation to the object or entity perceived as external to the field of language; second, in the subsequent relation between the object as sensed and thus functioning as referent within the event of language and the signified, now identified as a concept corresponding to or in some way in touch with that referred object.

Two concluding comments: first, the proposition spoken by A to B in the Forest Scene suggests strongly that at least in some contexts, the intended "transcendental signified" is perfectly and unambiguously identified and stable. If so, then it seems necessary to distinguish at the level of theory between transcendental signified with respect to *a given* context and transcendental signified with respect to *any* context. The argument is then that Derrida's denial of a stable transcendental signified may be applicable in some but not all contexts.

Second, recall Derrida's aim to erase the difference between sensible and intelligible. But the intended purpose of erasure surely fails, since referential difference must obtain in fact, in the world encompassing (a) language–articulated expression based on the intelligible as well as (b) entities referred to–their existence as extra-linguistic derived from sensible experience. The difference between (a) and (b) must already obtain to enable the advocate of deconstruction to claim that the difference between the two dimensions has indeed been erased. But if this former difference must exist in order to fulfill conditions necessary for the very attempt to erase the difference, then the difference has in fact not been erased. We return to these concerns in Chapters 8 and 9.

4.14 Signification and the History of Metaphysics

Derrida asserts that the distinction between sensibility and intelligibility controls "metaphysics in its totality." We interpret this broad claim: All metaphysicians sense the extra-mental world, then attempt to render that experience intelligible through language and concepts. But throwing into question the distinction between sensibility and intelligibility implies that regardless how metaphysicians have articulated that experience, regardless whatever words they use to evoke whatever concepts, all this material is now suspended by the force of questioning the distinction on which the totality of that discourse depends. Furthermore, the deconstructive strategy for coping with the sensible and intelligible as an especially privileged instance of opposition has comprehensively epochal consequences. This passage (the conclusion of which appears as **Passage 4H**) indicates their scope:

> **Passage 4I** The difference between signified and signifier belongs in a profound and implicit way to the totality of the great epoch covered by the history of metaphysics, and in a more expli-

> cit and more systematically articulated way to the narrower epoch of Christian creationism and infinitism when these appropriate the resources of Greek conceptuality. This appurtenance is essential and irreducible; one cannot retain the convenience or the "scientific truth" of the Stoic and later medieval opposition between signans and signatum without also bringing with it all its metaphysical-theological roots. To these roots adheres not only the distinction between the sensible and the intelligible–already a great deal–with all that it controls, namely, metaphysics in its totality (OG, 13).

Derrida emphasizes that the "difference between signified and signifier" belongs "to the totality of the great epoch covered by the history of metaphysics." This organic stretch of complex and variegated history displays a unity of primordial simplicity with respect to the distinction which Derrida here emphasizes. Note that Derrida qualified the mode in which this difference belongs to this totality as "profound and implicit." The overall import of "implicit" in this context is considerable since Derrida combines eras of Christianity, later medieval thought, and "the resources of Greek conceptuality" all within the ambit of metaphysics as it has been controlled by the distinction between the sensible and the intelligible, that is, "metaphysics in its totality." Whether it is possible that crucial differences defining the various systems of thought through these millennia could tell against Derrida's sweeping generalization is not considered, although a sufficiently sympathetic reading of "implicit" would apparently cover such questions of interpretation (along with all the scholarship devoted to these texts). Derrida also combines the metaphysical with the theological; all differences between and among the divergent systems of theology advanced throughout this epoch are no less affected by the same characterizations ascribed to the history of metaphysics according to narrower, strictly sectarian parameters.[55]

If we grant that close study of all the relevant texts falling in this epoch would in fact support this summary judgment concerning their implicit manifestation of the distinction under scrutiny, then we will also notice that Derrida parallels the level of constitutive implicitness with a commensurate measure of profundity. Derrida does not venture a reading of how this profundity becomes displayed but we recall from the reasoned conclusions of Part I that being as presence as a way of gathering the weight and force of the history of metaphysics under the collective organization of *différance* rests on a determinate sameness. The difficulty of eliciting this sameness can indeed be described as an exercise in profundity, whether or not the quest for this measure of sameness is successful. Thus the philosopher, if accepting this epoch-wide result of deconstructive inquiry, must think beyond the paired opposition of the sensible and the intelligible in order to specify this sameness, its content driving the intelligibility of *différance* insofar as *différance* has enabled the "great epoch"–the history of metaphysics.

55 Daylight, 24, provides background on the sources of Derrida's claim concerning medieval philosophy and theology.

This passage concludes with another undifferentiated generalization–conclusions which hold for analysis directed toward the sensible/intelligible distinction when components of this distinction are vehicles of metaphysical thought may be extended to include "all the concepts and sentences of metaphysics." The sameness in question must underlie not only differences between sensible and intelligible, but also how these differences extend themselves to cover each and every notion, argument and position asserted throughout the history of metaphysics. Thus all elements in the Forest Scene, for all their apparent obviousness and philosophical relevance, must be reexamined and reinterpreted as a simplified preliminary endeavor in approaching the sameness in question. The quest for such sameness is, again, an exercise in profundity of unique cast. And yet Derrida's understanding of signification as a fundamental component in the realization of language, even in the face of the fact that signification with regard to a necessarily transcendental signified must be questioned and rejected, leads to identical consequences concerning the pivotal place of sameness as the ultimate goal of reflective inquiry.

Finally, if "all the concepts and sentences of metaphysics" are regulated by the concepts and principles of formal logic, then the effective sway of Derrida's questioning of the distinction between sensibility and intelligibility not only extends to the many diverse positions which have been expressed by means of these "concepts and sentences," but also to the apparently "formal" structure of logic if its elements are transposed into articulated concepts. For such articulation is arguably no less metaphysical than the content of all that has appeared under the auspices of metaphysics throughout the western tradition, assuming that this content cannot be completely divorced from the regulating concepts and principles of "right" thinking–self-identity, difference, contradiction, the syllogism, modus ponens, tense logic, etc.–which allow that content to become accessible to rational sensibilities. The relevant question becomes whether these "logical" elements should be isolated, one by one, then subjected to close and intense deconstructive scrutiny in order to substantiate with the appropriate degree of methodological rigor the seemingly limitless implications that follow from the gambit of throwing into question the distinction between what is sensed and what is construed as intelligible based on what is sensed. We return to this crucial line of critical inquiry in Chapters 8 and 9.

4.15 Referentiality and Signification–the Question of Stability

Derrida insists that it is necessary to preserve "sign" to indicate a relation from the sign as such to something other than itself, i.e., a sign is necessarily and unavoidably a "sign-of" something. When this principle is developed within an actualized deconstructive setting, i.e., toward a given text, we may discern a relation comprising three distinct elements: (i) words constituting an interpretation of (ii) the tangible object read and referred to by these words–the text (here in the non-technical sense)–leading to

(iii) a conceptual response to what was read, i.e., the text's proposed meaning. This tripartite deployment implies a dependency relation–conceptual response depends on (a) text read in conjunction with (b) language formed as a response to that text.

Although this model is sufficiently articulated for present purposes, its simplicity must be diversified and amplified in order to accommodate the wide variety of concepts which characterize the interpretation of texts insofar as the surface meanings of those texts involve the lifeworld, as well as the modes of inquiry and expertise resident in that world. Despite this procedural qualification, however, the model yields two noteworthy senses of stability. Derrida must assume referentiality when he contends that the aspect of the system which is first and foremost to be questioned is "the opposition between the sensible and the intelligible." But note the important self-referential dimension which begins to emerge from this assumption: if there is no necessity in specifying a given transcendental signified–such specification is not possible for Derrida–then it follows that this same species of skepticism also applies to everything Derrida himself said and wrote. This inference, if sound, throws into question Derrida's own texts, since there is no compelling reason to connect the words of those texts as signs with determinate concepts that convey to the reader a certain stable or fixed semantic sense. That Derrida realized this kind of self-referential implication is illustrated by his assertion that the fundamental operator, i.e., *differance* "is neither a *word* nor a *concept*" (SP, 130–italics in text). But by right of consistency, Derrida should have made explicit that *none* of the verbal components, or marks, in *any* of his theoretical texts are or indeed can be concepts.

And yet these marks, both individually and collectively insofar as they animate Derrida's texts, certainly appear to convey...something. If therefore deconstruction can in fact be deployed as a series of assertions (a traditional way to name such ordered language) which have meaning, then at least one instance or range of instances of the transcendental signified must be preserved in order to establish deconstruction as an occurrence of language which an audience, especially one receptive to Derrida's message, can understand and, if such is their intent, apply to other linguistic events. Unless Derrida's own language, whether written or spoken, preserves signage and signification in this sense, his texts will suffer a unique kind of self-destruction in their very utterance or appearance, even though on the surface these texts are aimed at deconstructing, apparently with success, the traditional conceptualized structure of language itself.

How would this implosive result occur? Derrida's texts are language; language functions as signs; signs point to something other than themselves; the intended objects of these signs are themselves instances of language; thus language is pitted against language; but unless language intended to accomplish a critical end remains in some respects apart from its linguistic targets, the entire endeavor collapses in extended self-referentially nullified indeterminacy–the reader of this project when it has been so applied, sifting through these logically-invalidated heaps, receives no theoretically defensible results. The prospective student of deconstruction cannot put the strategies of deconstruction into practice because the very statement of these strategies is dissolved in its very utterance! To prevent this self-referential destruction, it appears (as already noted) that Derrida's writing must remain in some sense *apart from* what this very writing says insofar as it itself exists as a retinue of signs.

A second dimension of stability seemingly must obtain. Derrida restricts the import of sign to the concept indicated by the sign in, presumably, the mind of the speaker. But he does not mention in the passages analyzed here that a word used in a proposition has a simultaneously dual function–from itself to something outside itself, and in two respects. First, even if we were to give up "signifier" as, in Derrida's phraseology, a "metaphysical concept," ceding this theoretical perspective on language would in no way diminish or even affect the capacity of language, when used in the world for its myriad functions, to present vital connections to the speaker and to an audience. These connections derive from and contribute to actual contexts occurring in the moment–as in the Forest Scene–based on linkages between word or words spoken or written (or both) and entities in the world. Second, in a typical context defined by deconstructive concerns, interpretive words refer to an extra-mental text in a book or to writing of some sort (or, on a different plane of significance, to an extra-linguistic entity in the lifeworld). The same words then move in the opposite direction, so to speak, but again to something outside themselves, that is, to a concept inspired and produced by the mind attending to that extra-mental text (or object as referent) and determining the meaning of that text (or object as referent) on the basis of the language introduced referentially with respect to these intended entities. For without the intentional movement toward the extra-linguistic object, the mind would never be in a position to receive an impetus, any impetus, to formulate a concept.

Words as signs display dual, simultaneous directionality–between words functioning (a) extra-mentally, as referential indicator, and (b) with regard to generating concepts cognitively linked to that extra-mental entity insofar as the latter stands as an object of reference existing in a potential relation to the signified as concept. Thus even if the advocate of deconstruction denied the existence of transcendental signifieds, the question of describing the referential status of extra-linguistic entities would remain.

From a synoptic Derridean standpoint, lack of accessibility to a definite concept as an element in signification parallels lack of accessibility to particularity in the account of being as presence. Since the relation between word and concept has been severed or at least drastically compromised, it is not possible to articulate anything theoretical about, e.g., the metaphysical structure of the referential object which would, as a putative item of thoughtful concern, exist with sufficient stability to admit thoughtful analysis. Indeed, it is not clear that anything could be conceptualized about entities even if their existence coheres with the critique of the transcendental designated. How is it possible to form concepts at all if (a) language functions referentially but (b) the vector motion of language from a referent does not reach a stable element of signification–i.e., a given signified? What we have at hand is a world of entities resident referentially but existing without any apparent means available to the observer for analyzing their structure or even their appearance if, as Derrida insists, the sensory phase of human experience has also been cast into deconstructive suspension. The two individuals in the Forest Scene may be able to "perceive" (in quotes because this function is now under deconstructive scrutiny) the trees therein but they will be unable to speak about their experience with any degree of confidence that their discourse has anything to do with the forest in which they currently dwell.

The principle that nothing exists outside the text reinforces this seemingly counter-intuitive if not paradoxical consequence. This principle allows entities to exist extra-linguistically. However, these entities exist only in that their perceived "faces" or exteriors are and necessarily become involved in language–there is, in principle, nothing existing "outside the text." But the perceived exterior of an entity cannot theoretically make the transition from its status as (a) element in an articulated word within a proposition, as linguistic transcription of that experience, to (b) residence in the mind of a speaker in a mode of existence amenable to being transformed into a determinate conceptual vehicle. No stability obtains between any instance of signifying language and the purely conceptual correlate which allows an individual consciousness to codify awareness embodied in experience of language and thereby to initiate this codification into streams of discourse as reliable media for confronting the entirety of what is other than consciousness itself. Thus the conditions which hold in the deconstruction of a text as a purely linguistic event also hold for the dissolution of any stable metaphysical character capable of being ascribed to the referents of that text. In these respects then, Derrida's system is neatly coherent, at least internally, in that it issues in a form of rampant and seemingly unlimited skepticism with respect to both the theoretical structure of language as such as well as the metaphysical status of beings referred to by language as such.

Consider the following schematic summary: mark + referentiality + intentionality + concept + context = signification

This sequence of elements is pitched at a level of stability typically sufficient for feasible interpretations of texts or, on a different plane of human affairs, for purposive and productive linguistic activity in the lifeworld. Each of these areas so schematized must possess a measure of stability as well as the concurrent conjunction of all areas in order to establish the possibility of signification. Derrida will accept the stability of marks on a page as representations of words, but according to reasoning which animates deconstruction, all other components of this schematic lack the degree of stability which, as contended here, is necessary to generate signification as a factor for the efficacy of language in the lifeworld–as well as, presumably the activity of philosophy itself.

4.16 Summary and Prospect

Derrida intended to “shake” metaphysics by critically scrutinizing the language of metaphysics. The ultimate purpose of such treatment is to set in relief what being as presence has either concealed or distorted while the history of metaphysics has, so to speak, played itself out. The strategy for such revisionist analysis concentrates on a medley of questioning and, if necessary, rejecting standard concepts and distinctions as they have run through their history within the confines of metaphysical speculation. Taken as an integrated set of strategies, this questioning sanctions deconstruction’s function to “break apart” sclerotic ways of theorizing about language.

Derrida admitted that metaphysics must be employed in order to critique metaphysics; but if so, we may ask whether self-referential implications emerge from this necessity. Thus if a critique of metaphysics depends for its cogency on a stabilized configuration of thoughts, some (or all) of which themselves relying on metaphysical notions for their formulation, what effect does this self-referential reliance have on the intended point of such critique? This question becomes increasingly important as deconstruction progressively surveys the metaphysical conditions which characterize common concepts in the traditional understanding of the structure of language. As already noted, the question will be analyzed in Chapter 8.

The initial context of inquiry addressed in this chapter concerned the description and eventual rejection of signification as codified in the notion of a “transcendental signified.” Derrida has enlisted the support of a complex framework, itself inflected with necessarily transcendental notions and principles, in order to elicit the conceptual “shaking” he intends to procure. Thus the claim “there is nothing outside the text,” with a broad meaning ascribed to the referential range of “text,” leads eventually to collapsing a variety of distinctions fundamental to a number of standard philosophical issues. Once this exercise in deconstruction has been effected, it becomes possible to reexamine and rethink what the history of metaphysics has gradually but ineluctably concealed or distorted–to step beyond the signifier/signified and sensibility/intelligibility instances of opposition by throwing the entire relations established by these pairs of polar concepts into question.

This is a procedurally bold move. But although Derrida thereby establishes a deconstructively shaped openness, i.e., a plain of inquiry generally free, at least by intent, of the conceptual and theoretical assumptions embedded within the history of metaphysics, the question becomes whether this and related moves will compel his advocates to face consequences embodying even more pressing difficulties than anything which may have been derived from a static and seemingly repressed metaphysical residuum.

We may nonetheless assume for now, as a statement of purpose, that the relations and connections between and among three distinct elements–object, word, concept–must be rethought from, as it were, the ground up. Two areas of special importance regarding this trio of basic realities may be noted: the deconstructionist must, first, show the difference between (a) the extra-linguistic entity as an instance of what has been identified, in technical terminology applied within the realm of cognition, as the intended object of the transcendental signified and (b) that same entity as the referent in an utterance derived from and incorporating some aspect or dimension of the lifeworld. In short, what is the relation between the concept (as an approximation of the transcendental signified) and the object referred to by that concept? For if we return to the Forest Scene, it certainly seems self-evident to insist that the concept of "tree" is one thing; the tree itself is something else again.

A second, related problem concerns the connection between word referring to an extra-linguistic object and concept signifying that object. The seemingly common-sense assumption underlying this question is that words and concepts are not identical to one another. If the question emerging from that non-identity is granted legitimacy, we now appreciate more fully the tension between (a) the emphasis on Derrida's denial of the transcendental signified and the lack of attention to the extra-linguistic entity as referent in a linguistic event and (b) the parallel *de facto* loss of particularity in Derrida's treatment of being as presence (discussed in Chapter 3).

The extra-linguistic entity exists but, as argued here, cannot be securely referred to in language. Derrida does not deny referentiality as a theoretical necessity–various texts make its necessity explicit. What he denies is the capacity of referentiality to isolate a given referent for purposes of grounding a fixed and stable "signified." However, even if one accepted Derrida's denial of an existing transcendental signified, the problem of describing the referential status of an extra-linguistic entity remains. If texts refer to entities, how do they accomplish this end and to what do they refer? Even more fundamentally, if the distinctive character of particular entities cannot be approximated much less properly described metaphysically, then referentiality as a theoretical extension of particularity becomes at best moot. And without a stable dimension of referentiality, an extra-linguistic entity in effect ceases to exist as an object of concern within the lifeworld. Intimations of this conclusion appear in the effective dissolution of the foundational difference between word and object as embodied in Derrida's maxim that nothing exists outside the text. And yet it seems again clear that the tall tree exists apart from or independent of the phrase "tall tree." Is the philosopher helpless and without conceptual resources in the face of this difference?[56]

---

56 In "Signature Event Context" (MP, 318), Derrida asserts: "the absence of the referent is a possibility rather easily admitted today." But by contrast, in Derrida 1992, 48: "No doubt all language refers to something other than itself or to language as something other. One must not play around with this difficulty." Cf. also LI, 137 and 150, where Derrida insists that questioning reference does not mean denying that there are and should be instances of reference. As an aside, when Derrida is speaking directly during an interview, he tends to enunciate "common-sense" positions, e.g., on concepts such as reference, and to do so with the bluntness indicated in the comment from Derrida 1992 just cited. This personalized tendency moves John Caputo to write (Caputo 1997, 204): "The best introduc-

Chapters 4 through 7 (Part II of this essay) display a series of ongoing tensions between what Derrida is critiquing in language via deconstructive gambits and what he loses by so doing in terms of the subsequent effects on traditional philosophical concerns. We will also observe elements Derrida surreptitiously reintroduced from preexisting archives of metaphysical concepts and positions in order to be in a position to mount these critiques in the first place. There will be continuously growing evidence that however vigorous the attack on metaphysics may be when we read the calm (yet often playful) referential and semantic surface of Derrida's texts, the extent to which metaphysics has been disrupted and roiled depends precisely on the ability of metaphysics to erect theoretical structures in order merely to propose, much less execute, such "deconstructive" critical analysis. We have seen Derrida expressly admit this self-referential requirement even in his own

---

tion to Derrida, in my view, is his interviews, where he speaks for himself in a particularly clear and unencumbered way." For Derrida, however (P, 67): "The improvised speech of an interview cannot substitute for the textual work." As a rule, Derrida insists on the need for reference: in addition to the sources cited above, see Derrida 1992, 45, 47, 48; P, 66, 67; Kearney, 154.

Although Derrida's suspension of certitude regarding access to fixed and stable referents represents a suitable position for attaining certain theoretical ends, the question remains–what are the implications of these theoretical gestures when translated into the lifeworld? For discussion, see Stocker, 186. Norris 2001, 55, interprets Derrida's insistence on the referential dimension to justify identifying Derrida as a metaphysical "realist." Indeed for Norris 2012, 81, deconstruction has a "referential bearing" with "real-world object domains;" furthermore, 97, Derrida always subscribed to "a highly specific referential dimension…;" and finally, in his later writings, Derrida maintained "that in his earlier work the intent was…to 'complicate' the nature and workings of referential language rather than deny that language could ever achieve anything more than an endlessly deferred simulacrum of reference" (Norris 2012, 98–in support, Norris cites the "Afterword" in LI).Derrida's reaction to this reading as Norris proposed it in an earlier text is that it makes him "smile" (Derrida 2001, 113), with the implication of this affective response not immediately clear. Apparently for Derrida, interpretive distance obtains between (a) being labeled a realist and (b) what he believes he is doing substantively in his texts. We may note that if Norris is correct, then Derrida must have a much more determinate sense of reference in mind than many of the earlier texts seem to warrant. To hold language as undeniably referential is to presuppose a fundamental sense of difference if language can refer to distinct referents in such a way as to distinguish them from one another. But what then could Derrida have meant with the claim that the intent of his early work was to "complicate" how referential language functions? How could the referential language in the Forest Scene be complicated and still function within the parameters of its context? At Kearney, 156, Derrida observes that his thought "does not destroy" its subject; thus the preservation of the subject at least countenances the possibility of a realist reading of that thought. However, the relation between an undestroyed subject of discourse and a more complicated approach to referentiality must be explained given the demands endemic to language as it emerges within the lifeworld. For additional discussion, see Ch7n88.

deconstructive advances. The upshot–metaphysics is necessary, even and especially if the intention is to dismantle metaphysics!

What then should be done philosophically in the face of this necessity, assuming we want to maintain Derrida's initial strategy regarding the incessantly hovering and potentially harmful penumbra of being as presence? Or is this necessity merely a residual effect from the lingering quality that has accrued to metaphysics given its millennial duration in western thought and affairs? If the latter, then it seems that the most we can say regarding implications with respect to totality for this phase of Derrida's position is that they hold in suspension any connection between a linguistic utterance and the relation between referentiality commonly attributed to language and the possibility that such referentiality can, when vectored in the theoretically opposite direction, evoke a conceptual marker for purposes of cognition. But without the availability of some stable concepts–the phrase "stable concepts" being arguably redundant–how is philosophizing even possible as a meaningful activity?

These fundamental difficulties in connection and linkage become even more prominent when Derrida advances from the relatively insulated arena of transcendental signifieds to the public domain of context.

5 Context and Concept

The scene: a university classroom. An in-class examination has just concluded. The instructor announces: "Please turn in your examinations." The students comply, then they leave the room.

Consider the setting–where it happened, what was said, what was done given what was said. Someone philosophically inclined will appreciate that this setting is decidedly metaphysical (e.g., with respect to determinate spatio-temporal coordinates, displays of motion and rest, individual instances of substance and accident); as such, and with the additionally stipulated factor of "lived" language, it could effectively and accurately be identified as a "context." How often are these and similar contextual circumstances replicated in events involving language? How often does the speaker know exactly what he or she wants to say, the audience understand exactly what the speaker intends to say, and the audience choose (or, perhaps, ignore!) to do exactly what was conveyed by what the speaker has said?[57]

The occurrence described above has a beginning, middle, and conclusion, all happening uniformly within a determinate spatio-temporal and metaphysically-structured setting. Are such successful intersections and sequences of linguistic utterances and extra-linguistic circumstances rare, common, or intermittent? If common, how necessary is the notion of context to account theoretically for this fundamental frequency?

5.1 The Question of Contexts

This chapter concentrates on evaluating the critique (as well as seminal implications from this critique) Derrida directs against the theoretical reliability of the concept of

---

57 Aristotle makes the relevant point with customary energy (and pungency) at *Topics* 108^a22-29: "For as long as it is not clear in how many senses a term is used, it is possible that the answerer and the questioner are not directing their minds upon the same thing: whereas when once it has been made clear how many meanings there are, and also upon which of them the former directs his mind when he makes his assertion, the questioner would then look ridiculous if he failed to address his argument to this. It helps us also both to avoid being misled and to mislead by false reasoning…".Aristotle's appeal to the "ridiculousness" of not addressing the meaning present in the *minds of both participants* during an argument suggests that the clarity and conceptual precision in such contexts is much more rule than exception; thus *not* to proceed in light of this shared meaning is, in fact, "ridiculous." Derrida apparently denies that the mind, individually or conversationally plural, is capable in principle of such semantic determinacy and habitual accuracy. A contemporary and slightly foreshortened version of Aristotle's point (Gadamer, 99-100): "It seems to me a misunderstanding of the nature of philosophical concepts to think that a thematic usage of a well-defined word, whether in philosophical discussion or fixed in a philosophical text, somehow places hobbles on our thinking." (The reader may also recall the first motto of this essay, from Augustine's *Confessions*.)

context. This critique is based on the fact that, as Derrida saw it, many elements of a given linguistic event may be, indeed must be, repeated in other, metaphysically distinct circumstances in order for these elements to be meaningful in any given set of circumstances. As a result, any context if characterized according to rigidly specific dimensions (whether stipulated or implied) is in principle indeterminate with respect to "set" or "given" metaphysical coordinates.

Consider the following passages:

> **Passage 5A**...I would like to demonstrate why a context is never absolutely determined, or rather in what way its determination is never certain or saturated (MP, 310).

> **Passage $5A_1$** ... the finiteness of a context is never secured or simple, there is an indefinite opening of every context, an essential nontotalization... (LI, 137).

> **Passage 5B**...one must be able to recognize the identity, shall we say, of a signifying form. Why is this identity paradoxically the division or dissociation from itself which will make of this phonic sign a grapheme? It is because this unity of the signifying form is constituted only by its iterability, by the possibility of being repeated in the absence not only of its referent, which goes without saying, but of a determined signified or current intention of signification, as of every present intention or signified (and therefore from communication and its context) seems to me to make of every mark, even if oral, a grapheme in general, that is, as we have seen, the nonpresent *remaining* of a differential mark cut off from its alleged "production" or origin. And I will extend this law even to all "experience" in general, if it is granted that there is no experience of *pure* presence, but only chains of differential marks (MP, 318–italics in text).

The reference to context is not highlighted in **Passage 5B** but its muted status reflects the indirect manner in which Derrida often approaches his concerted rejection of the reliability of the notion.[58] Examination of the passage reveals that Derrida has surrounded the reference to context with many of his most fundamental elements of his thinking on language–signifying form, mark, iterability, referent, as well as phrases evoking "transcendental signified." The point is to show that if all these elements are variable, then it is not possible to arrive at a stable notion of context if "context" is intended as a foundational basis for the establishment of semantic certainty regarding an utterance in a text, whether philosophical or literary. A particular, concrete instance of language assumes what Derrida refers to in **Passage 5B** as a differential mark "cut off" from its apparent origin as a linguistic given, and as a result not completely determinable because of its location in what appears to be a given "context." The question that must be posed, however, is whether the concept of context is restricted to purely linguistic considerations or whether (as I argue below)

---

58 For additional sources on the inherent indeterminacy of contexts, see M, 310LI, 12, 79, 81; Derrida 1989, 42, 64. Derrida's general approach to the concept of a context is discussed in DFL, 12-4, 63-7, 129-30.

context includes as part of its conceptual structure an appeal to non-linguistic–indeed, metaphysical–considerations as necessary conditions for *any* linguistic event. For now, the reader is requested to consider this possible feature of the notion of context as viable and worth close attention.

The analysis of signification in Chapter 4 reviewed the extent to which the various component elements of that complex concept yield a kind of systemic indeterminacy. One of the most important implications of this position is the denial of a working and stable transcendental signified, a cardinal position in Derrida's deconstructive treatment of language.[59] Furthermore, the denial of a precisely identified and resilient transcendental signified is reinforced by Derrida's rejection of context as traditionally determined. For if no context is ever finally determined, there can be no guarantee that any notion which might appear assignable as a transcendentally signified concept putatively connected to a given word appearing within a certain setting can in fact be necessarily linked to that word.

This implication has additional consequences of even broader significance. Derrida's use of the phrase "transcendental signified" is reserved for primarily "mental" based entities. But philosophers have traditionally located or qualified concepts in or according to the same "mental" coordinates. Given this confluence of epistemic environments, the following question becomes unavoidable–what distinguishes something identifiable as "transcendentally signified" from something described as a "concept"? In fact, from a more foundational standpoint, just what is a concept in the first place?

To answer this question from a deconstructive perspective, it appears necessary to specify conditions guaranteeing the possibility of concepts. And one way to approach this question–surely an integral issue as far as the mechanics of philosophizing are concerned–involves tracing a series of implications regarding Derrida's explicit treatment of concepts. Furthermore, if this inquiry is then juxtaposed with Derrida's rejection of the reliability of contexts, the implications are vital for both concepts *and* contexts, especially when concepts and contexts as theoretical entities are viewed according to traditional specifications. The primary purpose of this chapter is to articulate these implications and to suggest that in the end, Derrida's acceptance

---

59 To supplement the sources pertaining to the general treatment of signification as cited in the notes for Chapter 4, see MP, xxiii, 319; OG, 7, 14; WD, 280; SP, 138; P, 30, 49-50, 82. For additional commentary on Derrida's approach to the notion of "transcendental signified" when taken in isolation from his general position on language, see Caputo 1997, 100; Daylight, 21-2; see also Alan Bass' useful account in his introduction to WD, xv. Close correlation exists between what Derrida construes as "transcendental signified" and the notion of a concept, in particular the sense in which Derrida's approach to the reality of concepts pertains to the problematic notion of a context. Critical analysis of this correlation is a primary concern of this chapter. Note that for Stocker, 67, "All concepts are contradictory for Derrida," an interpretive claim which, according to the strict dictates of formal logic, would render it impossible even to formulate a concept. For how then would Derrida himself have conveyed philosophical positions through the medium of language and the printed word?

of concepts, deployed in various modes and combinations, goes a long way toward undermining his avowed rejection of contexts. As we shall show, if concepts are possible and indeed accessible, then contexts are necessary.

5.2 Concepts: A Presentation of Context

Derrida's skepticism concerning the theoretical reliability of context as a framework for determining meaning and signification affects the very existence of the conceptual order. This consequence if carried far enough will undermine the viability of the philosophical enterprise assuming that philosophy deals with generalities and that generalities depend on concepts for their mobilization and expression. Note that this stark consequence is derived by implication from Derrida's texts, for according to the letter of these texts, concepts do indeed exist. Furthermore, the development of these implications and the resultant status of concepts in general involves an important correlation between being as presence and Derrida's deconstructive treatment of language, in particular the relation between language and concepts.

The pivotal features of this correlation become apparent through a close analysis of a passage from Derrida's seminal essay on *différance*. Commenting on a passage from Saussure's seminal work on linguistics, Derrida identifies and develops the crucial notion of a *chain*. What the passage asserts as well as what it implies and presupposes about the structural characteristics of a chain is sufficiently broad to justify an extended analysis.[60] Finally, it is worth repeating that implications derived from a critical consideration of this notion will invite a return to the currently discredited concept of a context.

After citing an extensive quotation from Saussure, Derrida asserts the following:

> **Passage 5C** The first consequence to be drawn...is that the signified concept is never present in itself, in an adequate presence that would refer only to itself. Every concept is necessarily and essentially inscribed in a chain or a system, within which it refers to another and to other concepts, by the systematic play of differences. Such a play, then–differance–is no longer simply a concept, but the possibility of conceptuality, of the conceptual system and process in general. For the same reason, differance, which is not a concept, is not a mere word; that is, it is not what we represent to ourselves as the calm and present self-referential unity of a concept and sound. We shall later discuss the consequences of this for the notion of a word (SP, 140).

The main thrust in this dense passage concerns key functions of *différance* and ways in which *différance* generates the "play" of elements leading to concepts. Within a given system of thought, *différance* generates differences thus allowing concepts to

60 Derrida appeals often to "chain," and does so in various theoretical contexts: OG, 150, 162-3; MP, 26n26, 39, 219-20, 317, 326; WD, 292; P, 40, 57, 81, 107n42; SP, 140; LI, 18, 21, 44, 59, 64, 71, 103. See also the discussion of "chain" in David Wills' Derrida glossary, 317.

emerge that define the limits of that system. Derrida also indicates crucial differences between *différance* as a technical term and traditional, metaphysical-inspired notions pertaining to the structure of language. The most dramatic differences are conveyed in assertions that *différance* "is not a concept" and indeed is "not a mere word." Our main interest here revolves around what is said and implied about concepts in relation to contexts; in this regard, the wording in Derrida's denial that *différance* is a concept is suggestive.

5.3 *Différance* and Conceptuality

Two notable consequences emerge from the structure of *différance* and we indicate them as a preamble to subsequent analysis: First, Derrida's appeal to negation in "not a concept," as if negation itself was a given, its logical import self-evidently clear in this context and presumably in many other contexts as well (we shall see in the following chapter that negation, especially in its modifying and adverbial guises, must be cautiously approached and carefully analyzed). Here, the phrase "not a concept" applies to *différance*, distinguishing it both from the more traditional "difference" as well as any other concept whether apposite to the scope of *différance* or not.[61] Second, the implications following Derrida's assertion that *différance* is "not a mere word; that is, it is not what we represent to ourselves as the calm and present self-referential unity of a concept and sound." What Derrida asserts as not belonging to the uniqueness of *différance* insofar as *différance* is "not a mere word" may, by implication, be positively ascribed *to* a word. Thus we may infer that a word possesses what *différance* does not–"the calm and present self-referential unity of a concept and sound."

The qualification "present self-referential unity" apparently asserts that the unity "of a concept and sound" exhibits a degree of presence which allows Derrida to refer to it without the same reservations he brings to bear regarding the concept taken in isolation. Consider this example, already introduced (in another context) in Chapter 4: (1) the word STOP as it appears on a sign; (2) the concept of stopping; (3) the sound of someone saying aloud "STOP!" if contextual circumstances required that this word be uttered in an urgent, exclamatory fashion.

If someone says "Stop" aloud, the concept and sound of this vocalized word constitute a "calm and present" phenomenon. The use of "present" brings out that aspect of being as presence which exists in the "here and now," i.e., in the present. This event is "calm" in that the inherent volatility and variability of the theoretical necessity for repetition of words is tempered by the sheer immediacy

61 Derrida once referred to *differance* as an "aconceptual concept" (LI, 117-8), suggesting that *différance* possesses at least vestiges of conceptual considerations. Recall (cited in Ch3n42) the extended discussion of *différance* in Ch. 4 of DFL.

of the spoken word. The concept of "stopping" and the sound of the word "stop" uttered aloud, when occurring simultaneously, constitute a "self-referential unity." The use of unity is intriguing given Derrida's systemic distrust of the concept, but its appearance seems to indicate that the conceptual component of "stop" is conveyed immediately and satisfactorily only when the word "stop" is actually being uttered. Since such an utterance need not occur "in the present," the implication is that words and concepts should be distinguished. A word could signify a concept but the word itself is not equivalent to the concept. But if the concept differs from the word representing the concept, the question of determining the constitution of a concept as such becomes paramount.

5.4 Concepts and Presence

In **Passage 5C**, Derrida makes clear that he fully subscribes to the existence of concepts, but only if the way they exist is strictly qualified. Concepts are said to refer to–in presumably an intuitive sense of "refer"–or are related to other concepts. Thus concepts are necessarily related to something other than themselves, i.e., "to another and to other concepts." We note Derrida's assumption that concepts admit plurality, perhaps a very large if indeterminate number of distinct instances of concepts. This plurality as a "given" is important to appreciate, as we shall see. In sum, each concept is inherently relational, exhibiting a plurality of relations to "other" concepts.

At this juncture, does Derrida presuppose a distinction between concept and *signified* concept (*concept signifié*)? If not, then "signified concept" becomes redundant since one could contend that a concept must exhibit signification in order to exist as a concept in the first place. Furthermore, in terms of logical priority, the account of signification discussed in Chapter 4 suggests strongly that "signified concept" would not be possible if signification is as inherently indeterminate as Derrida insists. We could, however, maintain the theoretical feasibility of "signified concept" for Derrida's overall position if a distinction were drawn between (a) the notion of signified concept as such and (b) the notion of transcendental signified–assuming that "transcendental signified" refers to concepts. Thus a concept could exist as signified without entailing–presumably a necessary entailment if Derrida is correct–that this concept also exists as a "transcendental signified." We will provisionally grant this distinction and thus assume that it is possible to establish with justification a viable "signified concept."

We are told that a "signified concept" is never "present in itself, in an adequate presence that would refer only to itself" as an entity capable of self-subsistence, thereby justifying its independent existence as a signifying vehicle. The appeal to an "adequate presence" has two important and readily recognizable implications: first, concepts as such possess presence (presumably in Derrida's technical sense). Thus each concept must have some degree of being as presence in order to exist even as a concept.

If, however, all concepts were to exist *only* as exemplars of being as presence, then it follows that any one concept must exist in relation to every other concept given that all concepts, to exist in the first place, become instantiated by virtue of being as presence. The alternative to this global consequence regarding concepts with respect to presence would be to demonstrate by argument (or, more bluntly, to assert *ad hoc*) that concepts in a respect other than that derived from being as presence concurrently occupy a *different* sector of existence. If such an eventuality were the case, concepts would not be governed solely by considerations pertaining to the range of reality controlled by being as presence.

Second, the presence of a given concept is not sufficient to confer an appropriate degree of self-reference unique to that concept. The presence proper to a concept once it is taken in isolation does not engender the fact that the concept is *this* concept and thus that it can then refer to itself. A caveat: Self-reference as Derrida uses it here functions in divergent senses and these senses are distinguished below. But Derrida can maintain this denial at all if and only if presence *admits of degrees*, given that gradations by degrees is not necessitated by the position concerning presence discussed in the previous paragraph. Thus the presence of a concept taken in isolation (*per impossibile* for Derrida) is in some sense less than the presence of that same concept when it exists as a member of a "chain or system" of concepts. Only such membership endows the given concept with its full stature *as* a concept existing with respect to the factor of presence in such a way that it displays its own self-reference.

Derrida claims that the presence of a signified concept is inadequate to sanction the concept existing in such a way that it would "refer only to itself." There are two senses in which a concept could refer "only" to itself. First, it seems at least plausible to maintain, against Derrida's claim, that a given concept would always refer only to itself if the concept was considered with respect to self-identity–i.e., minimally when regarding the word (representing the concept) as a mark, an assembly of letters with an assigned signifying status.Thus a concept can be related to itself as A = A (e.g., deconstruction = deconstruction) without that concept having any link whatsoever to anything other than itself.

This self-identity is purely formal, assuming that the principle of identity, A = A, stands as axiomatically necessary for the logical intelligibility of Derrida's position. But self-identity also functions in a substantive sense. Thus it is a question what goes into the constitution of a concept that makes it not only exist at all, but continue to exist as *this* concept, i.e., with a determinate identity. A concept continually fluctuating with respect to content will surely cease to exist *as* a concept. If so, then the dimension of substantive self-identity applies to concepts with as much relevance and force as it applies to extra-mental entities existing in the spatio-temporal world. Even if this distinction is granted, however, we observe that *formal* self-identity (A=A) holds at least from the perspective of traditional metaphysics even when *substantive* self-identity is subject to interpretation, i.e., what divergent systems of metaphysics contend constitutes the identity of a given concept (or a given entity).[62]

Derrida asserted that the principle of identity is the fundamental principle of being as presence (**Passage 1L**). Thus every concept exists and functions within what can be referred to as an "aura of presence" merely by virtue of the apparent necessity that this principle can be predicated of anything and everything, whether individually or as a member of a group. Furthermore, a concept is at least intuitively a very different sort of entity than an extramental object, but does it follow that the presence surrounding a concept of an object is the same or different from the presence surrounding that object itself? The use of "surrounding" is potentially misleading since in one sense the object under scrutiny simply *is* presence. But that qualification aside, we may nonetheless conclude that it remains unclear what exactly that aura is and how exactly it functions with respect to whatever exists within it–the concept as indicated by an orthographic mark as well as the entity seemingly referred to by that concept. Also, in light of the conclusions established in Part I, if as argued here being as presence does indeed admit degrees, then it is all the more essential to clarify the structure of being as presence–now taken as a unity–in order to understand how this unity can be differentiated for purposes of establishing not just *possible* but *actual* concepts. We therefore appreciate the theoretical need to clarify being as presence, both with respect to the principle of identity itself in the purely formal sense commented on in this section as well as the supposed application of that principle to concepts regarding their determination by being as presence.

5.5 The Structure of Chains

Derrida asserts axiomatically (**Passage 5C**): "Every concept is necessarily and essentially inscribed in a chain or a system, within which it refers to another and to other concepts, by the systematic play of differences." If Derrida's account developing the notion of a chain is itself a chain, then we may inquire whether assumptions are at work in the utterance of this position. If so, then these assumptions must be extracted

62 For Derrida on self-identity, see WD, 82, 296, 318; SP, 50, 521; MP, 278; PS, 222; Kearney, 148-9; Derrida 1992, 47, 65-6, 67-8 (i.e., regarding a work of literature as "singular"). Self-identity as a formal principle (A = A) must be distinguished from identity conditions predicated of entities or persons (Derrida commonly ignores the distinction–see the passages cited above for examples). For a useful introduction to positions available on the predication of the substantive sense of identity, see the accounts of "identity conditions" and "criteria of identity" in Chapter 2, "Some Tools of Metaphysics," in Cynthia Macdonald 2005, 36-76. It is fair to say that Derrida's various treatments of what he refers to as self-identity do not reach the level of specificity required to be relevant to the metaphysics of identity. For a more pointed critical analysis of Derrida on the formal character of the principle of identity, see DFL, 154-61, 244-5n1.

and brought into the open for critical appraisal with respect to both (a) deconstruction as an attempt to buffer the history of metaphysics and (b) metaphysics proper to the extent that the elements of a chain exhibit the same characteristics and bear the same names as a series of traditional terms in metaphysics.

The inscription "in a chain or a system" is not elaborated, so we assume that the disjunction differentiates between "chain" and "system." (This differentiation need not obtain if "or" expresses identity, i.e., that "chain" and "system" are denotationally equivalent.) In what follows, we also adopt an intuitive understanding of "chain" and apply this to "system" when appropriate. Thus we take a system to be a set of chains–e.g., deconstruction instantiates a system, with *différance* and iterability construed as chains belonging to that system. Also, iterability by itself is a system constituted by a number of chains (see Chapter 7 for details regarding the sets of concepts constituting these chains), each linked with one another within the limits of that system. Another example: the account of the structure of Forms in the *Phaedo* is a chain with respect to the articulated properties of the Forms as such, whereas the participation relation between particular entities and Forms constitutes a system.[63]

We now outline a series of factors–each possessing an historically extended metaphysical pedigree–all of which apparently pertain to the structure of a chain:

5.5.1 Self-identity

An existing chain requires self-identity in the formal sense to insure the integrity and unity of an individual concept so that concepts as elements in a chain (of concepts) can remain distinct from one another. Without establishing and preserving these differences, a chain ceases to be a chain in the sense that the various "links" constituting the chain would become indistinguishable from one another except nominally, i.e., each link identified only as a concept numerically distinct from another concept–$concept_1$, $concept_2$, etc. Although a metal chain used for prisoners, fences, animals, etc., comprises links typically identical to one another except for numerical individuation, concepts in the epistemic sense are presumably more complex by virtue of also being distinguished from one another with respect to content, i.e., what the concepts are concepts *of*. More will be said below concerning how this "of" functions but the disanalogy just indicated must be kept in mind.

63 The assumption is that a chain can be part of a system and that a system as a grouping unity constituted by a set of chains can itself be a chain as a part of a yet more inclusive system.

5.5.2 Relationality

What provides impetus for one concept to enter into a situation involving relation? Each concept must contain a sort of impetus to move toward something outside itself. If this approach to the phenomenon in question is phrased too animistically, we might tentatively accept the prior existence of *patterns of deployment* by means of which each concept necessarily functions in relation to at least one other concept–and, Derrida insists, to a number of other concepts. How then is such arrangement and configuration possible?

5.5.3 Concepts and Otherness

When a concept does relate to something other than itself, what are the "other concepts" to which Derrida refers? How many "other" concepts–*all* other concepts or only *some* other concepts?

Derrida needs being as presence as a premise in order to argue that every concept functions as an essential member of a chain of system–i.e., each concept must have some degree of being as presence in order to exist even as a concept. If therefore all concepts exist only as exemplars of being as presence, then it follows that any one concept exists in relation to every other concept at least with respect to the ultimate principle of Derrida's position, i.e., that all concepts, to exist in the first place, become instantiated in terms of being as presence.

However, Derrida presumably would not assert that *any* given concept must link to *every* other concept before the given concept can become a vehicle of specific signification. For if so, then it would follow that no one concept could be meaningful and exercise its own distinctive range of significance unless and until that concept was explicitly related to *all other concepts*. Furthermore, if each concept required all other concepts in order to constitute a chain just to realize the functionality of that one concept, then *all* concepts become equivalent to one another because each and every concept would have exactly the same links in its respective chain.

If these consequences are ultimately untenable, it follows that the number of elements in a given chain must be fewer than the number of possible concepts. The chain for a given concept must therefore contain a limited number of other concepts. In addition, if concepts differ among themselves, then so do their corresponding chains. Any appeal to a chain thus presupposes a coordinate appeal to a *limit* whereby that chain establishes itself as a chain and also distinguishes itself from all other chains (to preserve uniqueness of given concepts with respect to one another). Otherwise put, there are limits on the number of concepts necessary as a set to generate a sufficiently interrelated unity to allow any one concept to come into existence *as* a concept of distinctive scope.

5.5.4 Principles of Gathering

If a given concept can become related to multiple other concepts, what happens to a proximate but random juxtaposition of concepts so that this grouping becomes transformed into an ordered, unified sequence of concepts? If a number of individual links are heaped together, the result is a pile of links–not a chain. A *principle of unity* must therefore be present to the individual links so that their arrangement produces a chain. What provides the principle of unity for a chain?

It may also be noted that if the principle of unity for a given chain is distinguishable from that chain as such, then this abstracted principle of unity exists as an independent reality (even if only in a logical sense), apart from both the particular instance of the chain as a set of links as well as the links themselves taken discretely. If we make the quite plausible assumption that many such chains are both possible and actual, where and how do these unities as the ordering principles of chains originate?

5.5.5 *Différance* and the Origin of Concepts

For Derrida:

> **Passage 5C$_1$** Every concept is necessarily and essentially inscribed in a chain or a system, within which it refers to another and to other concepts, by the systematic play of differences.

But appeal to "the systematic play of differences" enters the explanatory picture only *after* individual constitutive concepts have already been established. Derrida adds that this "play" is "the possibility of conceptuality, of the conceptual system and process in general." Therefore, to assert that a concept *results from* the "systematic play of differences" is arguably inadequate in order to account for the distinct character of both a given concept as well as the set of concepts constituting a chain. In fact, "play of differences" denotes only, as Derrida explicitly says, "the *possibility* of conceptuality [my italics]" rather than a determinate concept or, in a chain, a set of concepts.

An existing set of chained-concepts will indeed generate a series of relations marked by sheer otherness. If concept A is other than concepts B, C, etc. and the same indeterminate "othernesses" hold between and among all other concepts in this grouping, then although this conjoint set of othernesses has engendered a series of "spaces" between and among the members of this set, the distinctive character of concepts A, B, C,–i.e., what concepts A, B, C represent–is not considered in Derrida's text. Thus the "play of differences" does not by itself effectively clarify the origin of the distinctive character of any given concept, much less any set of concepts insofar as elements of this set constitute a chain.

Furthermore, Derrida has prefaced the reference to "play of differences" with "systematic." And in view of the argument in the previous paragraph, this modifier does heavy and essential labor. Thus if the play of differences for a concept includes taxonomic ordering, then it would be possible to recognize that a given concept in terms of its distinctive content is related to all other concepts *as they have been ordered into a unity within the limits established by a given chain.* In turn, chains will be systemically deployed within the limits of a given system.[64] But according to this explanation, the specific content of each concept regarding its location within systematizing as an ordering principle of unification becomes just as important, if not more so, than the "play of differences" establishing the intervals of otherness separating each individual concept from all other concepts deployed within given chains and systems.

How then does *différance* establish the possibility of a concept? The "play" factor of *différance* must be stabilized in order to generate concepts in the first place, quite apart from and logically prior to their subsequent ordering within the limits of a given chain. If the system as a principle linking and coordinating concepts has greater specificity, then the only way to establish this characteristic is to indicate what the concepts are *of*, to what they refer. There is no obvious evidence from Derrida's texts that he has developed his position in that regard.[65]

In brief, where does the unity producing this stabilization originate? What is the principle of limitation restricting the relevant range of associated concepts insofar as one integrated set of concepts constitutes *the* chain under scrutiny?

5.5.6 The Extension of the Transcendental

If we assume that all the questions in this section of Chapter 5 have been addressed and answered thereby resulting in an account of an actually existing chain, then it can fairly be asserted that a "transcendental aura" characterizes Derrida's approach to the very notion of chain, at least in terms of what the appeal to a "chain" as the foundational component with respect to the existence of integrated concepts arguably presupposes. This aura is exhibited in two respects: (a) that a concept such as *chain* can be isolated and explained at all; (b) that *unity* as a notional necessity

---

64 I have employed "systemically" and not "systematically" on the assumption that the principle involved in ordering a system is in some respect (or respects) distinct from the principle ordering a chain. I have not located a text where Derrida addresses this hierarchical implication.

65 The notion of "play" is crucial here for Derrida. It will be discussed again from various perspectives later in this chapter, in the section "*Différance*, Chains and Possibility." For representative passages on play, see OG, 7; WD, 19-20, 62, 250, 260; 263, 278-9, 291-2; SP, 159; Derrida 1992, 65-6; LI, 116, 145, 149. See also DFL, 232n9 (with secondary sources cited).

underlying any and all such chains can be assumed as a structural element by itself, independently of its predication to any single given chain. The recognition of this extension underscores Derrida's implicit dependency on metaphysical notions of classic vintage even in the midst of an account which purportedly details an interpretive stance–incorporated within the confines of *différance*–enabling a proponent of deconstruction to move beyond the theoretical and doctrinal limits imposed by the content of the history of metaphysics.

5.6 Chains and Referentiality

If the deconstructive approach to language accepts as a given the position that propositions exhibit referentiality, then it becomes a question whether concepts, especially insofar as concepts are expressed through language, function in the same or at least similar way. In response, we will assume that concepts have some sort of referential function, that is, a concept refers to something outside itself. Of course, we are required by Derrida's position to accept that no one concept can function as a concept in isolation from a chain of other concepts. If, therefore, the concepts in a given chain differ among themselves, it follows that the referential functions of a chain of concepts will as a collective entity span considerable sectors of extra-conceptual reality.

A chain of concepts might therefore encompass vast differences in space and time even if these differences reside within a "mental" dimension of significance, that is, a dimension within or along which concepts are generated and related to other concepts. Consider referential instantiation of "rational animal:" Some animals are rational and some are not, so to predicate rationality is to envision rationality inhering in these animals and concurrently envisioning other animals to lack this property. But to do so at a level of instantiation is to situate occurrences of these concepts in spatio-temporal settings co-existing with other entities whether natural or inanimate (or both), etc. Therefore, if concepts have such referencing capabilities, the application of concepts presupposes conditions with a determinate metaphysical structure. This consequence is crucial, as we shall see.

5.7 Chains and Signification

Derrida asserts that a given concept can "refer" to another concept and to a series of concepts insofar as that series, taken as a unity, constitutes a chain. Thus in addition to its assumed referential capacity, a concept can get outside of itself, so to speak, and move toward another concept, another entity of presumably the same general type but also distinguished in some respect from the original "referring" concept. The question is whether a *chain* as such can also, analogously, get outside of itself in order to evoke something external to the chain, that is, something or things existing in a non-conceptual domain (where "non-conceptual" refers to entities existing in a mode other than whatever may be the appropriate mode of characterizing what is exclusively conceptual).

If chains do function in this regard, how is the process accomplished? Does the chain as a uniform set of distinct concepts have a distinctive transcendental signified? For if individual words even if set within a proposition never reach a transcendental signified, then what reason is there to believe that a chain of individual words will be any more successful? Concepts presumably refer to something extra-conceptual; otherwise, the traversal of a given chain as a set of concepts never moves beyond circumscribed dimensions of the chain and gets "to" something that a chain as a series of concepts evokes or represents. It thus seems necessary that a chain construed as a kind of organic unity must win access to something like a transcendental signified. This consequence is necessary to justify the conviction that discourse and thought insofar as they develop from and depend on chains (of concepts, following Derrida) are not self-negating and therefore self-stultifying processes. If the chain does indeed establish the possibility of variable meaning, then Derrida is again implicitly appealing to a notion of unity in order to guarantee that the individuality of distinct words can somehow be combined into something more than a linkage of autonomous and in principle unrelated–indeed, unrelatable–linguistic markers.

And yet, if Derrida denies that a transcendental signified can be assigned to a concept as individual element, it is difficult to understand how he could maintain that a chain–by definition, a set of concepts integrated with one another so as to establish significance of any one of those concepts–possesses a counterpart transcendental signified. As a matter of strict logical correctness, note that if the relation between concepts and their residence in a chain is of parts to whole, then to argue that what holds for the part (i.e., absence of a transcendental signified for any concept) must hold for the whole (absence of a transcendental signified for the set of all concepts constituting a given chain) constitutes a straightforward instance of the fallacy of composition. But the question remains–do chains as unities display a counterpart to the notion of a transcendental signified?

Earlier in this chapter, we assumed that "signified concept" was a feasible formulation, given Derrida's position on the notion of the transcendental signified. But now it becomes necessary to ask again:What is a concept for Derrida? Whatever it may be, a concept would not seem straightforwardly reducible to something existing the way a transcendental signified exists (a reduction also illicit on the grounds of internal inconsistency). For if concepts are not only possible but actual and the structure of a concept–its unity conferred by residence as a link in a chain–is not clarified with respect to the implied rejection of anything equivalent to a transcendental signified, then there is no accessible difference between what Derrida is implicitly *asserting* with respect to the existence of concepts (and their presumably precise referential content) and what he is *denying* with respect to the dimension of the transcendental signified.

The problem may be stated thus: if the denial of an existent transcendental signified encompasses all possible concepts as generated by language, then how is a chain of concepts moored, so to speak, so that any one link in the chain–any given concept–is relatable to any other concept in such a way as to engender the possibility of significance? Even if this succession of links is somehow produced, is there a connection between the resulting chain and anything "outside" the chain, e.g., anything which the language leading into the formation of the chain has referred to within the contexts of ordinary discourse? It certainly seems plausible to read the account of a chain as if chains are connected to the extra-conceptual world in some necessary way. But the question is whether a chain is related to some feature of the extra-conceptual world–the apparent "object" of the chain (as in the instances of language referred to by Derrida's positions on deconstruction when taken as examples of chains)–identically or differently than the way an individual concept is related to that world.

If the relation is identical, then there is no reason not to include a chain of concepts as affected by the same conclusions Derrida applies to concepts taken individually; if the relation is different, then it seems to follow that a chain can penetrate the penumbra of language and presence and somehow "reach" extra-conceptual realities without inherently affecting the metaphysical character of those realities. If we assume therefore that a concept to be effectively functional must (a) connect to something determinate in extra-conceptual dimensions of reality but due to deconstructive allegiance we also (b) do not want to endow a concept with illegitimate structure, we may refer to a "self-generative conceptual factor" as the theoretical desideratum. This designation distinguishes what makes a concept to be a concept from all the strictures laid against the notion of a concept being equivalent to or necessarily related to a transcendental signified.

A brief concluding note, it seems safe to say that the matter of signification requires additional analysis in all aspects specified in this section regarding the generation of concepts.

5.8 Chains: the Interplay of Unity and Limit

To appeal to the notion of chain in an account justifying the function of concepts presupposes the effectiveness of *unity* as a source of the chain's coordination and integration of elements. There are, in fact, multiple senses of unity at work in this theoretical setting. The different ways concepts are unified within chains in conjunction with the presumed referential content of concepts renders the unity exhibited by a chain to be an essential factor in this phase of Derrida's formulation of deconstruction. It is also necessary to recognize the integral connection between the content of a chain as a collective unity and the limit to that content as well as the relation between this limited content and what is other than or "not" that content.

We therefore continue to elicit presuppositions in Derrida's notion of "chains," with special attention to unity and its various deployments. As we shall see, this discussion of distinct types of unity initiates a return to the viability and theoretical necessity of contexts.

1. Unity of a concept as such: This measure of unity ensures integrity and self-subsistence of each concept. We will not speculate on possible "parts" of a concept (as Aristotle did regarding the parts of a definition, i.e., a matter of equivalent abstraction and importance given his project at this juncture of the *Metaphysics*).[66]
2. Unity of a chain: what provides the principle of linkage among the elements of a given chain? If chains are collections of concepts linked in some way, then any one chain (for a given set of concepts) differs from every other chain (for sets of all other concepts constituting chains which are not the given chain). These differences may take two forms, thereby requiring a different kind of unity (if we assume that different kinds of unities are necessary to account for different kinds of chains). The unity of a chain could be numerical or substantive (or both)–depending on the number and type of concepts in a chain, assuming that concepts as such differ among themselves in some substantial way.
3. If chains are hierarchically related to one another (e.g., genus/species arrangements in classical metaphysics), then the unity of a "higher" level of chain differs from unities of all "lower" levels of chain. Again, we seek clarification concerning the principles of unity marking these lower- and higher-level classes of chains.
4. If a system differs from a chain, then it is possible (we have assumed) that a system is a group of chains, e.g., the system of metaphysical concepts, principles, and arguments in Aristotelian metaphysics. If so, then the unity of a system qualitatively differs from all types of unity applying to each element of that system. This sense of unity captures the differences between and among, e.g., Aristotle, Leibniz and Whitehead as systematic metaphysicians. Each system differs from all other systems, so the unity which consolidates each distinct system A, B, C, etc., to be one system of coordinated components differs from all other types of unity.
5. If concepts can interlock in order to generate significance within the scope of any one given chain, then it may be assumed that (a) chains can interlock with other chains and (b) groups of chains can interlock with other groups of chains. Thus a multiplicity of chains characterized by differences in conceptual interaction within each chain presupposes a multiplicity of different types of unity–unities in commerce, rules of a game, sciences, subdivisions of science, e.g., sociology vs. physics, etc.

It is also evident that different kinds of chain will interlock–jargon, shop talk, trade talk, technical terms all exemplify chain-structures and it is possible that each of these

66 The parts of a definition are discussed in *Metaphysics* VII, 1037^b8–1038^a35.

types will intersect at least once (and doubtless multiple times) with other types. But what principles allow *different* chains and *different kinds* of chain to interlock? At issue are multiple senses of unity; indeed, Derrida's own texts represent chains of concepts, with different chains of concepts linking different thematic concerns circumscribed within the ambit of deconstruction.

Without investigating and articulating these senses of unity, the appeal to chains as the underlying source of connectivity establishing the reality and applicability of concepts remains ungrounded and arbitrary.

5.9 Deconstruction and the Question of Unity

Chains would not be chains without a unifying function. And unifying functions are what they are because they embody a unity. The question now is whether unity can be isolated and considered in its own right, as a kind of reality. The previous section described a number of clearly distinguishable senses of unity, all actively present (whether explicitly or implicitly) in Derrida's account of chains of concepts.

There are junctures in Derrida where he has explicitly rejected unity as a relevant feature of the philosophical landscape under consideration.[67] In addition, Derrida has effectively denied the notion of unity in his rejection of the concept of a context as justifying a single, inherently stable setting for the determination of significance during a linguistic occurrence. But can the ensemble of unities structuring a chain remain impervious to the stated goal of deconstruction to shake and disrupt all that is hallowed in the history of metaphysics? To answer this question, the advocate of deconstruction must address *the concept of unity as such*, as a formal concept or reality which imposes limits and stability on the inherent indeterminacy of thinking when it confronts the welter of the extra-linguistic world as filtered through the torrents of language available to describe and evoke that world. Both unity and limit perform transcendental functions in this regard; as such, both are decidedly metaphysical and it seems completely appropriate to anticipate that both will at some point be critically analyzed by deconstruction.

---

67 In the essay on *différance*, Derrida rejects the existence of an "organic unity" which can serve as the ground for an event (SP, 143–cf. **Passage 6D$_1$**). LI, 60 has a related observation on the connection between context and unity. See also WD, 38; LI, 75. In addition to the discussion in Chapter 4, Chapter 6 below develops a critical account of unity in Derrida. And for additional discussion of unity as a background concept for Derrida, see DFL 57-62, 127-30, 226n5.

5.10 *Différance*, Chains and Possibility

The notion of "play" appears frequently in Derrida as it does in **Passage 5C**'s reference to the "systematic play of differences," which Derrida immediately reidentifies as an equivalence for "*différance*." The qualification "systematic" apparently precludes taking "play" in a purely unstructured way; thus differences accruing from *différance* will cohere with one another within the confines of a given system. It also follows that play is an inherently futural perspective in this speculative context since all systems in the history of metaphysics are "set" in an empirically determined way, i.e., as texts. Although there is room–or "play" as the word is used to describe a rope that is not taut or a "loose" steering wheel–in a past metaphysical system for the generation of different readings of that system, the kind of novelty which would be capable of brandishing a radically distinct formulation of language for philosophical purposes awaits its protagonists. But this futural dimension of play is not completely open-ended; the following qualifications establish contours for the effective range of *possibility* regarding conceptuality. These considerations will become crucial in Part III of this essay.

5.10.1 Novelty

This characteristic pertains to systems not yet realized. However, does possibility include products which in their novelty can diminish or escape altogether the restrictive effects of being as presence?

It is not obvious that when a fully-enforced deconstructive treatment of a system has done its work, the results will qualitatively differ or be distinct from what great metaphysicians of past philosophy have produced. Will novelty include, for example, systems eschewing principles of unity or limit? If concepts of such radical novelty are possible, then the play of differences at a point prior to the formulation of these concepts must include the active agency of *imagination*, since at that juncture the proposed concept exists only potentially, that is, as an intersection of what could be (in the sense that something is logically possible, i.e., not contradictory) and what can be imagined to be. It appears that the only restriction on "play" and Derrida's use of "possibility" rests on the imagination and what it can conceive or "image" regarding the coordination of whatever may be given.[68]

---

68 The open-ended quality of the concept of possibility when integrated with Derrida's pivotal notion of "trace" becomes an important element in the argument of Chapter 9 below.

5.10.2 History

Can novelty arise in a self-contained expression fully sealed off from the history of metaphysics? Will residual effects continue to issue from the history of metaphysics on any and all new systems, systems now realizable through deconstructive efforts to shake the rigidity of that history? The status of the history of metaphysics in relation to the possibility of novelty must be addressed. (Chapter 10 contributes to this discussion.)

5.10.3 Givenness

However novel and distant a post-deconstruction system may be when juxtaposed with the entirety of the history of metaphysics, its "metaphysical" components will be seemingly contoured by the sheer facticity of what is "given," e.g., the presence of materiality in the natural order. Furthermore, the influence of what is given extends beyond the isolated instance of some particular given thing if we assume that possibility in this speculative and radical sense will continue to entail metaphysical presuppositions.

Consider Hume's example of the sun not rising tomorrow (*Enquiry*, Section 4). This possibility could eventuate only if a number of factors also changed as well as the distinctive mechanical action of the sun as one heavenly body among countless others, all subject to physical forces and circumstances. If the sun did not rise tomorrow, then the concept of gravity (and, doubtless, many other concepts) would have to be rethought. This example is taken from the lifeworld's material dimension but it illustrates that the post-deconstructionist must attend to the effects of the given on the scope of a novel, historically liberated and revamped metaphysical vision.

5.10.4 Logic

Any post-deconstructive instance of novelty will exhibit a unique sense of possibility from the perspective of *différance* in terms of its generative relevance to the history of metaphysics. But will the "formal" concepts found in traditional metaphysics be affected or, perhaps, displaced entirely? One example: will this dimension of possibility exhibit the property of self-identity when self-identity is predicated of it as well as any specification occurring within an emergent systematic expression? The dimension of "second-order" terminology–especially everything typically associated with formal logic–will have to be discussed in order that any post-deconstructive system of reflection can remain standing apposite to the traditional guidelines of intelligibility with respect to metaphysical inquiry.

5.10.5 Transcendental Dimension

Derrida appeals to possibility concerning generation of conceptual chains and systems via the underlying sway of *différance*. But how does *différance* establish from a realm of pure possibility the actualized existence of determinate chains of concepts? The conditions introduced in this section show that the notion of possibility must follow certain guidelines and parameters. If so, then in this respect we are justified in reading Derrida as continuing to think conceptuality according to a kind of transcendental formulation.

The argument is that it should be part of the goal of deconstructive interpretation to be as fully aware as possible of the conditions which must be met or at least taken into account in order to move beyond (1) the statement of deconstruction as an ensemble of interpretive techniques and strategies and (2) the application of deconstruction to given texts, whether philosophical or literary in their provenance, in order to realize (3) a positive expression of thought and language aimed at the lifeworld from which all philosophical activity ultimately originates. The "play of possibility" should therefore be a form of endeavor practiced to the fullest degree possible but, as argued in this section, with adherence to a set of limiting factors pertaining to its results.

5.11 Chains and Contexts Revisited

A metal chain viewed abstractly becomes an ensemble of part and whole, multiple parts and one whole or, depending on the type of chain, multiple parts and multiple wholes. For Derrida, a conceptual chain is a more involuted ensemble of whole and parts. In fact, it should now be evident that Derrida's position concerning concepts linked as a constituted "chain" carries with it presuppositions which, if examined, closely resemble structural considerations inherent in the traditional notion of a context. The problem: what distinguishes a chain from a context? The relevant conditions for considering and resolving this question are complex.

5.11.1 Metaphysical Assumptions for the Promulgation of Chains

If a concept includes within its structure viable connection to a group of extra-conceptual entities, then legitimacy of a concept depends on this connection being necessary and recognized as necessary. Without such necessity, a putative concept lacks any content, any means of making its presence felt, so to speak, as a vehicle of cognition (cognition understood very broadly).

All concepts exist by dependence on the universality of being as presence. If therefore a system as a mode for linking concepts has *qua* system greater specificity, the only way to establish this characteristic is to indicate what its concepts are *of*, what they refer to outside their own structural character as signs and independently of the commonality of being as presence which marks all entities, whether conceptual or extra-conceptual. Derrida's texts do not clearly indicate that he has developed his position in this regard, i.e., the content-specification dimension of concepts. But the following considerations appear necessary should such development be attempted.

If a concept has a measure of reality distinguishable from its necessary characterization by being as presence, then the possibility of concept-formation must connect the content of the concept with some sort of determinate extra-conceptual specifications. For example, the concept "moral action" would normally be connected to human beings, who exist within a spatio-temporal setting as a necessary condition for the possibility of pursuing moral action. This consequence can be generalized for virtually any concept.

If Derrida is committed to the existence of concepts as an element in human cognition (to use traditional terminology in order to identify the relevant referents), then he must either (a) posit a realm of chains lacking any tie with elements constituting the necessary regions of standard human experience (e.g., space and time, the existence of entities as, on occasion, referential objects of concepts) or (b) by accepting such ties in order to "realize" the epistemic and practical relevance of concepts, acknowledge the necessary metaphysical background as supplying contextual conditions underlying the possibility and effective range of concept-formation. Alternative (a) preserves "purity" of deconstructive description when its results are separated as much as possible from the pervasiveness of being as presence but alternative (b), although reintroducing elements derived from the history of metaphysics, allows an advocate of deconstruction to participate in the lifeworld as we conceive and interact with it during the epoch of metaphysical inquiry. Assume then that (b) is arguably preferable to (a).

The introduction of this background factor renders chains more metaphysically divergent than might appear if the "chain" image applied to linkage of concepts is taken literally, according to the visualized model of a chain tethering an animal. Thus the greater the number of concepts constituting a chain, the proportionately greater increase in metaphysical considerations underlying the possibility of referentiality for those concepts.

We have assumed that a concept has referential relevance to the realm of whatever is extra-conceptual; in this respect, concepts function like words in propositions with referential content. However, we have seen that Derrida's appeal to the chain factor in the establishment of a concept depends for its cogency on the much more abstract notion of *unity*, and unity deployed in a variety of circumstances with, presumably, a correlative variety of structural features. How is such a unity generated in order to establish the possibility of a chain?

The problem revolves in part on the question of evidence. When extra-linguistic objects and properties–plurality, individuality, sortal terms, space-time, etc.–constitute the elements linked in some way to one another in order to produce a chain, what provides the principle of connectivity and the foundational bases for the relevant senses of unity? Consider perception standing as the experiential function grounding unity–the controlling factor of a chain–when unity is predicated of a set of objects and properties, and compare that setting with a more extended set of objects when the evidence for the possible unity of this extension is provided by, e.g., imagination as a supplement to perception. Does the status of extra-linguistic objects as imagined entities provide the same kind of evidence with respect to the origin of unity as that of objects when their presence is experienced through perception? And in general, how are the elements of a chain correlated and gathered so that they coexist and display the unity requisite to exist as *one* chain?

5.11.2 Chains and Contexts–Isomorphic Assumptions

The analysis in this chapter of Derrida's appeal to and reliance on concepts has demonstrated the complexity and diversity in the array of background assumptions which ground that reliance. In fact, at a level of high generality, we recognize that the conditions required to establish the possibility of chains are isomorphic with the conditions required to accept the notion of a context if a context is construed as sufficiently stable to warrant inclusion as an essential component in the determination of referentiality and meaning. We are well aware that Derrida denies the purely theoretical reliability of this construal when applied to the notion of a context. But my thesis is that the minimal conditions for establishing the possibility that chains of different concepts can function with significance are of sufficient complexity that, taken as an integrated set, these conditions become virtually commensurate with if not parallel to conditions for maintaining the continued existence and relevance of contexts as necessary elements in any account of meaning and signification. Consider: spatio-temporal coordinates, physical objects, relations between and among these objects, the intentionality of consciousness with respect to these objects, discourse and/or the possibility of discourse–these factors are integral features of *both* the implementation of concepts in the lifeworld *and* the occurrence of a context with respect to a linguistic event.

This tight correlation between the underlying characteristics of chain and context implies that deconstructively denying the legitimacy of the latter must be clarified, since arguments aimed at rejecting or even questioning the reliability of individual contexts will have the same effect on chains if contexts and chains are virtually interchangeable regarding their respective actualizations. The other direction of interpretation is that if chains are legitimate bearers of significance regarding concepts, as Derrida forthrightly and frequently maintains, then so are contexts as structuring the interpretive accessibility of a wide variety of linguistic events. This congruence of conclusions has crucial consequences.

5.11.3 Concepts and Contexts–the Challenge of Inclusive Preconditions

Consider the example drawn from the lifeworld of one person asking another person for something. The setting which underlies the intelligibility and success of this utterance is complex–spatio-temporal intersections, natural substances or technology (or both), the formulation and utterance of a request, interpersonal relations, conceptual grasp by way of understanding, concrete and completed action. But it cannot be denied that *within the circumscribed limits* of this complexity, the language articulating the request is clear, set and meaningful (in the Forest Scene as well as the classroom scenario depicted at the outset of this chapter). Human action is intended, vocalized, established, understood and consummated. Questions of residual motivations and imagined consequences may be associated with this event as auxiliary considerations, but as such these considerations remain purely potential events relative to the actualized event at hand. And, given the event's realization, *none* of these potential-as-imagined events occur. The event as such, e. g., a request for a glass of water and the resulting acceptance or refusal, encompasses a complete scene, albeit one of small consequence in the grand drama of human affairs. Furthermore, I submit that the majority of human theater is, in fact, composed of such small scenes. The successful interrelationship of all these diverse elements is testified to by the number of linguistic events which fully and finally achieve their goal in the lifeworld.[69]

We introduce "threshold stability" to name the set of metaphysical concerns underlying the feasibility of a context as a theoretical necessity for signification. This notion refers to a segment of space-time circumscribing the immediate vicinity of the speaker (or speakers) of language. The entities (other than speaker or speakers) inhabiting this segment remain constant; temporal divisions and durations are continuous, not subject to sudden or volatile variations and changes. Beyond this threshold region lie, true enough, extensive sectors of space-time (theoretically

69 Consider the following from Dewey's *Experience and Nature*, 179-80: "Meaning is not indeed a psychic existence; it is primarily a property of behavior, and secondarily a property of objects. But the behavior of which it is a quality is a distinctive behavior; cooperative, in that response to another's act involves contemporaneous response to a thing as entering into the other's behavior, and this upon both sides. It is difficult to state the exact physiological mechanism which is involved. But about the fact there is no doubt. It constitutes the intelligibility of acts and things. Possession of the capacity to engage in such activity is intelligence." This eloquent passage elegantly codifies a theory of meaning by integrating speakers with each other and with the objects which concern them. Dewey's admission that "the exact physiological mechanism" constituting meaning is "difficult to state" would perhaps offer Derrida an opening to move past physiological opacity and into the sort of sustained philosophical skepticism he desired to promulgate. But Dewey remains adamant–"...about the fact [of meaning] there is no doubt." The reader will recall the second motto of this essay, from Henry James: "...the sentiment that words were acts and acts were steps in life."

without limit given our cosmic epoch), a locus encompassing a much broader sphere of possibility in terms of variations regarding entities, relations between and among entities, and configurations of temporality.

Now by contrast, it is evident that conditions for generating a concept must be more rigidly defined and specified than the counterpart conditions for situating a context. A context can be perfectly stable regarding its constituent elements as a precondition for establishing meaning and reference yet remain somewhat hazy around its periphery. But a concept–and *a fortiori* a chain of concepts–cannot afford to be so indeterminate as long as concepts function as the elemental components of knowledge and ordered experience. If concepts as such are structurally and inherently fuzzy, then it is difficult not to infer that any conclusions based on configurations of such concepts when they are deployed in propositions will be commensurately and necessarily fuzzy as well. The result–a continually teetering control of knowledge and an unstructured, disordered randomness in experience.

The complex arena of assumptions comprising metaphysical conditions necessary for implementation of concepts is in fact greater, i.e., more inclusive, than the counterpart set of assumptions for contexts. Therefore, if Derrida is a proponent of fully realized concepts, there is no reason why he cannot also support contexts.[70] More rigorously stated, Derrida is logically inconsistent if he advances the intelligibility of concepts but denies the relevance of contexts as a ground for interpretive stability. Indeed, the set of limiting concepts constituting a chain could be designated as a determining *context* for the meaning of each individual concept constituting that chain. Thus if every concept has a limited set of factors circumscribing the existence and application of that concept, then this measured segment of totality designates what could legitimately be called as its contextual relevance.

In effect then, Derrida implicitly supports the notion of "threshold stability" in order to maintain the functionality of chains regarding applicability of individual concepts as well as applicability of concepts when governed by a chain's unifying agency. Derrida's "chain" as consisting in individual concepts is merely another name for a series of interlocking contexts with the understanding that when a given concept links to its successor, the successor-concept does not instantly exist *ex nihilo* but emerges from a setting which possesses many of the same fundamental metaphysical

70 In *Limited Inc,* an extended defense against the criticisms of John Searle, Derrida insists on the stability of contexts, e. g., as found in his own writing. These exceptions and the resultant matter of consistency with Derrida's avowed theoretical position regarding contexts will be discussed in detail in Chapter 9 below. See also in this regard John Ellis 1989, 13n10. On the variability of contexts in the study of literature, see Derrida 1992, 41-2; see also MP, 310, 316, 320; PS, 144. For a critical assessment of Derrida's position on contexts from the perspective of formal logic, see DFL, 12-5, 48-9, 63-7, 73-4, 227-9n8, 230n15. And for discussion of the crucial relevance of context with respect to Derrida's reading (in *Dissemination*) of "*pharmakon*" in his interpretation of Plato's *Phaedrus*, see DFL, 228-9n9.

differentiations as those underlying the appearance of the original concept. Without this subtle allegiance to context, Derrida could not have developed the variety of deconstructive ploys which his advocates find so compelling when they use them in a variety of literary and philosophical...contexts.

5.12 Summary

Derrida's endorsement of concepts as viable elements in pursuit of the philosophical enterprise raises a series of fundamental issues–among the most prominent: what distinguishes a concept as epistemic vehicle from an instance of the "transcendental signified"? If concepts are in some way stable and apparent instances of the transcendental signified can never be stable, then it becomes problematic whether a concept exists as medium for knowledge of extra-linguistic reality. Indeed, how are concepts formed at all and, once formed, how can individual concepts relate to other concepts in ways which exhibit by such gathering modes of order and stability?

Various responses are available for these questions–responses drawn from the history of metaphysics. All such responses thus remain instances of being as presence (since everything that has been uttered or written throughout the history of metaphysics has been so characterized). Derrida's systemic reservations about being as presence and its consequences suggest the following dual resolution to the problem of concepts: either (a) jettison the very idea of concepts and remain consistent with deconstruction's goal of achieving a domain of results separate from the "shaken" history of metaphysics or (b) preserve concepts and either (i) rethink, if it is possible to do so, the traditional notions introduced to cope with the questions just raised so that they avoid the penumbra of presence or (ii) allow that at least some structural elements with a metaphysical heritage can be, perhaps must be, maintained in order to justify fundamental epistemic concerns. These concerns will, so it seems, receive their apparent theoretical–and conceptual–shape requisite for all their diverse functions within the lifeworld even though these elements remain unavoidably tinged with residues of being as presence. The alternatives just outlined (as well as related metaphysical issues) will be discussed in detail in Chapters 9 and 10.

An additional level of inquiry emerges based on second-order factors. Derrida's position concerning the feasibility and fact of concepts depends for its formulation on *self-identity*, *limit* and *unity* as notions ordering the direction and conclusions of that position. But self-identity, limit and unity are all quintessentially metaphysical notions. As a result, these appeals raise questions concerning seeming incompatibilities between assumptions Derrida introduced during the formation of his position concerning concepts, chains and systems when these assumptions run up against the stated goals of deconstructive investigation.

It is essential that deconstruction recognize these three formal and functional concepts; it is also essential that deconstruction not shy away from analyzing its dependence on these concepts. Deconstruction should bring these structural components to the fore, then "deconstruct" them if the metaphysical heritage of all such components warrants this treatment. When this analysis has been achieved, the exponent of deconstruction must regard with care the philosophical landscape once unity, limit and self-identity have been deconstructively transfigured and then must enumerate the consequences entailed by such treatment. Such a summary review and appraisal will (a) clarify the difference between the history of metaphysics and the results of deconstructive inquiry and (b) lay a foundation for future thought regarding emergent issues worth due philosophical attention.

These important issues must be addressed. But in terms of the long-range ramifications of Derrida's thinking, it is equally important to appreciate the fact that Derrida's exploitation of the notion of a chain along with all that such chains presuppose and imply indicates a subliminal recognition of the need to circumvent the skepticism of his position on the transcendental signified. The introduction and reflective treatment of chains is also a transcendental gesture in its own right since Derrida has extended his analytical and, indeed, speculative viewpoint beyond being as presence by embracing a variety of perspectives on unity. These perspectives are somehow privileged insofar as they ground the actualized linkage between separate instances of words so that these linkages establish the possibility of conceptual significance. Derrida reintroduced stability at a level of high generality by presupposing the relevant presence of unity as the ground of a chain. And in what can be construed as an assumption of even grander scope, Derrida also explicitly and implicitly takes advantage of an element with a uniquely broad and rich metaphysical heritage–that is, *possibility*.

Possibility can be seen as the notional culmination of the transcendental themes and motifs animating much of Derrida's thinking, both in promulgating the axiomatic character of being as presence as well as in developing the strategies of deconstruction. On one hand, possibility is an instrument of criticism insofar as the appeal to indeterminacy with regard to the notion of a context presupposes the existence and relevance of possibility as an omnipresent ground for "other" settings, i.e., extended and logically incompatible environments conflicting with the stated limits of the attempt to stipulate a given context. But on the other hand, possibility is also an endemic element in the formation of concepts through *différance*, the ever-present "play of possibility" that inserts gaps of otherness, individual differences, which for Derrida stand as the dominant feature in the deconstructive description of concept-formation.

In sum, possibility stands as the apex of the transcendental dimension during this phase of Derrida's thinking. The elements or "vectors" of this dimension move toward more explicit recognition of the need to specify preconditions underlying the possibility of a chain and thus to establish a transcendental setting for the possibility of knowledge, presumably as a way of continuing to counteract the skepticism built into the analytical and critical temper of deconstruction. In fact, increased appreciation for the subtle dominance of possibility emerges from considering both the letter and spirit of Derrida's approach to negation, as we shall now see.

6 Traces of Negation

No S are P. Some S are not P. Students of logic immediately recognize these standard form propositions, the modern-day E and O propositions on the Aristotelian square of opposition. The E and O have the same quality–both are negative. The "no" in the E proposition attaches to the subject, the class designated by S; the "not" in the O proposition characterizes the relation between subject and predicate, classes represented by S and P. All this is familiar, easily stated in early moments of basic logic textbooks with sections on the syllogism. But negation itself, what it "is" and what it "does," raises large questions for anyone wanting to move past the surface level of logical concepts and principles taken as "givens."

Alfred North Whitehead asserted: "The negative judgment is the peak of mentality" (PR, 5) and he adds, much later in *Process and Reality*: "The triumph of consciousness comes with the negative intuitive judgment" (PR, 273). This gilded realization of the unique stature of negation as it plays into how we think and speak was hard-won for the history of thought. Recall that for Parmenides, the impossibility of even holding a "negative judgment" in the mind, e.g., "No olive trees are deciduous" or "Some Athenians are not wise," drove him to deny the rational accessibility of negation and ultimately to posit as a consequence of what he thought conceptually impossible a perfect spherical One inflected only by appearance but not by anything negative. In general, how philosophers deal with negation reveals a great deal concerning their metaphysical tendencies and convictions, not to mention substantive features of their thoughts on reality, what it *is* and what it *is not*. Derrida is no exception. In Chapter 5, we saw Derrida exploit negation in his characterizations of *différance*–asserting, indeed emphasizing, in various theoretical contexts what *différance is not*. Negation instantiated through diversified applications in discourse is readily available as a device for thinkers and presumably just as readily understood by readers. But if negation as such becomes subjected to analytical scrutiny, the results are usually intriguing and important to a philosopher's overall teaching.

6.1 *Différance* and Language

We have been focusing at various junctures in this essay on the notion of *différance* in its broader, more metaphysical context of operation and we need to do so as well with respect to negation. But for Derrida, *différance* has a more immediate range of import:

> **Passage 6A** ...we shall designate by the term *differance* the movement by which language, or any code, any system of reference in general, becomes "historically" constituted as a fabric of differences (SP, 142).

Derrida uses scare quotes around "historically" in order to alert the reader that the texts constituting the history of metaphysics as a "system of reference" will not be considered a straightforward linear sequence distinguished only by customary dating but rather as a function of *différance* in concert with all the technical features Derrida has built into that notion. Furthermore, language itself–regardless of its exalted purpose with respect to philosophy or its mundane ends addressing concerns of the lifeworld–is also characterized as activity inherently beholden to the workings of *différance*. We cannot overlook Derrida's direct appeal to reference; he is not dealing with language as a self-contained activity but rather with language as referentially related to something other than itself. It is perhaps tempting to separate philosophizing about being, i.e., metaphysics, and philosophizing about the language used in this form of inquiry, i.e., a branch of philosophy of language. However, it is not clear whether Derrida's position would allow this distinction. We examined in Chapter 4 the dictum "There is nothing outside the text" and then discussed the reasons offered in various passages and contexts to undercut the conceptual neatness of that distinction.[71] **Passage 6A** makes clear that *différance* informs language; thus *différance* is commensurately functional with respect to all the beings to which language can refer, especially the language marking the history of metaphysics. The role negation plays in this theoretical context is decisive.

6.2 Signification and *Différance*

The core of Derrida on negation emerges in this section. **Passages 6B, 6C,** and **6D** appear as one continuous text (SP, 142-3) and are separated into three sections to facilitate commentary. The relation between negation and Derrida's important term–trace–takes shape here. The account of these elements in Chapter 6 will be complemented by discussion of trace in Chapter 9 where the principal line of analysis addresses an application of trace to deconstructive possibilities relevant to metaphysical concerns. Chapter 6 develops critical consequences on key issues concerning negation as a foundation for the later, more intricate and speculative inquiry.

In the following passage, Derrida initiates a series of observations concerning the function of *différance* with respect to signification. This account includes several focal points relevant to the phenomenon of language and, as noted, leads directly (**Passage 6C**) into a description of trace proper and its connection with negation.

> **Passage 6B** *Differance* is what makes the movement of signification possible only if each element that is said to be "present," appearing on the stage of presence, is related to something

71 The relevant section in Chapter 4 is "Reasoning to the Roots of Language" as well as analysis of the subsequent inferences derived from that argument.

other than itself but retains the mark of a past element and already lets itself be hollowed out by the mark of its relation to a future element.

When Derrida identifies "what makes the movement of signification possible," immediately adding to this statement of intent the important logical qualifier "only if" (followed by identification of conditions), he continues to pursue a transcendental position, i.e., he is stating conditions which establish the possibility that something will eventuate and be duly experienced according to these specifications. Derrida thereby steps outside the realized limits of the phenomena he is purporting to explain in order to elicit conditions enabling the existence of those phenomena insofar as they stand as an object of philosophical inquiry. It follows, as it would for any philosophical treatment which adopts this procedure, that what Derrida assumes conceptually in his articulation of such transcendental considerations is open to identification and critical review. What might be described as the "transcendental dimension" of deconstructive inquiry will emerge often in the following discussions.

We may begin by asking: does "the movement of signification" refer just to reaction in the cognitive agent, i.e., the process whereby something initially linguistic becomes signified in the mind of an observer, or does it also include a metaphysical dimension, i.e., the conditions occurring within extra-linguistic realms as prerequisites underlying the possibility of signification? Does the conjoint appeal to "element" refer to (a) words in a language and then ultimately to conceptual cognition, (b) the referents of words in a language, or (c) both, words and referents, insofar as the conjunction of the two elements generates a relation? Derrida addresses aspects of this basic question below, but at this juncture much of what he could have intended is not obvious.[72] If, however, *différance* is taken as a determinant at the most fundamental level and the image of "the stage of presence" in **Passage 6B** establishes

---

72 Russell Daylight, discussing a paragraph in "Structure, Sign and Play," contends that it "never actually becomes clear...whether it is the *concept* of the sign, or the *name* of the sign [i.e., a purely linguistic component] or *both* elements which have continuity...." Daylight concludes, 54, that in this paragraph, "there is evidence for all positions" (all italics in text). Daylight then appears to mitigate the force of this observation, 54: "Derrida's slippage from signs to signifiers, and from concepts to words and back again, can be interpreted as either lack of care with the elements of the theory he is critiquing, or as simply another instance of how the epoch of the metaphysics of presence subsumes all such distinctions within its operant conditions...." Two observations: (a) if the "epoch of the metaphysics of presence" does indeed subsume "all such distinctions within its operant conditions," then there is all the more reason to subject "the metaphysics of presence" to close scrutiny (as in the present essay) assuming that "concept," "name," and "sign" are terms representing fundamentally distinct components in cogent theorizing about language and knowledge; (b) bluntly stated, perhaps Derrida did not sufficiently think through his position, thereby enabling him to play imprecisely with the terminology of a criticized theory relative to the scrutinized concepts as Derrida believes they function in that theory. Such "slippage" also appears later in this chapter and in Chapter 7.

a setting in which the manifestation of presence is a mode of performance, which in a way it is, then the "movement" in question engenders signification as a kind of *relational realism* between language as a set phenomenon and extra-linguistic entities capable of interacting with and entering into language in a variety of ways.[73] This interpretation effectively extends the implications of the reading of **Passage 4E** regarding Derrida's intention to reason by means of deconstructive considerations to the roots of language.

Différance collaborates with the notion identified as "trace" to make signification possible. The immediate context of specification is temporal–past, future, present–but it would be a mistake to restrict the effect of trace, or tracing as it might be named, to verbs as vehicles of tense. *Any* word as an element in a proposition is governed in terms of the possibility of conveying signification through temporality by means of the functioning of trace.

This element, verb or noun, exists by itself as a mark but as vehicle for signification only insofar as it is "related to something other than itself." First, the openness referred to is the entry point to negation as a kind of fundamental "otherness." And second, the appeal to "something other than itself" should be connected to the notion of chain (discussed in Chapter 5). Element in this context is necessarily relational and thus functions within a chain-like configuration. Derrida's claim about relation involves minimally a given word linked necessarily with other words; whether this relation also includes referenced entities linked to other entities (referenced or not) is not clear from this passage alone, although we have argued that Derrida's overall position commits him to such relational realism. Thus for reasons stated above, the relevant context of investigation includes both linguistic (words related to other words) and metaphysical (entities related to other entities) components.

Past, present, future are integral phases of "the stage of presence." I take this specification to mean that when something is past, then it is related to something other than itself but so that the "pastness" of this relation comes to be and remains apparent as past. Similarly, when something is futural then it is related to something other than itself but in such a way that the "futurity" is this relation is and also remains apparent as futural.

In preserving the classic past/present/future trichotomy at this juncture, Derrida's position rests on grounding notions which invite questioning. For example, subsequent mention of being "hollowed out" (*creuser*) by the mark of its relation to a future element suggests that this hollowing is not intended to refer in parallel fashion to an entity, existing in the present, undergoing a metaphysical process of being "hollowed out" in order to continue that existence into the future. Rather, this sense of "hollowing" exists at a linguistic level, the image conveying that existence

73 Cf. Ch4n56, the reference to Norris insisting that Derrida be read as a realist.

in the present must yield its fullness in this respect by allowing itself to "stretch" from its present modality and then to "dip," as it were, to what it will be in the future, in conjunction with the corollary that its degree of reality at that point in the future remains nugatory in the present. So interpreted, this allowance for "hollowing out" seems to work more efficiently, or at least more intuitively, in the context of language than it does in the context of envisioning entities characterized by temporality somehow losing their temporal shape with respect to their mode of existence in the present in order to fulfill a possible mode of futural existence.

If this reading is accurate, unavoidable questions arise even if they seem redolent of traditional metaphysical concerns–for example, identifying the "indicators" of temporality however pastness and futurity are denoted. Past and future appear "real" here, or at least distinguishable from temporality commonly referred to as the present. Thus the past has left a mark on the entity currently existing as present but the future exists only as a relation to something other than whatever constitutes the trace when it exists as present. But how is the pastness of the element *qua* past indicated? And the same question holds for the futurity of the future element *qua* future. *Différance* presumably functions at a level where common life (Hume's phrase) distinctions between past and future originate and contour experience, not just an attenuated past and future as structural components in a theory of temporality inscribed by a metaphysician. Our senses of the past and the future may be vague at the level of intuition but they are structurally undeniable in terms of temporality both in the lifeworld and, apparently, in the more rarified concerns of the philosopher. But on Derrida's own deconstructive principles, how are these temporal distinctions determined so that they can be at least provisionally trusted in a philosophical context?

6.3 Trace, Temporality, Negation

This tentative analysis of the dimensionality of time with respect to *différance* and its function regarding signification is complemented by broader speculative considerations.

> **Passage 6C** This trace relates no less to what is called the future than to what is called the past, and it constitutes what is called the present by this very relationship to what it is not, to what it absolutely is not; that is, not even to a past or future considered as a modified present.

We confront Derrida's notion of negation in this passage, and the diversification of doctrine will justify the following thematically differentiated commentary.

6.3.1 The Patrimony of Trace

The element under scrutiny above has now been explicitly named "trace" with the initial development of that element in **Passage 6C** concentrating on temporal features extending the account of time advanced in the previous section.

The ordinary sense of "trace" is inherently relational, as in "trace (or traces) *of* something;" this relationality also applies to Derrida's technical usage (i.e., of "element" as an introductory mark to represent trace). We observe that in some contexts "trace" can also exhibit nuances of considerable subtlety, as in Emily Dickinson's evocative description tracing the flight of a hummingbird as a "route of evanescence." And yet the imagistic vividness adhering to the phrase "route of evanescence" does not seem sufficiently supple and sinewy to accomplish the various functions which Derrida's texts impose on the notion of trace.

What can be said about a trace when it "is"? I suggest that it is reasonable to make the following assumptions about, as it were, the structure of a trace:

i. Refers to something (understanding the pronoun in a very broad sense) "leaving a trace," a non-vanishing residue.
ii. Leaving a trace is a process and thus presupposes some measure of continuity.
iii. This continuity as ongoing allows a trace to come into a relation with something other than itself.
iv. Assumption iii presupposes that trace admits of plurality–there are many traces (presumably in some or perhaps several respects distinguished from one another).
v. The continuity of a given trace is at least temporal (and it may have other attributes, as we shall see).
vi. The temporality of trace has a vector factor, i.e., it possesses directionality.
vii. The elements of continuity are, initially, linguistic. But if these elements included at the moment of their origin *in* language encompass entities as referred to, then trace also includes necessarily a dimension of referentiality. Thus (cf. Assumpion i) *something* leaves a trace; even in the technical Derridean sense, a trace does not just appear *ex nihilo*, mysteriously but effectively emerge in a linguistic setting, then just as mysteriously vanish into a limitless void.
viii. If trace is of something, then trace presupposes sameness (in a sense perhaps related to Derrida's undeveloped sense of sameness in being as presence) in order to guarantee that a given trace maintains itself, i.e., its self-identity. Establishing uniqueness of this sort is necessary in order to distinguish any one trace from other traces.

This complex slate of properties is not overly zealous interpretation but represents characteristics readily extracted from the diverse dicta Derrida provides in texts on trace and on notions immediately related to trace. We will attempt to elaborate a selection of these properties as they appear in neighboring passages.[74]

6.3.2 Temporality

In **Passage 6C**, Derrida describes trace initially as what "relates no less to what is called the future than to what is called the past...." At least two readings are possible: first, the suggestion that past and future may be something other than what they appear to be, since trace relates to "what is called the future" (*ce qu'on appelle le futur*) in the same way that it relates to "what is called the past" (*ce qu'on appelle le passé*). This circumlocution calls attention to the deconstructively established fact that "past" and "future" are and can only be modes of presence (since in principle everything metaphysical is a mode of being as presence). From this perspective, "we call" the past the past only because we need at least for practical purposes a contrasting dimension to what "we call" the future. But additionally and again from the perspective of deconstruction, these modes of calling or naming are in a real sense misnomers since they posit distinct temporalities which require some mode of differentiation in order to justify any claim based on their supposed distinction over and above the fact that "past" and "future" are, as metaphysical gambits, both exemplifications of being as presence.

The second reading posits that Derrida is contending more pointedly that the past, taken as a distinct dimension of temporality, achieves that status only because the verbal phases of natural languages are tensed with built-in indicators of pastness and futurity. What we call "the past" exists only because language has evolved in certain ways, originally with respect to development of its "action" words. Derrida's position seems to be moving toward a subjectivist approach to temporality in that differences pertaining to time derive solely from linguistic considerations rather than from external, non-linguistic factors in the natural order. The temporal dimensions of past and future lack any kind of independent existence; rather they function as perspectives derived entirely from a human source, molded in some way with respect to temporality by means of linguistic considerations.

---

74 See "Conclusions" in Paola Marrati 2005, 177-97 for a summary of trace in two contexts: Derrida on, respectively, Husserl and Heidegger. Marrati's final sentence of this account reads, 197: "Neither trace nor *différance* can be gathered." However, I will argue in Chapters 9 and 10 that Derrida's texts leave open the possibility of a less skeptical and more substantive application of the notion of trace (cf., e.g., **Passage 6E**). Derrida uses "trace" frequently and in a wide variety of senses. For representative passages on trace, see OG, 60-1; MP, 65-6, 313; WD, 265, 292; SP, 142-3, 152-3, 154, 156-7; P, 7. For early usages of trace in apparently a non-technical sense, see WD, 68, 247. For discussion, see Thomas Baldwin, 110; Harvey 118, 122; Rorty 1982, 177; G. B. Madison 1993, 198; H-G. Gadamer 1989, 183. The account in Chapter 6 is only an introductory exegesis of this pivotal notion.

In order to contest this approach to time and challenge its apparent subjectivist bent, it is necessary to get beyond language, so to speak, in order to access, apparently through extra-linguistic means, temporal dimensions other than those available in the immediate present. This possibility assumes that verbal functions of language can *refer* accurately and legitimately to regions of temporality existing in different dimensions than those limited to the present; it also assumes that linguistic formulations can circumvent the epochal effects of being as presence.

In general, even at the level of language it seems fair to assert that there must be a way to discriminate between presence when it informs what is called past and presence when it informs what is called future. Without the possibility of such temporal discrimination, the present-called-past and the present-called-future become indistinguishable from one another. Some determinate feature is required in order to distinguish past from future–is it in the language we use, some empirical aspect of the referents of language, or a complex interaction between language and referential (i.e., extra-linguistic) reality?[75]

This fundamental question regarding time puts additional pressure on understanding how Derrida evokes the present. As we shall see, the characterization of the present involves negation construed as what is "absolutely not now." Thus the temporality of an entity as well as the temporal considerations animating the language used to signify or refer to that entity engages perspectives on negation, a notion of the utmost generality and a notion, as Parmenides taught us, with an inherently puzzling surface and structure.

6.3.3 The Question of Negation

Derrida contends that trace generates "the present" by its relation to "what it is not, to what it absolutely is not; that is, not even to a past or future considered as a modified present." For Derrida, the present does not exist *qua* present merely by

---

75 In *Confessions* Book XI, Section XX, Augustine maintains that instead of dividing time into past, present, future, it would be more correct to say that "there are three times, a present of things past, a present of things present, a present of things future." Augustine then situates each of these three temporal designations: "the present of things past is memory, the present of things present is sight, the present of things future is expectation." Although this approach clearly differentiates between and among the classic temporal taxonomy of past, present, future, Derrida could hardly subscribe to this approach since it is (a) a quintessential instance of "being as presence" and (b) derived from dimensions of consciousness. Given the axiomatic necessity that consciousness as a concept must be deconstructed, it seems fair to infer that any phase of consciousness, such as the "psychological" aspects Augustine introduces, would also have to be similarly "shaken." Thus for all the apparent relevance of linking temporality to facets of common experience, Augustine on time seems to be a "vulgar" (following Heideggerian parlance) approach to this concept. For a systematic treatment of Derrida on time concentrating on Derrida's own terminology, see Joanna Hodge 2007; see also DFL 231n16.

being "other than" apparently distinct dimensions of temporality, even when–his vigorous insistence on this point should be noted–these dimensions, past and future, are qualified as modes of presence. The present and by extension temporality in all its dimensions becomes intimately and necessarily related to "what it is not"–in a word, negation.

It therefore becomes crucial to specify or at least approximate the metaphysical scope of "what it is not" since it is axiomatic for Derrida that "what is called the present" separates trace–assuming that some determinate content can be assigned to it–from the identity of "what it is not." It is clear then that within this highly theoretical configuration, "not" does not function merely to negate a proposition (e.g., "S is not P"), i.e., as a purely formal operator. Now we recall that according to traditional formulations, "S is not P" is a negative proposition made so by the negation "not,"; by contrast, "S is non-P" is an affirmative proposition with a predicate which is the complement of the class P, i.e., non-P. This distinction is relevant with regard to Derrida's approach to negation. For I suggest that his phrase "what it is not" implies that "not" represents the notion of a complementary class, commonly rendered as "non-" and attached to the class on the basis of which the complementary class is generated. Thus when Derrida writes that "trace relates...to what it is not...", trace relates to what is "non-trace." What then does the "non-" of "non-trace" refer to? What is included in the scope of that which is "non-trace"?

For comparison's sake, consider the concept of non-rational. Strictly speaking, the "non-rational" names a class which includes everything existing, whether extra-mentally or imaginatively real, which does not belong to the class representing whatever has the property of being rational. This complementary class is very large, of course, since it encompasses the totality of reality except for rationality and whatever beings possess that quality. Therefore, if non-trace functions identically to non-rational, then non-trace refers to *everything other* than trace (recall that the precise reality of trace remains indeterminate at this point).

It might seem plausible to characterize the present–i.e., an entity insofar as its trace constitutes "the present" of that entity–by exclusion when situated with respect to past and future; thus if an entity is not in the past and not in the future, then it exists in the present. But Derrida denies this possibility; if he did maintain this relational position regarding trace, then he would be defining it solely in terms of how past was present to an entity as well as how future was present to that entity. But then trace would become a function of presence, and presumably indistinguishable from the content of any position occurring within the history of western metaphysics since for Derrida metaphysics regardless of its specific formulation (cf. **Passage 1C**) implies being as presence.

To avoid this inconsistency, Derrida asserts that "what is called the present" comes into existence not by a relationship characterized by temporality but by the "very relationship to what it is not, to what it absolutely is not...." Negation as a class-complement of a given entity is not sheer, undifferentiated absence; rather, negation–

what is not–is the sum total of all beings but only insofar as this totality exists apart from, or *other than*, a given entity. It becomes evident then that what makes trace present is everything, the *complete totality other than* whatever a given trace is trace *of*. However, if negation is not sheer absence but can accurately be restated as non-X and this complementary class encompasses totality, then it follows that what makes X, a given trace, to be what it is *qua* trace encompasses everything other than whatever may be the self-identity of that X. But the question remains, what is X *qua* trace?

6.3.4 Trace, Presence, Negation

Derrida derives two advantages from this position: first, it obviates addressing what "sheer absence" taken all by itself might mean as a matter of conceptual content (Parmenides' poem casts a long shadow!); second, any single trace now bears a concretizing relationship to *every* entity which falls under the complementary class of non-X, where X represents as a variable any single determinate trace. This consequence becomes crucial if we then ask about the connection between "the present" so defined and "presence" in the canonic and crucial notion of "being as presence."

Later in the essay on *différance*, Derrida opens a paragraph with this:

> **Passage 6C$_1$** The trace is not a presence but is rather the simulacrum of a presence that dislocates, displaces, and refers beyond itself (SP, 156).

This passage extends the account of the relation between trace and negation (in the sense specified above). If trace is a "simulacrum" (*simulacre*) of presence, then trace is in some way *like* or *similar to* presence. But what does likeness between trace and being as presence indicate in this context? Derrida stipulates that trace "dislocates, displaces, and refers beyond itself." Note that dislocation, displacement and extra-referential process have not been underscored as integral elements of being as presence except perhaps in senses which remain inchoate in Derrida's texts. We will develop one possible reading of these properties, especially with respect to possibility, in Chapter 8. But the relevant point with respect to trace and negation is that if presence exhibits the three functions Derrida has identified then so does trace to the extent that trace stands as a "simulacrum" of presence. Furthermore, if metaphysics throughout its history is characterized by being as presence then the complement class, non-trace, will include *everything* other than trace-as-simulacrum insofar as that "otherness," including everything found in all metaphysical systems, suffuses the totality of western metaphysics.

Trace exists "by this very relationship to what it is not," and then, apparently to emphasize the importance of negation, Derrida immediately adds, "to what it absolutely is not" (... *ce qui n'est pas lui*: *absolument pas lui*). The second phase of this locution is naturally read as intensive restatement of the more neutrally expressed first phase and thus to lack significance regarding any qualitative shift in theoretical import. However, as we shall show in Chapter 9, Derrida at this juncture in **Passage 6C** is laying a foundation for a speculative venture in "fundamental ontology," as long as we keep the referents of this name apart from their Heideggerian antecedents.

The identification of trace as a "simulacrum" of presence has two important implications: (a) to the extent that "simulacrum" entails features of similarity, then trace is in some respects (which must be discussed) *like* the products of *différance* and being as presence; however, (b) to the extent that "simulacrum" maintains its uniqueness *apart from* being as presence, the displacement and extra-referentiality Derrida mentions will engender vocabularies and conceptualizations which may be, to use again the word Derrida cites in *Of Grammatology*, "monstrous" in significance when juxtaposed with the traditional terminologies of western metaphysics. A relation of similarity to some phase of being as presence implies that these characteristics would not differ *in toto* from properties or characteristics locatable within being as presence (i.e., within the history of metaphysics). Therefore, it may be inferred that "non-" must also point to something which exists in a way or ways completely other than any of the moments of the history of metaphysics as well as all those moments taken as a unified set of positions. It will be necessary to traverse the domain of (a) above in order to be in a position to catch a glimpse of the content of (b), a sequential strategy that will come into play in the argument of Chapters 9 and 10.

6.3.5 The Factor of Sameness

In order to situate a theoretical locus for subsequent inquiry, we consider several ways to clarify the notion of trace with respect to the past/present/future deployment. Although Derrida maintains that trace is derived from its relation to that which is not ("non-trace" in our formulation), the classical tripartite divisions of time remain constitutive of trace (cf. **Passage 6C**). Assume that the aspect of trace capturing the past factor also inheres in trace insofar as trace is present. If so, then sameness obtains between trace-when-past and trace-when-present–at least in this respect, trace is the same as dimensions of time "move" through a given trace.

A contrasting possibility: the distinction between past and present (as well as between present and future) presupposes an entitative commitment in that such temporal differences could not arise unless *something existed* from which trace as a purely linguistic phenomenon has been derived and this mode of existence is construed in such a way as to generate the possibility of temporal differentiations of the past/present/future sort. The temporality of a trace thus would depend on a relation between the givenness of a particular trace and something other than that givenness. This hypothesis implies that trace must be understood as envisioned in tandem with an object, i.e., trace is always trace *of*, with the next question the equally obvious query into what can and should be said about the structural character pertaining to how this "of" is constituted. In this case, the relevant sameness is "in" the entity which yields its trace, not "in" the trace proper. And as we see in the next section, Derrida's appeals to *division* and *interval* as components of trace will also be affected by this problematic sense of sameness.

However this substantive issue may be resolved, we must note the implicit appeals to unity and sameness as embedded in the notion of "trace." These are second-order or formal properties and they apply here with respect to establishing and preserving the integrity of given traces just as necessarily as the same formal properties pertain to *différance* as the pivotal element in the theoretical formulation of deconstruction. It follows then that at some point the advocate of Derrida's notion of trace must distinguish and discuss the difference between, e.g., (a) the sameness that characterizes *différance* and (b) the apparently more limited instances of sameness which constitute the inner workings of a given trace. For if more than one trace exists, a plurality which is intuitively certain, then the question of determining the distinct samenesses inhering in each trace taken individually as well as the conceptual techniques for distinguishing between them becomes serious and problematic.

6.4 The Duality of Interval

Derrida covers considerable theoretical ground in this final excerpt from the extended passage in the essay on *différance*. Discussion of this excerpt will identify and elucidate elements rooted in traditional metaphysical considerations. The primary purpose is to suggest how far Derrida must go to distance his position, consequent upon the application of deconstructionist principles, from these considerations. It should be obvious that Derrida's approach to negation is speculative, including the introduction of the notion of interval and the process of division.

> **Passage 6D** In order for it [trace] to be, an interval must separate it from what it is not; but the interval that constitutes it in the present must also, and by the same token, divide the present in itself, thus dividing, along with the present, everything that can be conceived on its basis, that is, every being–in particular, for our metaphysical language, the substance or subject. Constituting itself, dynamically dividing itself, this interval is what could be called *spacing*; time's becoming spatial or space's becoming temporal (*temporalizing*) (SP, 143–italics in text).

Since "an interval" (*un intervalle*) is necessary for the existence of a trace, we must determine, if possible, what properties Derrida assigns to this mediating notion. The interval (a) "constitutes it [trace] in the present" but (b) the interval also, and apparently with necessity ("by the same token"), divides "the present in itself...." The ellipsis, what Derrida then adds to this characterization in **Passage 6D**, is crucial and requires separate discussion.

In its usual usage, "interval" refers to a break or gap between two determinate elements (points on a line, musical notes, periods of time–"interval" in Britain denotes the span of time between two halves of a soccer match). In this respect then, trace becomes constituted "in" or "as" present by virtue of the "interval" between (a) the trace as such (recall that at this point, precisely what trace "is" remains, as an X, indeterminate) and (b) non-trace X, that is, everything which is not X (the totality, defined by negation, excluding X). Now if this were all that the notion of interval did, then it would be a question of envisioning an interval, a gap of some sort, *between* the unique X to be constituted as present and everything other than that X. It would be consonant with this account to understand interval in this sense as a *static* gap, a determinate segment measuring something (as an interval of specified duration measures the amount of time *between* the halves of a soccer match). This gap would be open to revision depending on circumstances but once established by due authority, then it is and remains fixed.

However, the next segment of Derrida's text indicates that to envision the notion of interval to function only as a static gap is incorrect. The interval is, in fact, *dynamic* and in conjunction with its function of establishing a given trace as such, one and the same interval, "... by the same token, divide[s] the present in itself, thus dividing, along with the present, everything that can be conceived on its basis, that is, every being–in particular, for our metaphysical language, the substance or subject."

We suggested initially that Derrida's interval might be taken as a static gap between two distinct entities or states of affairs–the X to be constituted as a trace and non-X, the totality other than that singularity represented by X. But now we know that interval in fact interacts with the present that it establishes for a given trace, dividing that present and, presumably, continuing to divide that present apparently without limit (more on this possibility below). Interval thus displays a complex function: interval both (a) distances the X, the potential trace, from everything other than that X *and* (b) penetrates the present which it has established, the trace of X, and acts upon that present, dividing it (according to specifications which Derrida does not indicate in this passage).

Derrida does stipulate that the relevant object of division is, apparently, an entity. We infer this because the interval divides everything–such as, in terminology of metaphysics, a substance–which can be conceived according to the present. The assumption is then that "substance" represents a standard metaphysical concept for an entity of some sort. Although there appears to be a defensible difference between substance as such and substance insofar as it is present, Derrida's claim is that since the latter is generated by the interval in the separation of a given X from non-X, the former also becomes subject to the same kind of effect, presumably because substance as a metaphysical concept must be present in order to exist as a metaphysical concept in the first place.

Even if Derrida's use of negation cogently presupposes a completely indeterminate absence, sheer otherness, it must still be open to continual division in order to engender something completely other than the original subject of division. The classical notion of class-complementarity remains in effect but with decided emphasis on the fact that whatever is "non-X," where X is the given interval and "non-" refers to the negation to which Derrida is appealing, can refer to *anything*, however "monstrous" in form (or formlessness) that something may be. But it is difficult to understand how one concept–interval–can simultaneously separate and divide in the senses Derrida specifies. When interval separates, it constitutes a trace *qua* trace; when it divides, it acts on what it has constituted by dividing that trace, apparently without limit. I propose to discuss this question by examining a series of assumptions and implications derived from what Derrida writes about division in **Passage 6D.**

6.5 Interval and Unity

It seems a fair question to ask what this interval represents or encompasses. Is its content strictly temporal or are we to understand the "stretch" of the interval to include not just temporality as such, i.e., abstracted from something else, but also the entity which exists "in" or "during" this stretch of temporality?

Relevant to this question is Derrida's position on *unity*. Shortly after the extended passage we have been analyzing, Derrida discusses why he coined *différance* rather than use the more common (and already existent) "differentiation." Derrida rejected using "differentiation" because it would suggest

> **Passage $6D_1$** ...some organic unity, some primordial and homogenous unity, that would eventually come to be divided up and take on difference as an event (SP, 143).

The apparent inference is that no such "primordial and homogenous" unities exist. Furthermore, the word "unity" becomes suspect as one of the residues from the history of metaphysics which must be either rejected as a construct of misapplied metaphysical zeal or subjected to the closest possible scrutiny in order to determine what "unity" *really* is and does insofar as that reality emerges under the more percipient strains of deconstructive analysis.

The unity Derrida has in view functions at the highest level of generality, i.e., as counterpart to "differentiation." Thus the unity in question is proper to totality and would classically be appealed to as a foundational factor underlying any subsequent speculative split introduced for metaphysical purposes. The question is whether Derrida also intends to deny *any* unity whatsoever, e.g., the unity of a given entity insofar as it has been experienced as stable, or the unity of a proposition uttered or written about that entity. The proposition "The glass contains water" is *one* proposition about *one* entity; it is also a true proposition if indeed the glass contains water. Thus perception (e.g.) as a means of providing evidence will ratify the rightness–the truth–of that proposition when it is juxtaposed with the entity in question. But the truth of this proposition depends on the independent existence of a set of unities–the glass, water as a determinate liquid. It would be the fallacy of division to argue that the denial of unity when predicated of a whole–indeed, *the* whole!–implies an equivalent denial of any part of that whole, that is, any particular unity as resident within the whole taken as a unity. Derrida explicitly denies unity in the former sense; the question is whether he would deny the latter sense as well.

6.6 Interval: Division of the Present and Beyond...

The status of unity with respect to the interval in question is crucial not just by itself with respect to clarity regarding fundamental concepts, but also because the character of unity will determine, at least in part, what transpires when division in this technical sense is in operation. The scope of unity becomes additionally problematic when we recall that Derrida wrote: "...thus dividing, along with the present, everything that *can be conceived* on its basis, that is, *every being*–in particular, for our metaphysical language, the substance or subject" (my italics). Not only can (and, for Derrida, must) "the present" itself be divided but also whatever can be conceived as existing in the present–"every being"–must be subject to such divisibility.

I suggest that this passage admits two readings: if we emphasize the factor "can be conceived," then what is being divided is not the being itself existing apart from the activity of becoming the subject of a concept but rather the product of this being conceived, i.e., the concept ascribed to the extra-linguistic being. However, if we emphasize the isolation and attention given to "every being," presumably every being as such and in itself, then what is being divided is indeed the being itself, not just the conceptualizations introduced to render the being accessible to cognition and practicality. In sum: unity applies either to words about being or to the beings themselves, or perhaps to both.

It seems fair to assume then that the unity of a word used in a proposition to refer to an entity is a different kind of unity than the unity of the entity so treated. The ramifications of each possibility are explored in the following section. But as a prelude consider these questions: what substantiates Derrida's introduction of the fact of division, where does this division originate, and does division as a process have determinate structure? The answers to these questions are not evident at this juncture. We proceed therefore with inferences based on assumptions derived from metaphysics. The burden of proof is on Derrida's defender to demonstrate by argument why these inferences and assumptions should not be introduced into this interpretive context.

Is the interval to be distinguished from Aristotle's "now"? If so, should we construe this interval as a point or a continuum? If neither model is appropriate, then it is difficult to entertain any sense to the assignment of "constitution" to the interval in question. The question is not inappropriate since this interval is what separates "the present" from the totality of that which is *not*. Thus if trace refers not just to a dimension of temporality but also to the content of what has emerged within this dimension of temporality, then the specification of how most properly to understand the constitution of the interval will bear directly on what factors come into play regarding the content of the interval with respect to actualizing signification.

If we do take the fact that the interval constitutes the trace "in the present" to be "the present" as construed on the model of an Aristotelian "now," then there are irresolvable difficulties. Aristotle's now as evocative of the present separating past from future is equivalent to a point as a unit along a continuum and thus is not subject to divisibility (from which it follows, interestingly, that on Aristotelian grounds it is impossible to live "in the moment" if a moment lacks any duration).[76]

We proposed that interval be understood as either a point or a continuum. If "to divide" retains standard metaphysical design, then to divide the "present in itself" means presumably that the present is a continuum; if so, then Derrida has replicated metaphysicians such as Aristotle and Leibniz, both of whom employ division as a factor in the structure of a continuum.[77] To divide the present means presumably that the present is always capable of division never reaching a point on the continuum which, insofar as a given segment even of infinitesimal extension is continuous, exhausts the possibility of further division. If Derrida intends a limit to divisibility of the present, it is not indicated in this or similar texts. In a related vein, does "dynamically" mean that division is ongoing continually or is it only a qualifying expression to indicate that when division occurs, it does so dynamically, leaving open the possibility that division is not always continual and that temporal segments exist when what is being divided remains stable and intact? We develop these and related questions in the next section.

6.7 Division: Words and Objects

We have interpreted Derrida's position with respect to trace and negation as a foundational gesture itself based on a kind of relational realism. Thus the notion of division which Derrida has incorporated into his thinking concerning signification encompasses both language and referents insofar as the combination of each element constitutes signification. But at this point, for the sake of clarity of understanding, we separate language itself, whether spoken or written, from what in Part I we identified as referentiality, extra-linguistic entities as they became the referents of linguistic

76 For Aristotle on the "now," see *Physics* Book IV, Chapters 11 and 13. Derrida discusses time in Aristotle in MP, 52-6.

77 For Aristotle on division and the continuum, see *Physics* 207^{a}8, 231^{b}15; for time as continuous and thus subject to unlimited division, see *Physics* 219^{a}10-14; for Leibniz on division as it pertains to the inherent divisibility of monads, see *Monadology*, 65-7.

activity. The question becomes whether the relation between (a) trace as such and what it is *of* differs from the relation between (b) trace and negation construed broadly as "something other than itself." Derrida has specifically addressed the latter (although, as argued above, without sufficient development regarding the configuration of negation) but it is not obvious that the former is considered at all.

6.7.1 Language and Concepts

If we emphasize the linguistic dimension of **Passage 6D**, then the divisions to which Derrida refers here are limited to (a) language about beings rather than to (b) the beings themselves. This distinction between the two frames of reference is or at least seems obvious. After all, even if it were true that every concept, proposition, distinction, and theoretical proposal as instances of metaphysical language are all subject to instantaneous and ongoing divisibility, it certainly seems plausible to maintain that none of these divided elements would have any effect on the "metaphysical" character of the being so described and any structural considerations pertaining to that being. If we perceive a tree and declare it as an entity in order to instantiate the concept of substance, the percipient remains a tree even if the concept applied to the tree divides itself so that it becomes something other than "substance." Trees and concepts metaphysically denoting trees are different from and irreducible to one another. However, this neatly drawn distinction is not so obvious from the standpoint of Derrida's principle that nothing exists outside the text. Although we will adopt the distinction, we do so tentatively, an offshoot of what appears "obvious" to the sensibilities of western metaphysics; we also qualify the distinction by assuming that its import is more a matter of emphasis than a difference marked by crystalline clarity.

Yet even with this distinction and associated reservations in hand, an important problem now becomes evident. If the divisibility generated by the interval through which trace comes into existence pertains only to language itself and not to the entities to which that language refers, how is it possible for students of "metaphysics" to extricate their inquiries from such continually dividing language in order to re-experience the being in question and then attempt to restate the "metaphysical" character of that being? For even if the relation between language and concepts is mysterious in many respects, it is reasonable to approach the phenomenon of the conceptual realm with the understanding that the very existence of concepts is in some respects dependent on language. But if language as an activity occurring in the present is always being divided, we must question whether it follows that the very possibility of conceptualization is thrown into an unstable flux. We will return to this consequence below.

Our interpretation of Derrida's position is that it involves a relational realism. Now consider implications from the extra-linguistic dimension, the perspective of referentiality.

6.7.2 Space, Time, Transcendentality

If the interval "constitutes itself," then trace includes a referent, an entity capable of generating such constitution; if so, this referent must have an extra-linguistic mode of existence. Furthermore, the functions of (a) ongoing division and (b) time's becoming spatial or space's becoming temporal suggest that Derrida has explicitly extended the domain of trace beyond language itself, whether spoken or written, and into the *referential range* of language when it is employed by speakers in determinate or limited contexts.

The introduction of time and space complicates the context. Is the reference to time's becoming spatial or space's becoming temporal intended in a metaphorical sense, i.e., that an assigned meaning to a given instance of language has space to alter that meaning depending on a shift in contextual circumstances? Or is the intent to claim that trace functions metaphysically as well as linguistically, that is, that trace generates an opening, as it were, in the fabric of space such that a being can come to exist in space and therefore assume a mode of existence capable of being perceived and then depicted or evoked linguistically?

If the latter metaphysical line of interpretation is feasible, then "spacing" may refer not only to the range of meaning a given word or phrase generates from context to context but also to referents themselves insofar as their inclusion within signification encompasses entities referred to by the purely linguistic components involved in signification. Derrida would then have "spacing" incorporate the extra-linguistic spatial settings of entities as these entities have entered into an event of signification (e.g., our hypothetical request for a glass of water). Thus a given component of referentiality (an entity as referred to) might occupy very different "objective" (i.e., non-linguistic) sectors of space depending on the overall situations which surround that component of referentiality.

It is not clear, however, how Derrida is justified in introducing space and time in the first place if neither of these fundamental realities can be completely divorced from metaphysically set, i.e., "being as presence," considerations. When Derrida asserts (**Passage 6D**) that the "interval is what could be called *spacing*; time's becoming spatial or space's becoming temporal (*temporalizing*)", perhaps he intends to "shake" the pristine givenness of time *as different from* space and space *as different from* time by having each sector "become" the other insofar as we are pursuing a process (note "temporaliz*ing*"and "spac*ing*"–my italics). If the result approximates space-time as an organic unity, then the point may be that the theory of relativity is philosophically congenial as arrived at or at least pointed to through deconstructive means.

By contrast, Kant made space and time into necessary conditions for the very possibility of receiving perceptual data and he devoted considerable energy to stating properties belonging to space and time respectively, properties arrived at by close argumentation. Derrida introduces "spacing" and "temporalizing" here without providing much if any supporting description and argumentation regarding their structure. If Derrida wanted to distance himself deconstructively from Kantian metaphysics but to retain space and time as a hybrid unity, then the problem is creating terminology which will recount this newly fledged unity in appropriate discourse. This suggestion is for now, of course, thoroughly conjectural.

Derrida's appeal to a conjoint spacing and temporalizing has a direct effect on the content of division. The process of division, in addition to instituting the present (of trace), also divides "everything that can be conceived on its basis, that is, every being–in particular, for our metaphysical language, the substance or subject...." Furthermore, this process is one of "dynamically dividing" itself, presumably without accessible limit. Derrida's claim thus extends the sway of division from a purely temporal setting–the present as subject of division–to, as he puts it, "everything that can be conceived on its [i.e., trace's] basis." Just as the interval construed as a continuous present can always be subdivided insofar as the subdivisions are emerging from a continuum (implying that the continuum will permit unlimited divisions of this sort), so also everything conceived of as occupying this present–such as a given "substance or subject"–will, indeed must, also be continually or "dynamically" divided and subdivided.

What then of "substance"? One possibility is that the process of dynamic division will preserve the identity of a substance but refer to that instance of substance as existing in a different context. In this case, it is entirely possible to understand the conjoint spacing-temporalizing as inhering in each substance uniquely; space-time becomes reducible to place, in the Aristotelian sense (and similarly for Leibniz, if construed relationally).[78]

A second possibility is that the notion of substance itself is divided so that its "deep" structure becomes something other than what it has been taken to be by the history of metaphysics, i.e., divided dynamically into a notion other than substance. In other words, whatever explanatory value the notion of substance may have commanded will be deconstructed into other, perhaps less familiar ways of expressing the "present being" of an entity. But if this process is dynamic to the point of lacking limits, then the very possibility of formulating a notion as an element of a metaphysical theory to account for "reality" seems obviated. Derrida's rejection of a transcendental signified also receives metaphysical support from this reading. For a concept can never be necessarily linked to a given entity if that entity construed according to a certain configuration, i.e., as "substance," can (and, for Derrida, will) also be construable according to other naming notions of a different structural sort.

A final scenario illustrates the potential breadth of Derrida's use of division in **Passage 6D**. Consider a tree (our example of a substance). Is the tree, as a present

78 For Aristotle on place, see *Physics*, Book IV, Chapters 1 and 2. Leibniz adumbrates the relational view of space in Section 57 of the *Monadology* (and more substantively in the Fifth letter to Clarke).

substance (which, for Derrida, is redundant), *continually* being divided? Two implications: first, from a standpoint of traditional metaphysics–i.e., any theoretical claim or position advanced is necessarily bifurcated (with each resulting segment of a theory itself subjected to and bifurcated on to infinity). Although this position may appear counter-intuitive, perhaps even to the point of manifest absurdity, recall Leibniz on monads; he proclaims in the *Monadology* (Section 65) that there are an unlimited number of monadic substances capable of being divided from any given instance of matter. One potentially relevant difference between Leibniz and Derrida is that Derrida never asserts directly that divisibility of a given substance produces other self-subsistent substances; he maintains only that the process of divisibility indeed obtains, leaving it open what this process produces in terms of a recognizable metaphysical component.

The second implication: is Derrida maintaining that this divisibility as a functional component of trace occurs *in fact* in the face of evidence which suggests otherwise, at least at a level of immediacy?

We may construe Derrida's position to evoke the ultimate purpose of the factor of divisibility as a phase in a transcendental strategy–if the latter, then it would be open to take the divisibility in question to function as possibility rather than as actualization. If the division is not accomplished in fact, then it is an implication residing only in the realm of possibility, whether in the past or future. Thus a theoretical element in a given metaphysics exists in the present but always has the self-governing capacity of being redefined, restated, even rejected outright in the face of additional evidence or argumentation. The relevant revision in this reading is that this element nonetheless has sufficient stability to maintain existence in the present until it is dislodged from an external source either by the protagonist of this element or external criticism. The divisibility Derrida intends does not occur before our very eyes; rather, it can occur independently of our perceiving the entity in question at some point in the future, and only in the context of some linguistic event pertaining to the extra-linguistic existence of this entity.

Again, the transcendental strategy is apparent. The implications of division in this context require a future in order to be realized (and presuppose a relevant difference between what a substance exists as at present in contrast to what it might exist as in the future). Derrida reinforces the transcendental character of his thinking but he also faces a problem–what to do in theory with language stated at present about an entity enjoying what appears to be full existence in that present? We return to this question in Chapter 8.

6.8 The Question of Individuation

Derrida's general account of trace describes its function in relation to a determinate signification (e.g., the concerted references to "it" in **Passage 6D**). Furthermore, if pluralities of entities and language for these entities are a given, then we may infer corresponding plurality of traces to account for these "data," individually and collectively, as they exist against a totality background named by the notion of negation. This plurality of traces raises an important question: if each trace indicates the present by a relation to "what it is not, to what it absolutely is not...", then how are different traces distinguished from one another?

If Derrida on trace can be read according to transcendental parameters, then this account stands as a structural consideration for trace regardless what a trace is trace *of*. Furthermore, if trace is connected with something being present even given Derrida's severely restricted approach to the present, it must still be possible to distinguish between and among traces in order to preserve differences between and among what traces are traces *of*. But how to distinguish between traces of a given A and B remains problematic. For if A and B as elements in experience differ from one another, then surely the trace for A differs from the trace for B. But if distinguishing factors in such instances cannot be discerned from the assigned properties of trace, then they can be elicited only from A and B insofar as A and B possess factors other than those pertaining to participation in processes integral to the notion of trace. However, these factors will derive from the fact, the metaphysical fact, of A being A and B being B–considerations reinforcing the conclusion argued for in this chapter that trace must include (in order to account for individuated traces) an analysis of structures pertaining to what traces are *of*.

If, however, we inquire concerning the structure of what a trace is a trace of, then since all traces involve the totality of what is not (**Passage 6C**), the individuation of *distinct* traces becomes problematic. Assume that Entity A and Entity B exist in such a way as to generate Trace A and Trace B. The difference between Trace A and Trace B will therefore be decided not by considerations pertaining to totality, since *what is not* Trace A is exactly the same as *what is not* Trace B, except for, respectively, Trace B (which falls into the domain of what is non-trace A) and Trace A (which, in parallel fashion, falls into the domain of what is non-trace B). But it apparently follows that the desired differentiation can be secured only by describing the difference between (a) *any one* trace and (b) *every other* trace.

We see here again in another context the importance of preserving not only the referential function of language but also the actualization of a determinate "metaphysical" character as this determinacy arises from and adheres to referred entities. If traces can indeed be described as *of* some sort of determinate object ("object" understood as widely as possible), then it seems feasible–if not necessary–to follow out such designations to the point where the intended object of the trace becomes realized in some sort of differentiated manner. Without such specification, it is not obvious how a given trace can be informed and defined as *this* trace (of Entity A) as opposed to *that* trace (of Entity B). Therefore, a metaphysical component emerges as essential to a theoretically complete process of differentiation, a component requisite as a necessary condition for the possibility of establishing connections between the appearance and utterance of words or marks and extra-linguistic entities and states of affairs. Any mode of resolution applicable to the question of individuation as we have outlined that question here is thereby committed to the possibility of determining "metaphysical" considerations over and above the extent to which a given trace or a set of traces exhibits being as presence.

A second problematic dimension of individuation also emerges. For if a given trace exhibits self-identity over time, it becomes necessary to distinguish between the sameness underlying being as presence (the problematic notion of sameness as argued in Part I) and the derivative sameness integral to the uniqueness of Derrida's notion of trace when that notion is instantiated by means of a given entity. The argument is that the sameness which underlies the enduring existence of a trace over time, indeed that in fact justifies the temporality displayed by a trace, is not equivalent in all respects (or, perhaps, in any respect) with the sameness which grounds being as presence as the determinant of an entire epoch. The predication of the common name–"trace"–commits an advocate of this decidedly metaphysical move to develop the necessary properties and characteristics to respond to these questions, classical in heritage yet intuitively sound and indeed unavoidable in their relevance.

6.9 Trace: the Transcendental Character of Possibility

Trace initially names the principle of continuity which presents and justifies the possible significance of language over time. This dimension of trace grounds all actualized instances of language for whatever purpose, whether prosaic or profound. But another dimension of trace arises from Derrida's position on the "closure" of the history of metaphysics and it is important to appreciate these additional ramifications built into the notion of trace. Thus:

> **Passage 6E** I have tried to indicate a way out of the closure imposed by this system, namely, by means of the "trace" (SP, 141).

We show in Chapter 9 that Derrida's notion of trace also includes a specialized route toward the possibility of recognizing what has been hidden or suppressed throughout the history of metaphysics. This dimension of trace is apparently unique in terms of its intended speculative results, results which presumably may or may not be actualized even by the most perceptive, incisive, and eloquent practitioners of deconstruction.

Such speculation and the notion of negation here intersect. In this chapter, we suggested that negation should be understood as a class complement–expressed formally by "non-" connected to a class term. According to this approach, negation then encompasses totality, i.e., everything existent *other than* whatever is designated by the given class term. However, the totality denoted by this sense of "non-" remains fully determined by the history of metaphysics. Thus everything which could exist that would fall within the scope of this dimension of negation will retain the effects of being as presence as these effects encompass all traditional concepts and visions of reality engendered throughout the history of metaphysics. It would seem then that anything substantive derivable on the basis of trace, i.e., that which is "other" than trace, must come from a region of reality completely determined by the entirety of western metaphysics.

But as we shall see (in Chapter 9), the epoch of metaphysics has been "closed." And "trace," when it becomes functional in this trans-epochal sense, can lead *beyond* this epoch. Furthermore, this radical dimension of trace retroactively bestows new meaning to the account of negation in the *Différance* essay. When Derrida spoke of the connection between trace and negation as a relation to "what it [i.e., trace] is not, to what it absolutely is not...," the qualifier "absolutely" may be understood to refer, not to what is "not" trace when this otherness remains *within the totality of western metaphysics* but rather to what is "absolutely" other than trace, i.e., whatever trace is capable of opening up *beyond* that totality, given the closure of the totality of western metaphysics.

In this respect then, Derrida's "non-" anticipates his sense of *possibility*. However, possibility must be extended from anything which might appear within a determinate totality to a region parallel to the history of metaphysics but transcending that history in terms of novelty, its content irreducible to anything which has or will ever appear from the history of metaphysics. With this new ontological dimension in play, trace bears similarities to the arena of transcendentality since only if western metaphysics is surrounded, so to speak, will it be possible for whatever trace allows to emerge relative to that totality to surge into existence, or perhaps better, into life.

What initiates novelty? Does it happen, if it happens at all, randomly, by accident or can it be discovered by sustained speculative and analytically precise thought? We address these and related questions in Chapter 9.

6.10 Summary

When employed symbolically in the language of formal logic, negation is typically taken as a "primitive," its applications (as in the more traditionally-framed E and O propositions) assumed to be obvious and immediately accessible to modes of common perception and understanding. By contrast, Derrida's approach to negation as a pivotal element in his notions of *différance* and signification is more theoretically divergent when examined both by itself and also with respect to its implications. As a key component in his comprehensive approach to being as presence and to language, negation for Derrida is as rich and convoluted as is the counterpart employment of negation in Hegel, Heidegger and Sartre.

The notion of trace emergent from and defined in conjunction with negation represents the primary entry into Derrida's understanding of language realized theoretically as a mode of expression to some important degree independent from the calcifying tendencies of being as presence. Thus what a trace consists in depends on how trace relates to negation, as "absolutely what is not," a factor implying how to understand the distinctive character of the trace *qua* trace with respect to its revelatory function. The appeal to the totality of "what is not" invokes the parallel importance of totality in the formulation and application of being as presence (Part I), and, as we shall see, anticipates crucial phases incorporated in the critical assessment of deconstruction (Part III).

The pivotal connection is between particularity denoted by Derrida's use of signification when it functions in living language and the "not," or negation, which allows trace for that particularity to exist at all. At issue: the status of particular entities insofar as they are capable of being named by individual words and referred to by individual propositions about those entities. For Derrida, the "interval" between the content of a trace as such and the totality-dimension of reality as established by negation generates "the present" as a mode of existence for trace. But this chapter has shown that interval lacks sufficient theoretical development to stand by itself in this regard–for example, it is not clear whether Derrida's intends an interval to be (a) subject to continuous division in a manner analogous to that of a continuum and (b), if so, whether the interval in question encompasses particular entities. If such division, which for Derrida obtains in fundamental contexts such as temporality and spatialization, indeed functions in the conceptualization of a given "subject," then it is difficult to see how the unity and stability of both entities *per se* as well as language referring to those entities can be preserved.

The import of these questions should be combined with the fact that traces exhibit additional residual properties of metaphysical cast. Each trace *endures*, displaying continuity throughout the history of metaphysics; each trace also exhibits self-identity. These properties are ascribable to trace as a fundamental element of language and presumably as generic descriptor of entities referred to by language. The foundational character of trace–or, more accurately, traces–entails acceptance of these substantive and formal properties. In fact, the entire approach to trace reflects transcendental concern for itemizing and describing conditions for the possibility of language functioning as it has traditionally been conceived to function. Furthermore, Derrida surrounds the formulation of trace with negation, thereby generating a complex metaphysical context which serves in the advance of deconstructive agendas as a foundational device for the various ways in which trace becomes a carrier of epochal novelty. As noted in the previous section, this phase in the structure of traces becomes evident in Chapter 9.

7 Iterability

Part II of this essay began by analyzing an excerpt (**Passage 4A**) which referred to presence and its importance for Derrida's thought on language. Now, at the culmination of our review of seminal concepts in the deconstructive approach to language, we return to this passage but from a different perspective. The point of concern is the notion of *iterability*. For Derrida, iterability is a necessary component in an adequate philosophical account of language–any signifying utterance must be capable of being iterated, that is, being repeated in presumably an unlimited variety of contexts (whether spoken or written) in order to exist *as* such an event in the present or in any given set of circumstances. Iterability will therefore have direct or implied connections with all philosophical considerations involving the phenomenon of language. The structural facets Derrida builds into his depiction of iterability as an element in the theoretical composition of deconstruction are of vital importance, especially when the applications of deconstruction are examined for their cogency and explanatory power.

This passage opens discussion:

> **Passage7A** (cf. **Passage4B**) This nonworldliness is not another worldliness, this ideality is not an existent that has fallen from the sky; its origin will always be the possible repetition of a productive act. In order that the possibility of this repetition may be open, *ideally* to infinity, one ideal form must assure this unity of the *indefinite* and the *ideal*: this is the present, or rather the presence of the *living present*. The ultimate form of ideality, the ideality of ideality, that in which in the last instance one may anticipate or recall all repetition, is the *living presence*, the self-presence of transcendental life. Presence has always been and will always, forever, be the form in which, we can say apodictically, the infinite diversity of contents is produced.The opposition between form and matter–which inaugurates metaphysics–finds in the concrete ideality of the living present its ultimate and radical justification (SP, 6–italics in text).

7.1 Iterability and Presence–The Transcendental Mirror

The passage begins by emphasizing the ongoing necessity for possible repetition of a word–i.e., iterability–in order to enable both the emergence as well as subsequent uniformity of linguistic meaning, and concludes by drawing a moral concerning the inauguration of metaphysics. In both contexts, language and metaphysics, presence serves a uniquely fundamental function. In the initial section of Part II (Chapter 4), we offered an exegesis of presence insofar as it pertains to the notion of signification. Signification and iterability are related for Derrida–broadly stated, the latter is a necessary condition for the former–but for purposes of critical assessment, we distinguish them in Part II. The reader is reminded, however, that concentrating on a single line of interpretation pertaining to a pivotal concept, a procedure we adopt in this chapter, presupposes that being as presence and all its incumbent implications continue to control the content of any concepts introduced in order to explain and clarify Derrida's notion of iterability.[79]

Derrida employs the locution "ideality of ideality," and asserts that this double phrasing emphasizes the necessity of appealing to "living presence" as the ultimate, foundational reality grounding the very possibility of "transcendental life." To state and indeed to emphasize the status of living presence is, in effect, to isolate and identify the condition which must be granted in order to generate even the possibility of meaningful discourse. We note again the crucial importance of assigning specific content to the notion of presence, especially in the qualified sense of "living presence" introduced here. The suggestion is that presence as such is broader than "living presence," i.e., that presence could exist in some way other than qualified by the property of "living." However we have argued that presence (if taken as shorthand for being as presence) necessarily includes a factor of consciousness; if so, then "living presence" is not so much a redundancy but a phrasing which emphasizes that being as presence is necessarily "living" to the extent that a factor denoting life is part of its constitution. We shall investigate the extent to which this notion becomes associated with or even part of Derrida's own position.

When we "recall" instances of repetition or iterability, we do so within the more general experience of living. But Derrida identifies what is being experienced as "transcendental" life and insofar as this condition evokes Husserl's unexamined reliance on being as presence, Derrida finds it a target of deconstructive scrutiny. Although he still has a critical eye on Husserl in this context, the question that arises is whether the subsequent account of iterability will itself be developed according to transcendental modes of analysis. It is important to determine the procedural status of transcendentality given Derrida's predilection for stating his position in a transcendental manner–consider, e.g., the claim that the self-presence of transcendental life "has always been and will always, forever, be the form in which, we can say apodictically, the infinite diversity of contents is produced." The conditions governing iterability must be stated according to conceptualizations which are classical in their universality, necessity and omnipresence. In these respects, Derrida continues to maintain a transcendental dimension in developing approaches to his own position even if from another related perspective, he is concurrently subjecting Husserl's appeals to transcendental elements to deconstructive scrutiny.

---

79 Derrida's most concentrated treatments of iterability are in "Signature Event Context" (MP, 307-30; LI, 1-23) and in the defense and clarification of his position against John Searle's critique in the "Afterword" of LI, 111-54. See also SP, 50-2; WD, 246, 296-7; Derrida 1992, 41-2, 64, 65-6, 67, 68-9; LI, 12, 48, 53, 56, 63, 70-1, 76, 100; Derrida 2002b, 27. As far as I know, existing scholarship does not take into account the connections between the conceptual configuration Derrida ascribes to iterability and implications derivable from being as presence insofar as they involve iterability. See, e.g., Culler, 110-34; Gasché, 212-7; Hill, 27, 28, 94; Stocker, 60; Farrell 1998; Wills, 318; Hodge, 204-5; Norris 2012, 131.

The ramifications from Derrida's treatment of iterability will progressively constitute a uniform but speculative apex encompassing all the major elements analyzed in Part II–signification, context, negation. We will include an expanded analysis of these elements in this chapter. But at the moment we offer an implication from Derrida's initial approach to iterability which will contour much of what follows in both this chapter as well as the rest of the essay. Since the repetition of a linguistic element never occurs in a vacuum, the universality and necessity for such repetition entail a set of minimal background conditions–including, presumably, an all-encompassing spatio-temporality–commensurately requisite to guarantee the possibility of iterability in Derrida's sense of the term. The resulting transcendental mirror will frame and reflect this set of conditions, the core of which will be unavoidably metaphysical in character. Identifying the principal elements of this core embodies a project which will move beyond the letter of Derrida's texts. And suggesting avenues for amplifying these elements will be the primary concern of Chapter 10, the concluding chapter in this essay.

7.2 Derrida on Possibility and Necessity: A Critical Review

The notion of possibility is crucial to Derrida's formulation of iterability and we will examine its function closely. In fact, possibility as Derrida deploys it ramifies into all major elements of the transcendental approach Derrida directed toward the notion of iterability.

Possibility has a venerable heritage in the history of metaphysics. In *Metaphysics* Book V, for example, Aristotle says that the possible "means that which is not of necessity false; in one, that which is true; in one, that which may be true" (1019^{b}30). Thus "that a man should be seated is possible, for that he is not seated is not of necessity false." Even if a person never took a seat, it remains possible that this constantly vertical individual could always take a seat. This approach to possibility illustrated by Aristotle's generous treatment of the concept should inspire caution when Derrida appeals to possibility. As we shall see, possibility, like being (as Aristotle famously observed), can be said in a number of ways, and it is crucial to appreciate distinctions regarding the structural complexity of possibility and the conclusions Derrida bases on selected phases of this structure.

The primary text for understanding Derrida's notion of iterability is the essay "Signature Event Context" (published in 1973), the work inspiring John Searle's strongly-worded response. The immediate context is Derrida's defense against Searle's criticism of Derrida's critique of John Austin's concept of speech acts. In this defense, Derrida expands and clarifies the ways in which possibility and iterability relate to one another given the position maintained in "Sec" (Derrida's abbreviation of the original essay, "Signature Event Context"). In order to appreciate how Derrida situates this relation, we examine two passages from Derrida's reply to Searle, which Derrida entitled *Limited Inc* (published in 1977).

> **Passage 7B** If I insist here, it is because this is indispensable to the demonstration and even to the minimal intelligibility of *Sec*. I repeat, therefore, since it can never be repeated too often: if one admits that writing (and the mark in general) *must be able* to function in the absence of the sender, the receiver, the context of production, etc., that implies that this power, this *being able*, this *possibility* is *always* inscribed, hence *necessarily* inscribed *as possibility* in the functioning or the functional structure of the mark. Once the mark *is able* to function, once it is possible for it to function, once it is possible for it to function in case of an absence, etc., it follows that this possibility is a *necessary* part of its structure, then the latter must *necessarily be such that* this functioning is possible; and hence, that this must be taken into account in any attempt to analyze or to describe, in terms of necessary laws, such a structure (LI, 48–italics in text).

We concentrate on the structure and ramifications of possibility as Derrida formulates that concept. However, note in the core development of his defense Derrida's clearly demarcated acceptance of modal differences, i.e., between possibility and necessity ("...must be able to function...") as well as the logical function of implication ("... that implies that this power..."). This point is procedural in terms of the immediate context of inquiry but the fact that Derrida has so broached his defense has crucial implications for the internal coherence and ultimate feasibility of deconstruction. The distinction between possibility and necessity is at least as old as Aristotle and in traditional formulations it is also quintessentially metaphysical, as are all the logical concepts and principles implicitly embodied in implication as Derrida has here employed the concept. But what theoretical justification allows Derrida to fashion his defense by adopting these conceptual programs and simultaneously, as an advocate of deconstruction, claim to throw into question–apparently without exception–*all* the most fundamental elements of western metaphysics?

This inconsistency as it emerges from an intersection of deconstructive theory and expositional practice certainly appears blatant; furthermore, this problem arises quite apart from whether or not the letter and thrust of his defense effectively blunts the force of Searle's criticisms. The relevant question does not concern the argumentative strength of the defense–rather, the question is how Derrida can mount such a defense in the first place without falling prey to an especially virulent instance of logical inconsistency.

From what source, for example, does Derrida derive his notion of necessity? At precisely this point, the transcendental dimension of his position becomes even more palpable, indeed undeniable. The necessity in question is not that of a formally logical kind–e.g., as derived from identity or contradiction; rather, its context is language and its origin is based on claims pertaining to how language will always and necessarily function. It seems legitimate therefore to ask whether conditions must be in place in order that the subject of this necessity–language–will in point of philosophical and metaphysical fact *always* so function as Derrida claims. The contrast between deconstruction's stated goals and Derrida's immediate defensive strategy along with the seemingly unavoidable logical tension establishes an issue of interpretation which compels examination. We will discuss this matter in detail in Chapter 8.

Derrida argues (**Passage 7B**, in the context of writing and its marks) that if an eventuality is necessarily possible for something, then such a possibility becomes an indelible component of that something. However, we must proceed with care in determining the relevant senses of "possible" in this regard, especially in contexts involving language. Consider Hume. In Section 3 of the *Enquiry concerning Human Understanding*, Hume characterizes a matter of fact as something the contrary to which is always possible.[80] Hume uses several examples to illustrate this approach to possibility, one of which is the connection between fire and heat.

It is a matter of fact that fire is hot. But it is possible that (1) fire could not be hot. However, at this point a distinction must be drawn. It is possible that the sun in our solar system will incinerate earlier than astronomers predict; that possibility is entirely within the realm of the empirical constitution of the sun. However, although this possibility does indeed belong to the sun, the sun may never in fact actualize this possibility. To return to the example of fire, it is also possible that (2) fire may in fact *never* exist as anything other than hot. In other words, the proposition "if fire exists, then fire is hot" may always and necessarily be the case, a necessity conferred ultimately by virtue of the nature of fire itself.

We must therefore distinguish between material possibility which may be realized and material possibility which may not, in fact, ever be realized. A material possibility which for whatever reason is never realized still necessarily exists as part of a given entity or situation–it is just a possibility that remains, since it will never be actualized, more appropriately in the realm of what is typically referred to as "logical" possibility.[81] To summarize: both cases (1 and 2 above) with respect to fire and heat embody material possibility but the first case includes a dimension of actualizability not applicable to the second case.

In general, empirical possibility extends to the outermost extremes of what could in fact be the case, but empirical possibility cannot extend beyond logical possibility. In other words, it is empirically possible for fire not to be hot but it is not even logically possible for fire to be both hot and not hot at one and the same time and in the same respect; therefore, it is necessarily not an empirical possibility that such a state of

80 The difference between Hume's approach to possibility compared with that of Aristotle in the *Metaphysics* cited above is the addition of the "contrary" formulation along with the understanding that if this contrary does not contradict the given matter of fact, its realization is "possible;" otherwise, the two positions are similar to one another.

81 We may distinguish between (a) logical possibility as a state of affairs not inherently contradictory and (b) a parallel state of affairs also not inherently contradictory but which may not, in fact, be possible if its realization is incompatible with existing physical laws. In this case, a change in physical laws might result in the realization of the state of affairs in question.

affairs could ever occur. But it is entirely possible, both logically and empirically, for fire to cease to be hot (although, again, as an empirical matter fire may never cease to be hot).

A second example with coordinate implications reinforces this preparatory response to Derrida's position: If I regularly play the lottery, there is a possibility I will win. Indeed, there is *always* a possibility that I will win (on condition, obviously, that I continue to play). But in fact I may *never* win the lottery (whether I play the lottery once or many times). Does the fact that the possibility of winning the lottery–a possibility which is necessary to the act of actually betting on the lottery–is never realized tell against the ongoing reality of this possibility *qua* possible? Not in the slightest. The point is that the reality and necessity of the possibility attaching itself to the act of betting must be kept apart from *actualizing* that possibility. Possibilities can be real without ever being transformed from (a) the reality of necessarily being possible to (b) the reality of possibility insofar as that reality has been reduced to actuality.

The types of possibility pertinent to the current discussion may be designated *logical* and *material.* And the above analysis suggests that material possibility then be subdivided into two distinct states of affairs (a) *materially actualized possibility*–a possibility which may or may not be actualized in fact (e.g., fire becoming in fact either hot or not hot) and (b) *materially non-actualized possibility*–a possibility which will *never* be actualized in fact (e.g., fire never becoming not hot although retaining necessarily the possibility to do so, given certain physical laws). In these states of affairs, (a) includes (b) but given that the underlying context is possibility and necessity, I submit that for the sake of clarity concerning possibility in relation to its various outcomes, the empirical difference between the consequence of (b) and the possible consequences of (a) justifies the distinction.

We will discuss in more detail the application of this distinction to Derrida's notion of iterability after additional investigation into the subtleties of the concept of possibility. But if we return for a moment to Derrida on iterability and repetition with respect to possibility, it becomes evident that iterability as defined must fall under materially actualized possibility, not under materially non-actualized possibility. The example from Hume detailing materially actualized possibility emphasized the limit case of fire becoming not hot because nothing contradictory precludes that factual possibility. But materially actualized possibility also includes what would in fact be expected in the case of fire, i.e., that the next experience with a flame will produce the sensation of heat. Shift the context to language; iterability then not only falls under material actualized possibility, but Derrida's position also necessitates that what exists as a possibility for a given signifier *must in fact be actualized.*

In **Passage 7B**, Derrida's "absence" stands for *something* in that the absence under scrutiny functions as a kind of indeterminate but necessarily existing place-holder for the range of possible instances of a given signifier, in the process enabling that signifier to exist in the first place. In other words, if a signifier had never, not even once, been in fact repeated, then this signifier would not have had the metaphysical weight, so to speak, to function as a signifier on the strength of the sole occasion in which it did indeed appear in a given context. For if a signifier had retained all actualizable instances of its realization in a condition of non-existence as resident within the boundaries of possibility, then this signifier would in fact have ceased to exist even as a putative signifier.

The context of inquiry thus advances to absence, a notion evoking what is other than–or, more abstractly stated, *what is not*–the given mark as a candidate for signification if the mark as word did in fact become iterable.

7.3 Iterability and Absence: the Question of Negation

The following inference might appear initially justified for Derrida: if the possibility of iterable meaning belongs necessarily to a given instance of a word as signifier (more precisely, as a *potential* signifier), then this necessity somehow guarantees that what is only possible must at some point be actualized. In this case, Derrida would move from the necessity of possibility (which in general the present approach to the concept of possibility accepts) to the necessity of *actualized* possibility, i.e., an existence in actuality (which the present approach to the concept of possibility denies as without warrant, given the distinction drawn above between the two types of material possibility). This apparently plausible inference concerning Derrida's position would, however, be incorrect. But in order to reject this inference with due cause, we must consider the complex relation between presence and absence (taken as a variant of Derrida's use of negation) insofar as both notions pertain to possibility.

Presence continues as an explicit element in Derrida's thinking on iterability but in the following passage its function is indirect, setting up the point Derrida intends to make by way of contrast along with an appeal to absence. A remarkable feature of this passage occurs when Derrida refers to "structural law." One wonders what Derrida means by "structural law" and how it can be established given deconstruction's concerted interest in dismantling the kind of conceptual conditions necessary in order to install a law, any law, in the first place.

We return to this point at the close of this chapter. For now, we examine iterability from the perspective of possibility as the latter notion is deployed in the following passage and in conjunction with the distinctions drawn in the previous section. The resulting interpretive position will allow the demonstration of complexities in the concept of possibility which Derrida did not seem to recognize, at least explicitly.

> **Passage 7C** Even if it is sometimes the case that the mark, in fact, functions *in-the-presence-of*, this does not change the structural law in the slightest.... Such iterability is inseparable from the structural possibility in which it is necessarily inscribed. To object by citing cases where absence *appears in fact* not to be observable is like objecting that a mark is not essentially *iterable* because *here and there* it has not *in fact* been repeated (LI, 48–italics in text).

The notion of "absence" appears three times in **Passages 7B** and **7C**. Both passages are thus stellar examples of the force of context, since in context "absence" refers to a specific mark (or word) which was present in a given textual setting but is now absent, i.e., no longer present. But for Derrida, the "absence" circumscribing what is necessary to the notion of iterability represents the possibility that any given instance of a signifier can be and for Derrida must be an essential component of the theoretical understanding of signification as applied to that and any particular signifier. Otherwise put, what a signifier "is not" is just as necessary to the reality of that signifier as what "it is" at its initial appearance on a linguistic stage.

But now the crucial relevance of the meaning of negation for Derrida becomes paramount, especially when negation–or absence as it is named in this context–is connected to the analysis of types of possibility which we have just established. Consider that the description "no longer present" can apply to the potentially signifying mark insofar as it can become present in some determinate sense *anywhere and everywhere*. Thus a given mark no longer present (or, as Derrida writes in **Passage 7C**, mark as it "functions *in-the-presence*-of"...) can appear–i.e., satisfy the function of being iterable–at any given point throughout the totality of possibility as that totality is determined by the gamut of reality described as "no longer present." This expansion of "absence" thus becomes denotatively equivalent to negation as that concept was characterized in Chapter 6. In other words, the relation between the mark and the totality within which the mark is no longer (but could be) present parallels the relation between X, the variable for a trace, and the negation by which X is determined to be what it is as a trace. In context, the references to "absence" in these two passages may be taken initially to refer only to the absence of the given marks. But this absence does not characterize a metaphysically empty environment–rather the absence in question occurs within a setting encompassing totality in the sense that the absence of a given mark will become an actualized presence of that mark *at any point within the totality of possible locations for that mark.*

The reemergence of totality, now as an element in the situation of iterability, has crucial consequences. In order to render these consequences with due force, I offer an example of the repetition of a mark the circumstances of which necessitate modifications and restrictions in Derrida's purely–and, as we shall see, inaccurately stated–formal characterization of iterability.

Francis Bacon's *The New Organon* appeared in 1611. When Bacon wrote in the account of the Idol of the Market Place that the meanings of words are assigned "according to the capacity of the vulgar,"[82] it was not possible at this moment in

82 See Section LIX of *The New Organon* where Bacon uses "vulgar" several times. We shall see that the discussion of change of meaning over time as it pertains to the word "vulgar" represents an important phase in the critical analysis of unstated assumptions in Derrida's notion of iterability; it also serves as a prelude to *possibility* relative to shifts in meaning in words and, more broadly stated, to the formation of new terminology, especially in philosophical contexts. See discussion of the relevant sense of possibility in Chapters 9 and 10.

history to anticipate that as time passed, "vulgar" would become significantly different in meaning. In the 17th century (and well into the 18th), Hume uses "vulgar" in the same way–cf. Section VIII of the *Enquiry concerning Human Understanding*), "vulgar" meant language spoken or written according to common or ordinary usage of common and ordinary people. The first OED entry for "vulgar" reads: "employed in common or ordinary reckoning of time, distance, etc…;" the second entry: "in common or general use; common, customary, or ordinary, as a matter of use or practice." By contrast, in contemporary usage, "vulgar" frequently if not typically means "offensive in language" or "lewdly or profanely indecent" (Merriam-Webster online dictionary). The OED does list "coarsely commonplace" for "vulgar"–but not until the 13th entry. Furthermore, "vulgar" when used in the 17th and 18th century regarding language did not exhibit any vestige of the normative or prescriptive aura that the word currently possesses. As written in 1611, "vulgar" does not (yet) have the meaning that it will have three centuries later. Thus the meaning of "vulgar" in 1611 is "not" what it will be.

It was indeed possible that "vulgar" could change from its meaning in Bacon's day to its current sense. But the relevant question is whether, at a given point in history, the possibility that "vulgar" could significantly change its meaning *in the future* will affect the actual meaning of "vulgar" *in the present.* It would appear implausible (if not patently ludicrous) to argue that because the 1611 meaning of "vulgar" will eventually change, it follows that no definite signifying sense can be attributed to "vulgar" when used *in the present* (1611 and for some time after that date) precisely because the reality of "vulgar," even and (with Derrida) especially when used in 1611, includes the *possibility* of a significant if not radical shift in meaning *in the future.* Thus what in this case "vulgar" *is not* when "vulgar" is used in 1611 has no bearing on what "vulgar" *is* when "vulgar" is used in 1611.

As a result, the concept of possibility with respect to iterability must be recalibrated and made more precise; if so, then the above distinctions between types of possibilities become crucial. Although it was always logically possible that "vulgar" could shift its meaning (because nothing involving contradiction bars such a shift), it was not always the case that "vulgar" would *necessarily* shift its meaning. After "vulgar" had assumed the meaning it had in 1611 and on through Hume's time, it was always possible that "vulgar" would have maintained precisely the same meaning, unchanged, into the present and beyond. In this scenario, the history of "vulgar" as a term of signification would have been guided by non-actualized material possibility; that is, it was possible, necessarily so, that "vulgar" could have changed its meaning but in point of lexical fact, "vulgar" did not do so.

These considerations also pertain to iterability in terms of theoretical completeness. Consider that in the Bacon example, the same mark, "vulgar," is repeated at two distinct periods of time (the early 17th century and the present). But the advocate of Derrida's notion of iterability and the possibility of repetition will object that iterability in the intended theoretical sense does not refer to the repetition of the same mark when the later instance of that mark *differs in meaning* from the earlier instance. Yet precisely this kind of difference characterizes the two instances of the mark "vulgar" in the Bacon example just described. Rather, what Derrida intended was the repetition of the same mark when the signifier in question *retained the same meaning* in both instances.

As a defense (albeit temporary), this response is legitimate. However, Bacon's use of "vulgar" reveals a feature of repetition which must be taken into account as far as general theory is concerned. The Bacon example illustrates the first of the two types of material possibility. However, although the mark remains the same, i.e., "vulgar," the meaning changes from its sense in Bacon's time to include the more contemporary signification. But this alteration in meaning does not occur in a terminological vacuum. External circumstances changed, doubtless a variety of different external circumstances; as a result, the difference in meaning must be understood in relation to what might be termed a contextual setting, that is, a set of minimal conditions which underlie the possibility that a signifier such as "vulgar" can alter its sense. After all, nothing in the original meaning of "vulgar" requires (a) that its sense will necessarily be something other than what it was at a prior period of time–as noted earlier, it is entirely possible that "vulgar" could have meant and will continue to mean what it meant when Francis Bacon was philosophizing; and (b) that a change in sense, if one occurs, would become *a discernibly determinate signification*, i.e., what "vulgar" did and indeed does mean, especially with respect to the use of coarse language.

The point generalized is that instances illustrating the repetition proper to iterability function necessarily *within contexts*. Thus to understand Bacon's use of "vulgar," the original meaning of "vulgar" must be maintained and reestablished in the current present, i.e., the common contemporary associations attached to "vulgar" must be kept apart from "vulgar" as this mark was used in Bacon's *The New Organon*. The present context characterizing the signification of "vulgar" weighs on the reading and understanding of that mark–there is, in fact, an intersection of contexts; thus, when we read Bacon in the 21st century and assign the original meaning to the use of "vulgar" in his text, we restore if only momentarily a cross-section of Bacon's own world. We may assume that a complex set of factors determined "vulgar" to mean what it meant at that time four centuries ago, just as it was doubtless an equally complex set of factors which went into the transformation of the signification of "vulgar" from its predominantly descriptive evocation to its contemporary normative and negatively charged sense.[83]

A more extensive and vital conclusion may also be drawn concerning iterability. The metaphysical necessity underlying the possibility that a word can change meaning over time points to the fact that the same kind of contextual necessity underlies the possibility that iterability can in fact be realized *at all*. Otherwise put, the same underlying conditions which account for the possibility of words changing

---

83 The online Merriam-Webster also restates the original meaning of vulgar as "common," or "ordinary," etc.

their meaning also apply to the iterability of a word when it does not change its meaning. It is essential then to connect the notion of negation and the necessity for contextual considerations. As an instance of iterability moves from a present mark into the future, into what is "not" in a sense characterized by totality, the instance which substantiates the appeal to iterability occurs *in a certain set of extra-linguistic circumstances*. Furthermore, every such instance will also be characterized in accordance with the same qualification. The enumeration and description of the circumstances will differ, perhaps greatly, from this instance of iterability to the next instance of iterability. But *all* instances of iterability will realize themselves within the same general framework of fundamental metaphysical components.

This argument demonstrates that Derrida's notion of "absence," which we have taken in this context to be synonymous with negation, must be carefully examined. The relevant point pitched in metaphysical terminology is that the element of possibility as it pertains to the notion of iterability has *limits* which must be identified and described. The inference is not to deny the relevance of the notion of iterability, i.e., that a signifier must have the capacity to be repeated in order to be meaningful; rather, it is to focus attention on the fact that *how* such repetition occurs, with respect to temporality, contexts and circumstances, is a necessary feature of a cogent and empirically sensitive theoretical treatment of iterability.[84]

In general then, the appeal to iterability necessarily encompasses commitment to extensive and serious metaphysical situations, circumstances and conditions–in a word, to contexts. In fact, the cogency of iterability as a theoretical component in Derrida's treatment of signification rests on a set of complex metaphysical parameters. Indeed, the theory remains incomplete without recognizing and enumerating these parameters. This implication has serious repercussions if deconstruction is advanced as a theoretically cohesive set of gestures for reading and interpreting texts. We now examine one of these repercussions.

7.4 Iterability and the Possibility of Parody

Words change in meaning. But words do not so change by themselves alone, from internal principles of self-motion seeking nuanced or extensive variety of sense. *Contextual givens and parameters* underlie the possibility of a word having a certain meaning; whereupon a shift in those givens collectively engenders and contours an alteration of the original meaning, changing it over time. Absence, or broadly stated, negation names the field within which the actualization of iterability for all

84 Chapter 8 demonstrates how three distinct transcendental–and metaphysical–dimensions underlie Derrida's position on iterability as well as the panoply of gestures which characterize and specify the various agendas of applied deconstruction.

significatory events occurs, whether it be the mere repetition of a word with the same meaning or the appearance of a mark with a new meaning. To generalize: iterability presupposes a configuration of contextual givens when over time a word either changes in meaning, does not change meaning, or the word (e.g., "vulgar") retains the original meaning as well as gathers a new meaning–the difference determined by the combination of the speaker's intention and context. Thus Derrida's notion of possibility when applied to linguistic phenomena is necessarily metaphysical in order to account for (a) a possible change in meaning of a given word and to ensure that (b) a given word with a certain meaning at a certain point *in the world* will preserve that meaning when it is used at a subsequent point *in the world*.

When iterability is considered with respect to certain specific contexts understood as extensions of the concept of possibility, the implications are crucial with respect to what will limit the material circumstances which can and cannot establish the possibility of iterability. Here is an example. A noted commentator on Derrida, Jonathan Culler, summarizes Derrida on iterability as follows:

> **Passage 7D** Something can be a signifying sequence only if it is iterable, only if it can be repeated in various serious and nonserious contexts, cited, and parodied.[85]

Culler gives the language of a play as an example of a "nonserious context." Consider the act of making a promise. For Derrida, no priority regarding philosophically relevant aspects of language exists between two quite distinct states of affairs: a "nonserious" promise made, e.g., during a play and a promise made "seriously" in, say, a contract arrived at in a business context located outside a theater (or, of course, anywhere). The iterability of each instance of a promise allows for–and for Derrida requires necessarily–the iterability of *any* instance of a promise, and Derrida's approach to an undifferentiated region of possibility sanctions this conclusion.

This extension of the scope of iterability presents circumstances where the distinction established earlier between types of material possibility becomes critical. Readily available arguments show that if the linguistic sequence and context involved in making or, more accurately, simulating a promise as actually performed in a play had equal weight with the same sequence executed in a non-theatrical or, as it might be put, serious context, it would not always be possible to determine what elements of the sequence of promise-making could be extracted and then treated as the subject of parody (cf. **Passage 7D**). It follows then that only *certain instances* of making a contract are such that a parody of a contract becomes possible.

This incommensurability between serious and non-serious contracts obtains under various circumstances:

---

85 Culler, 120. This formulation neatly encapsulates Derrida's position on the equivalent status of plays with respect to iterability and the possibility of parody. Cf. MP, 324-6; LI, 34, 89-90.

Given the present social order, an agreement to fulfill a contract as a formal, legalized promise is an important if not essential type of conduct in the complex and variegated circumstances defining human interaction. But it is entirely possible that a promise of this sort–doing something with words–would *never* have been introduced as an element in any play throughout the history of drama. Although a promise could have become such an element, nothing in the nature of dramatic license and limitless imaginative possibilities open to playwrights requires that it must ever have been selected to exercise this role. Had lack of interest on the part of the world's dramatists been the case, then speech acts involved in making promises and contracts would illustrate the category of material non-actualized possibility. It would necessarily always have been possible for dramatists to introduce contracts and promises into the action of a drama, but in fact they never did so. In this case, only serious contracts could be parodied for the simple reason that only these instances of such speech ever existed.

Assume that contracts were employed in some dramas. Even in this case, improperly formulated contracts could have been inserted into the dramatic action. For example, if the intention to keep a contract is a necessary condition for a legitimate contract, then it is possible that every contract that ever appeared in the history of drama was entered into by characters who never had the intention of keeping these supposed agreements. If the intention to keep the promise is necessary to the words articulating the promise in order that those words constitute a *de facto* promise, then an actor could say "promise-words" without intending to keep the promise and the words as such would not qualify as a "serious" promise (even if they sounded serious to the character so addressed, as well as to the audience, because the words fit the dramatic circumstances). The plots of these dramas might have been variations on the putative contractor ostensibly entering a contract but then reneging on the promise to fulfill it because not to do so would advance that character's self-interest. Thus contracts were used as a plot device but were not used accurately. The *idea* and external wording involved in making a contract would then exist in a dramatic setting, but not a contract in the legally canonic sense even within the stylized setting of a play.

A final possibility: contracts in their authentic formal structure in fact could have been used *accurately* as a plot device but none of these contracts were ever *actualized or fulfilled*. Unexpected events interrupt the completion of a contract (these events might mesh perfectly with the overall plot) so that an audience would witness the advancement of a legitimate contract in all requisite details but would never see what happens when in fact a contract is met. Thus only if a promise-sequence (leading to a contract) occurring in a drama were staged to its completion, in straightforward imitation of a promise made in "real life," would it be possible to determine what elements in the entire process could be subjected to parody and how the parody could be exhibited. But in this eventuality no formal difference would obtain between an authentic and a staged contract; it follows that it would be otiose to insist that the latter instance of a contract is materially relevant to the concept of iterability.

The first two examples above illustrate material actualized possibilities–they could or could not have been utilized throughout the history of drama. Thus if any one, pair or all three of the possibilities just enumerated and described had been the case, then contracts could never be parodied because a real, i.e., serious contract had never existed in a dramatic setting to stand as a model for purposes of duplicating an actual contract from inception to completion. To summarize: nothing in the necessity of the *possibility* that playwrights would at some point in the history of theater employ contracts, whether complete or incomplete (as legally defined), entailed the necessary *actualization* of this possibility. Thus a contract can be parodied if and only if at least one contract had been fully and seriously (i.e., in a non-theatrical context) formulated and realized by the participants of that contract.

We conclude that a promise could not be iterable *as a promise* in the defined sense unless it had been carried to completion. Call this the serious *realized* promise. But as just argued, such a promise as a fully-fledged linguistic event need never have happened in the context of a play. Nothing would have been lost in terms of the structural necessities involved in a promise if all the promises ever made–whether or not they involved the legalized machinery of contracts–were in fact serious. Would a promise made by two individuals each fully aware of what it meant to "make" a promise be any less of a promise if this particular arena of proposed iterability–the repetition of a promise in a theater–were eliminated? In a word, no.

We now more fully appreciate the importance of the generalized analysis of structural considerations pertaining to possibility which opened this section. Derrida (as well as advocates such as Culler) has conflated *materially actualized possibility* and *materially non-actualized possibility* and asserted, without justification, that instances properly belonging to the latter–promises which could possibly have been made in plays but in fact never were–must necessarily be included in possibilities that belong only to the former, i.e., when those possibilities are (or are not) fully actualized.

In order for iterability to be a feasible element in determining the possibility of signification, it is essential to introduce metaphysical stipulations and commitments. We have argued that it is a mistake to claim that serious and non-serious settings are equally relevant to the notion of possibility with respect to the implementation of iterability. But to show this error presupposes that (a) appeals can be made to contextual situations and, more precisely, that (b) these appeals will contain explicit references to descriptions which introduce metaphysical factors. Therefore, not only does iterability presuppose a contextual setting with definite metaphysical dimensions but we now appreciate that these settings have limits, i.e., not all possible settings will in fact actualize the iterability of a given mark in the requisite sense. As a result, a telling feature in the denial that a promise made during a play is relevant to the notion of iterability is precisely the fact that the words constituting the promise are spoken by actors in a theater, i.e., a spatio-temporal setting reserved primarily for timed events which involve by their very nature "playing" at reality and without the participants of those events enduring the consequences, whether beneficial, neutral or harmful, of their participation, in particular whatever they may have said.

In sum, the notion of iterability must be construed in relation to contexts as well as transcendental conditions of a metaphysical character, e.g., places existing in certain spaces and characterized by limited stretches of temporality. We now consider another approach to iterability which will further refine the relevant notion of possibility, especially from the standpoint of its associated metaphysical structure.

7.5 Possibility and the Concept of Supplement

The previous analysis with its focus on and development of the multi-dimensional constitution of possibility has sounded an alert regarding Derrida's treatment of concepts allied with other integral elements in the makeup of deconstruction. Additional analysis will reveal the root explanation for the imprecise condition of the notion of iterability and how this imprecision dictates implications regarding the use of possibility with respect to fundamental metaphysical concerns.

We examine Derrida on *supplement*, the name in the basic vocabulary of deconstruction for Rousseau's approach to writing, a dismissive approach for Derrida, which sees writing as a "supplement" to language as spoken and which depends for its discursive and theoretical support on the notion of possibility. The following summary from the *Derrida Dictionary* is a useful introduction to this term:

> In each case, the supplement adds itself to an ostensibly ideal or original presence in the form of exposing the lack and self-difference at the very origin. For Derrida, in other words, supplementarity is at the always divided 'origin' of presence. It is, precisely, that without which the **metaphysical** tradition could not constitute itself. But the supplement is also that which metaphysics projects as an inessential and privative attribute, something it could well do without.[86]

And Derrida himself describes the seemingly quixotic, self-erasing nature of anything which functions as a supplement in the prescriptive sense underlying Derrida's use of the term:

> **Passage 7E** It is the strange essence of the supplement not to have essentiality: it may always not have taken place. Moreover, literally, it has never taken place: it is never present, here and now. If it were, it would not be what it is, a supplement, taking and keeping the place of the other. What alters for the worse the living nerve of language ("Writing, which would seem to crystalize

86 See Wortham, 2010. This summary appears on 204 of the *Derrida Dictionary* as part of the article on "supplement." The bolded word is in text.

> language, is precisely what alters it; it changes not the words but the spirit of language...") has therefore above all not taken place. Less than nothing and yet, to judge by its effects, much more than nothing. The supplement is neither a presence nor an absence. No ontology can think its operation (OG, 314).

This passage may be read as an argument describing the metaphysical character of supplement. Derrida identifies the essence of a supplement as "strange," asserting this qualifier because a given supplement "may always not have taken place." Let us evaluate what Derrida has written by reflecting on a common instance of a supplement, then extrapolating conditions based on that instance. We shall see that the concept of possibility becomes crucial in these reflections (including the two types of material possibility distinguished earlier in this chapter), as well as several related notions of equivalent generality.

If a household has a vitamin supplement on hand to use in case of need, it is entirely possible that this supplement would remain on a shelf for the duration of the household if all the nutritional requirements of the members of that household were met by the regular pattern of their meals. If so, the vitamins, existing as an intended supplement to the family's diet, "may always not have taken place" as functioning in this regard. But would we then say of these vitamins that they were "neither a presence nor an absence?"

Derrida contends that a supplement "literally...has never taken place: it is never present, here and now." He then adds, as a reason:

> **Passage 7E$_1$** If it [i.e., the supplement] were [i.e., present], it would not be what it is, a supplement, taking and keeping the place of the other.

Consider Derrida's assertion: if the supplement were indeed present, it would not be a supplement, "taking and keeping the place of the other." *Two* senses of present must be distinguished. If the supplement were present, not in the sense that the container of vitamins sits before someone with a vitamin deficiency but in the sense that the vitamins had in fact been consumed by this individual, and if in *this sense* of presence the consumed vitamins did nothing other than take and keep "the place of the other," then the vitamins would *not* have fulfilled their resident potentiality to function as a dietary supplement. It would have been as if the vitamins had never been taken at all, given that consumption of the vitamins had no effect on their consumer. However, a consumed ineffective vitamin *continues to exist* as a vitamin–but only as an *ineffective* vitamin, not a vitamin which "has never taken place" and been "never present, here and now." Derrida's reason in **Passage 7E$_1$** conflates (a) the *potentiality* of the supplement to function this way (assuming it were taken), when as such, as purely potential, the supplement is not present "here and now" for the simple reason that the supplement (the vitamins) has been consumed with (b) the *actualization* of the supplement when the supplement has ceased to exist since it has in fact fulfilled its function *as supplementary* to a pre-existing physical condition which can be altered (and improved) by its consumption.

Would we then conclude with Derrida's decided air of finality that "No ontology can think its operation"? We would not. All that is needed to "think its operation" are the following: first, recognition that vitamins exist by themselves, as items in a container, whether or not they are ever taken by anyone; second, that the reality of the vitamins *relative to their prospective users* includes a dimension of potentiality. The vitamins exist as *actual* tablets, with measurable size and ingredients. However, the vitamins also exist as *potential* supplements to the diet of individuals who may require such additional sustenance. Finally, the vitamins possess both types of reality *simultaneously*. Derrida conflates actuality (of the vitamins existing in the container) with potentiality (that they may or, as he emphasizes, may not be used); he also conflates the present (when the vitamins rest in their container in the cupboard) with the future (when the vitamins are or are not taken by those who may or may not need them). In sum, all that is ontologically required to analyze fully the "essentiality" of a supplement given what Derrida has asserted about it at this juncture are (a) concepts of potency and actuality and (b) a notion of time with divisions into past, present and future. Many such ontologies are available throughout the history of metaphysics.

7.6 The "Logic of Supplementarity"

The critical analysis in the previous section reacts to the argument Derrida employed at the conclusion of his treatment of supplement in *Of Grammatology*. However, what Derrida intends supplement to accomplish from the standpoint of deconstruction trades on additional considerations pertaining to possibility as well as an unexamined and (as shown in Chapter 6) loose sense of negation. The connections between the more speculative aspects of supplement and the structure of iterability are worth exploring, especially with regard to the dependency on possibility in both contexts.

Christopher Norris has been a prolific and energetic defender of Derrida as the primary proponent of deconstruction. In a chapter from Norris 2012 entitled "Of Supplementarity: Derrida on Truth, Language and Deviant Logic," Norris develops a reading of Derrida's notion of supplement which may serve as a convenient platform for discussion. We divide Norris' account into two main areas–logic and metaphysics.

7.6.1 Supplementarity: Logical Structure

The phrase "logic of supplementarity" is Derrida's (OG, 215) and it occurs in his discussion of Rousseau in *Of Grammatology*. For Norris, "...it is Derrida's claim that Rousseau's overt (intentional) meaning is contradicted by certain other, strikingly discrepant formulations whose logic runs athwart the manifest sense of his argument" (125). But if this were all the "logic of supplementarity" (hereafter LS) encompassed, it becomes merely a disguised form of contradiction. Norris seems to reinforce this equivalence: "Yet this system (or logic) cannot be ignored if one is to take account of the objections that rise against Rousseau's thesis *by his own admission elsewhere* and which constitute a standing refutation of his claims with respect to the order of priorities between nature and culture, speech and writing, or origin and supplement" (126, italics in text).

However, contradiction as classically defined is not all LS entails. For Rousseau uses "supplement" so that "...it has to function *both* in a privative, derogatory sense ('supplement' = that which subtracts and corrupts under the guise of adding and improving) *and also*–despite Rousseau's intention–in the positive sense 'supplement' = that which fills a lack or makes good an existing defect" (126, italics in text; cf. OG, 144-5). The analysis of supplement developed in the previous section reveals that the distinguishing element of the LS is this conjoint positive/negative factor. The standard usage of supplement (as analyzed here) does not include a special, technical semantic dimension encompassing something which "subtracts and corrupts under the guise of adding and improving."

To illustrate, consider Rousseau on music. For Rousseau, the "harmony, musical notation or the calculus of intervals..." had "somehow *befallen*..." music "through an accident of 'progress' that need not–should not–have happened yet which also (by a certain perverse compulsion) marked their development from the outset" (134–italics in text; cf OG, 198-200). The result is "...the difference that Derrida constantly remarks between that which Rousseau *expressly wishes to say* and that which he is nonetheless *compelled to describe* by a logic that results, contradicts or countermands his avowed meaning" (144–italics in text).

Furthermore, the LS is not a narrowly idiomatic phenomenon circumscribed by Rousseau's theoretical quirks. For Norris, that "...Rousseau is unable to sustain this thesis against certain powerful objections that arise from the logic of his own discourse is a fact that should interest logicians as much as musicologists and cultural historians" (140). For Rousseau's thinking is "...caught between two logics–that of classical (bivalent) truth/falsehood and the logic of supplementarity" (148). Two dimensions characterize these logics–contradiction and modality–and the relevance of the LS for iterability emerges from their intersection.

7.6.1.1 *Contradiction*

The standard expressions denoting contradiction in a formal sense are–"p. ~p" in symbolic logic and a "material mode" version, e.g., Aristotle's *Metaphysics* 1005^b18-20: """...the same attribute cannot at the same time belong and not belong to the same subject and in the same respect." Consider this example which illustrates both formulations of contradiction:

A. Harmony is part of melody and harmony is not part of melody.

Although Example A is a blunt contradiction, its present form does not adequately reflect the structure of LS since in Norris' restatement of Rousseau's basic approach to the nature of music, harmony "...need not, *should not* have happened..." (134–my italics). The valuational element–"should not"–is an integral component in LS. Therefore, Example A may be written:

> B. Harmony is part of melody but [logically equivalent to "and"] harmony *should not* be part of melody (my italics).

If Example B adequately states the normative component of "should not" and if B and A are juxtaposed and compared, then B is more accurately described as an incompatibility rather than a contradiction. All contradictions involve juxtaposition of incompatible propositions but not everything which is incompatible is contradictory. The modal quality of "should not" indicates that the author intends to reject a certain property in relation to the subject of the proposition in question because this property is for some reason inappropriate to the subject. This modal quality establishes a complex normative relation between predicate and subject. This relation differs fundamentally from the straightforward predication characterizing the proposition in Example A. A parallel example: "War is a fact of life and war is not a fact of life" is a contradiction immediately recognized as such; the relation between war and life is completely indeterminate if all we knew about both referents was uttered in this single proposition. However, "War is a fact of life and war should not be a fact of life" is not false by its form alone (as are all contradictions)–indeed, the whole statement stands as an avowal of hope for the human condition.

This normative dimension so analyzed removes the generalized statement of the LS from the arena of contradiction as classically characterized. On the basis of Norris' versions of Derrida's position, the status of contradiction in the LS is involuted and formally complex. However, one apparent aspect of this complexity can be readily resolved. At one point, Norris speaks in a potentially misleading way about contradiction and this sense must be clarified. Derrida's deconstructive reading of Rousseau shows "...how Rousseau's discourse is compelled to undergo such 'supplementary' swerves from manifest or overt expressive intent in order to avoid *more blatant instances of self-contradiction* and thereby preserve at least some semblance of coherent sense" (138–my italics). Strictly speaking, instances of self-contradiction do not admit of degrees, i.e., being "more" or "less" blatant. All contradictions are identical in terms of their formal representation. What Norris perhaps has in mind is the difference between a contradiction such as Example A above which is "blatantly" contradictory because it can be immediately recognized as such and an instance such as Example B where the elements producing the (putative in this case) contradiction may be separated by pages of text and couched in language initially concealing the logical character of what has been asserted. The "blatancy factor" might affect the capacity of a contradiction to be recognized but it has no bearing on the logical character of the contradiction as such.

7.6.1.2 *Modality*

Norris asserts that "Derrida arrives at these claims through a reading of Rousseau that undoubtedly places considerable strain on the precepts of classical (bivalent) logic..." (143). Again, "...Derrida is no doubt pressing beyond any order of statement that might be acceptable in terms of those various modal or tense-logics that philosophers have lately proposed by way of extending and refining the resources of the first-order propositional and predicate calculus" (137). Norris cites "...*always already* present... (136–italics in text) to illustrate this logical variance. Later (147), Norris similarly refers to "the complex array of tensed and modal constructions ('had already', 'must [should] have been', 'would come from', 'would be simply', 'would have it that', 'should be already', etc)."

Tense logic concerns the past and future (what "was" and what "will be"), with a kind of necessity conferred by the verbal phrase "always will be." Derrida's "always already present" is presumably a case in point. Modal logic in the narrow sense encompasses possibility and necessity but a broader application can include deontic logic, i.e., what is permissible and what is obligatory. If therefore obligation is expressed in terms of "ought" or "should," then Derrida's introduction of "should not" as the appropriate rendering of the place of harmony in melody for Rousseau falls under the rubric of deontic logic. Deontic logic is not mentioned directly in Norris' discussion but it may be what Norris intended. The need to be clear on the scope of modal logic becomes evident in the next section.

7.6.1.3 *Validity*

Norris offers a pair of global impressions of the LS, and with crucial implications. First, "...the Derridean 'logic of supplementarity' has this much in common with other, more 'classical' modes of logic: that while laying claim to its own kind of formal rigor and validity conditions, it must also correspond to the way things stand with respect to some given subject domain or specific area of discourse" (139). Thus Norris sees the LS as having its "own kind of formal rigor and validity conditions." Second, Norris refers to the "...extreme complexities of modal, subjunctive or counterfactual reasoning..." [which Derrida resorted to in order to capture Rousseau's texts] "whose gist can be paraphrased (albeit very often at tortuous length) and whose logical form can sometimes be captured in a suitably refined symbolic notation but which serve above all to indicate the aberrant (logically anomalous) character of Rousseau's discourse" (149). Norris contends that not only does the LS have formal rigor and validity conditions (139), but also that its logical form "can sometimes be captured in a suitably refined symbolic notation" (149).

This phase of Norris' approach to LS requires a series of comments:

a. Derrida's restatement of Rousseau involves at least three distinct types of variant logic–tense logic, modal logic (in the narrow sense) and deontic logic. But each of these types of logic involves different operators. Therefore, it will be a decided challenge to produce, as Norris claims, a "suitably refined symbolic notation" encompassing *all three* types of operator into a single formalized notation. Furthermore, the modally diversified expression of the principle of contradiction (as shown above) will exponentially complicate this project.

b. The formalization Norris asserts as "sometimes" possible must be developed and laid out explicitly in order to determine the relevant samenesses and differences to existing formalized positions in modal and tense logic. To claim that such formalization is possible remains vacuous until it has in fact been done.Also Norris' qualifier "sometimes" suggests that the formalization will necessarily be incomplete, which of course weakens the extent to which the LS substantiates itself as a viable candidate for independent status in logical theory.
c. Norris describes the LS as a "quasi-logic" (123; see also 138) and later as "a mode of paradoxical pseudo-logic" (138). But if the LS has these characteristics when applied as Rousseau has done in the contexts of music and culture and the LS as such is as inherently contradictory as Norris maintains, then how will the formalization be possible to state at all?
d. Whatever rules are introduced in order to formalize LS, if abstracted from their application in this context and then examined in their own right, would be statements of highly abstract generalities. These rules would then be equivalent to rules performing the same function for bivalent logic, that is, fully metaphysical in scope and structure when translated into an appropriate material mode. If so, then they must also become a necessary target of deconstruction (an inference which will be developed in similar contexts in Chapter 8).

These consequences are of such magnitude that Derrida on supplementarity "offers some of the best, most searching and perceptive commentary anywhere to be found in the recent literature on philosophical semantics and philosophy of logic" (149); however, we wonder whether such high praise indeed involves a structurally significant "strain" on bivalent logic or whether Norris is following Derrida on a line of interpretation that involves–as we now show–a truncated approach to the explanatory relevance of several basic metaphysical concepts.[87]

---

87 In *Of Grammatology*'s concluding section, "The Supplement of (at) the Origin," Derrida writes; "... yes above all, in as much as we designate the impossibility of formulating the movement of supplementarity within the classical logos, within the logic of identity, within ontology, within the opposition of presence and absence, positive and negative, and even within dialectics...." (314). If the formalization of the logic of supplementarity Norris promises will preserve the formal contrast between true and false (i.e., "bivalent logic") as basic truth values of propositions, and "positive and negative" captures that contrast, then Derrida explicitly says that it is impossible to "formulate" supplementarity in this way. Indeed, the passage as a whole removes "the movement of supplementarity" from *any* mode of formulation "within the classical logos," a removal which presumably includes at the very

7.6.1.4 *The Heuristic Dimension*

Two additional points concerning heuristic possibilities derivable from the logical aspects built into supplementarity: First, if Rousseau's position is contradictory in the classical sense, then the conjunction and consequent contrast between what is asserted and what is denied compels us to determine what proposition (or propositions) should be accepted and rejected. This response holds for any text deemed a contradiction (assuming the reader's conviction that the text merits additional scrutiny). Norris approximates this point when he asserts that the "quasi-logic" of LS engenders "...a process of diagnostic reasoning that questions the premises (the 'unthought axiomatics') which can be shown to have produced that dilemma" (149). The "unthought axiomatics" is Derrida's phrasing and refers to whatever conceptual considerations underlie the generation of texts describable as supplemental in his technical, and problematic, sense.

Second, Norris' development of Derrida's use of "logic" ends up closer to certain modes of ordinary language analysis in which attention is paid to the "logic" of how a word is used. In this case, Derrida employs "supplement" as a technical term to advance a certain interpretive end, i.e., a way into determining meanings of texts which is controlled by the linguistic trappings of the texts in question. These texts include putative contradictions more or less obvious in terms of what has been explicitly written. As just noted, this textual characteristic will engender additional thought simply by virtue of the need to escape the logical incompatibility a contradiction produces. But Derrida's "supplement" exhibits a dimension of negativity in the sense that the contrast producing the logical incompatibilities includes an element described in language which indicates the author's unfavorable attitude toward that element.

The basic conflict so presented encompasses a problem–i.e., for the Rousseau texts under scrutiny, how to integrate "nature" and "civilization" and, on a smaller scale, "melody" and "harmony" in music. But the perspective of inquiry inspired by the application of supplementarity involves the extent to which the supposed negative element–civilization or harmony–is indeed as negative as the author would apparently maintain. The reader is asked to feel the negativity arising from the original formulation, then think against this aura in order to determine whether the negativity is justified at all, marginally, or whether it is completely out of bounds given that the supposedly negative element introduced into the discussion is, in fact, necessary to the phenomenon or entity under scrutiny.This type of heuristic incentive may be useful in helping to cope with problems involving complex or emotionally-laden features subject to vigorous criticism or straightforward rejection.

---

outset the stark and rigid contrast between true and false. Norris does not discuss this passage from *Of Grammatology*, and it seems incumbent to do so if he wishes to maintain that the logic of supplementarity can be rigorously formalized–even if only "sometimes"–given, as we have just seen, Derrida's outright denial that such meta-analysis is possible.

7.6.2 Supplementarity and Metaphysics

Derrida’s deconstructive reading of Rousseau “...has to do not only with certain curious blind-spots or logical anomalies in Rousseau’s text but also with the plain *impossibility* that things could ever have been as Rousseau describes them.” (139–italics in text). Norris also insists that Rousseau’s results are “*strictly unthinkable*” (138–italics in text). Why impossible and unthinkable? Consider the Barber Paradox. Bertrand Russell’s well-known description of the small town barber is self-contradictory; whether that barber shaves or does not shave himself cannot be determined because no such barber can exist. In general, a contradiction proposes states of affairs which issue in a logical impossibility and since no such entity (supposedly) referred to by these stipulations can exist, the attempt to think about that (supposed) entity is nullified.

A contradiction involves a conjoint true-and-false proposition about the same state of affairs. Now Norris maintains that Rousseau uses demonstrably false premises: “For indeed it is the case–empirically so, as a matter of acoustics and the overtone series, and phenomenologically speaking, as concerns the ubiquitous role of harmony in our perceptions of melodic contour–that what *ought* (for Rousseau) to figure as a mere ‘supplement’ turns out to be the very *condition of possibility* for music and musical experience in general” (139–italics in text). Therefore, it seems Rousseau could have avoided contradicting himself, thereby skirting the logic of supplementarity entirely, if only he would not have asserted that false premises were true. But Norris explicitly denies that Rousseau had this option: “...this is not merely a matter of Rousseau’s having fallen prey to conceptual confusions which he might have avoided with a bit more care in framing his proposals or clarity in thinking through their implications. For the logic of supplementarity is both indispensable to Rousseau’s argument–the only form in which he is able to articulate its various propositions–and also (as Derrida shows) the main point of leverage for a reading that effectively subverts all its governing premises” (147).

But if we consider Norris’ appeal to “the only form in which he [Rousseau] is able to articulate its various propositions...”, then it becomes evident that the prerequisite for establishing the LS as an “indispensable” theoretical construct is the *fact and progression of Rousseau’s text itself*. In a recapitulation emphasizing the fundamental source of Rousseau’s need to resort to supplementarity, Norris writes: “...what should by all rights have been a self-sufficient entity requiring (or admitting) no such addition turns out–by the logic of Rousseau’s arguments–to have harboured a certain incompleteness *at source* which belies that claim and thus complicates his argument despite and against its manifest intent” (134–italics in text). The “swerves” Norris notes (149) derive from the underlying pressure of a metaphysical assumption, i.e., that it is reasonable to assume that for Rousseau, a self-sufficient unity is capable of discursive and articulated analysis and description. However, Norris’ phrase “a self-sufficient entity” refers to (a) the nature of music as (b) Rousseau describes that nature. Thus it does not follow that “a certain incompleteness *at source*” when the source refers to the letter of Rousseau's text concerning music also includes "a certain incompleteness *at source*" in the nature under scrutiny, i.e., music in this case.

The phrase "self-sufficient entity" is roundly metaphysical and seemingly promises to embrace more abstract notions such as unity and nature as worthy grounds to justify the proposed self-sufficiency. Of course, Norris' "self-sufficient entity" is not intended to assert metaphysical fact; it only grounds what Norris, following Derrida, takes to be the approximate status of the subject under Rousseau's scrutiny. Thus in terms of the context at hand, we will not require Norris to fulfill traditional expectations regarding theoretical matters merely by adopting a certain turn of phrase. However, the process of inquiry Rousseau has pursued appears to presuppose something (e.g., music) existing in a way which, given due modes of inquiry, will yield positive and defensible results. For example, if an entity–culture, nature, music–is self-sufficient, then it should be analyzable in ways which are coherent. The "sufficiency" to which Norris refers would therefore include logical compatibility as a minimal condition but it would also include, in terms of content, a comprehensive account of the "entity" under scrutiny. And finally, the propositions comprising this account must be true, using as criteria for truth the specifications Norris himself provided–empirical evidence and testimony of phenomenological experience.

One might object that these concerns arise from placing excessive weight on a delimited textual origin of the LS, i.e., Derrida's reading of Rousseau. But Norris insists that the LS has a much more prevalent and pervasive foundation: "This particular form of deviant ('supplementary') logic is sure to emerge whenever it is a question of fixing–or attempting to fix–some notional point of origin for language or society that would *not yet* partake of the defining traits (articulation, difference, structure, hierarchical relationship, etc.) in the absence of which no language or society could possibly have come into being" (144–italics in text).

This generalization of LS concerns claims about "points of origin" in accounts of certain fundamental activities. If Norris is correct in so extending the scope of the LS, then origins of such entities or activities are either impervious to analysis or when analyzed, care must be taken to avoid dual causal or explanatory elements displaying hierarchy or dependency, for such structural configurations become the ready target of deconstructive scrutiny and consequent upheaval. By implication then, any potentially feasible analysis advanced must identify common and shared qualities. However, this procedural inference concerning content involves substantive metaphysics, e.g., regarding unity and the interplay of whole and part. But surely these attributed considerations are equivalently unwarranted additions and assumptions given analysis within a rigorous deconstructive setting.

Perhaps not. Norris insists that applied deconstruction is not philosophical criticism exercised for its own sake. For Norris, Derrida is claiming "…that textual close reading of this kind is able to bring out certain logical complications which also have much to tell us concerning the real (as distinct from the mythic or idealized) conditions of emergence for language and society" (145). But if the "real" conditions of emergence for “language and society” encompass activities such as music, then whether or not music as one activity among many others should be approached as a “self-sufficient entity” represents a decidedly metaphysical commitment. After all, music, etc., may not be such an entity or even if there are such entities, they may (following implications from later Wittgenstein) not be open to linguistic evocation at a level of determinacy adequate to reflect that deep metaphysical fact. The only effective way to undercut this reasoning is to analyze certain realities from certain directions and observe the results.

7.6.3 Supplementarity and Possibility

For Rousseau, melody should exist without harmony. But Norris (armed with data) contends that melody includes harmony as integral to melody itself. If, however, Rousseau’s text can be read to assume that these integral unities were already and truly “present,” then Rousseau did not recognize this metaphysical fact for the simple reason that he was wrong about the nature of music. More fundamentally, Rousseau was wrong about nature itself insofar as a process (or entity, to use Norris’ word) can exist and function with seemingly disparate elements integral to its “natural” unity. Thus for that entity which is music, harmony is not a contrived adjunct to melody; rather, melody *is* harmony in a sense which must be properly experienced and duly expressed in coherent theoretical discourse.After confronting Rousseau’s texts on music, it becomes necessary for the philosopher not only to rethink the nature of music but also to rethink the notion of nature itself insofar as certain entities constituted as wholes and parts exhibit “a” nature and collectively constitute “nature.”

If Norris is correct concerning the substantive contributions deconstruction can and presumably will make about “real life” issues and phenomena, where then do we turn? Norris quotes Derrida sparingly in this account of LS but one brief passage from *Of Grammatology* is especially pertinent in this regard. Derrida says his deconstructive efforts with Rousseau and writers generally concentrate on “...a certain relationship, unperceived by the writer, between what he commands and what he does not command of the patterns of the language that he uses” (125, quoting OG, 158). If the relationship is “unperceived,” then the writer presumably does not have the option of rectifying what the writer has written if the resulting text illustrates the “logic of supplementarity.” Thus the intersection of what the writer intends and what is “present” but unintended by the writer seems unavoidable, although presumably there are degrees of this hiddenness since not all writers are (a) contradictory at least at a surface level of analysis or (b) modally mixed in their assertions. As a result, their work can convey meaning without falling under the ambit of supplementarity in the virulent sense that Rousseau did.

Derrida clarifies this powerful hiddenness when he refers to a contrast between what a writer commands and "does not command of the patterns of the language" the writer employs. Language possesses and is controlled by patterns encompassing a writer's work irrespective of the writer's intentions; these patterns presumably affect whatever has been written as marks on a page. If we allow a theoretical gap to define the difference between what a writer commands and does not command, then Derrida is appealing to a *dimension (or dimensions) of possibility* underlying the difference between a perceived text and that text's "unperceived" patterned background. In fact, Derrida is pointing out a web of metaphysical intersections permeating the language Rousseau–and presumably any author–will employ. If, for example, the patterns in question include the capacity of words to refer to different things in different contexts, then the possibility in question may extend even beyond the structure of the logic of supplementarity and include any and every linguistic configuration, especially with respect to reference. This interpretation implies that Derrida has introduced a transcendental aspect to supplementarity which is crucial to his overall project regarding the philosophical analysis of language, even and including deconstruction itself.

Identifying and describing the primary metaphysical components of this prerequisite is a major concern of Chapter 8. There we see that the connection between considerations pertaining to the logic of supplementarity detailed in this section and the notion of iterability includes a more complete explanation of the range of, using Derrida's terminology, what is under the writer's command and what is not. This explanation will include a series of explorations into the potential relevance of the notion of possibility–an inquiry necessary if Norris' conviction concerning the capacity of deconstruction to disclose truths about reality should be given justified weight. Thus the metaphysics of possibility and potentiality, properly interpreted given Rousseau's construal of the original form of music, covers what is for Norris (and of course Derrida) the emergence of an apparently paradoxical confluence of opposed qualities or states–the "logic of supplementarity." Iterability and Derrida's use of supplement jointly culminate in appeals to the effective reality of what is possible as evocable in language.[88]

---

88 The following sources illustrate the more speculative aspects built into the notion of supplement: LI, 50; WD, 290; MP, 312-3; P, 100n5; Derrida 1987, 259. For additional discussion, see Culler 103. Supplement is one of Derrida's wedge concepts for eliciting, via deconstructive gestures, concealed significance in written texts. 20th century philosophy offers other approaches for realizing the same end. Thus John Dewey in *Experience and Nature*: "But the realm of meanings is wider than that of true-and-false meanings; it is more urgent and more fertile" (410). However, Dewey then adds immediately that when "the claim of meanings to truth enters in, then truth is indeed preeminent" (410-11). And in the first chapter of *Process and Reality*, A. N. Whitehead notes that for a philosopher who intends to advance thoughts about reality: "The only possible procedure is to start from verbal expressions which, when taken by themselves with the current meaning of their words, are ill-defined and ambiguous." Whitehead then observes (13) that such "meanings are incapable of accurate apprehension apart from

7.7 Derrida and the Fact of Writing

For Derrida, writing construed as a supplement is "never present, here and now." In fact, supplement is "neither a presence nor an absence." But writing is indeed present here and now in a way analogous to vitamins being present here and now when the mode of existence of writing is presence defined by a state of actuality–i.e., the vitamins as such, prior to the empirical fact of their being actually consumed. The words constituting the supplement exist by themselves as words but without being "added to" the original text, just as the vitamins exist by themselves as entities but without being "added to" the diet of their potential consumer. Consider writing in relation to speech with this perspective on supplement in mind. Thus writing can be understood as a potential adjunct to spoken language with the additional property of preserving or adding to what had already been uttered in spoken language, which occurs in the present, for future consultation or reference.

Writing is an actuality; speaking is a different actuality. If writing identified as supplemental is approached as a *type of actuality*, it should be possible to determine its reality compared to the reality of speaking. In the *Phaedrus*, Socrates observes that writing says the same thing forever (275d). The static stability of writing results in a uniformity and sameness of content which lasts over time. These abstract characteristics should be juxtaposed with parallel circumstances proper to spoken speech and the results generalized. The sense in which writing could legitimately be construed as "supplemental" to spoken discourse then becomes apparent.

In the same dialogue, Socrates also says about writing that "...anyone who leaves behind him a written manual, and likewise anyone who takes it over from him, on the supposition that such writing will provide something reliable and permanent must be

---

a correspondingly accurate apprehension of the metaphysical background which the universe provides for them. But no language can be anything but elliptical, requiring a leap of the imagination to understand its meaning in its relevance to immediate experience." Dewey and Whitehead explicitly allow for the possibility that meanings emergent from language will be more inclusive than what according to conventional standards may initially appear semantically assignable to that language. But both philosophers make this claim from the vantage point of adherence to and residence within their own versions of systematic philosophical thought. This consideration and its potential for future insights must be kept in mind as an available contemporary alternative when Derrida appears to overreach regarding the possibilities inherent in language as "supplemental." Finally, note Whitehead's concluding appeal to the need for imaginative leaps regarding ranges of meaning–on its face, a most Derridean thought–but leaps which for Whitehead must be tied to the "relevance of immediate experience," what in the present essay we have been referring to as the lifeworld. The preeminence of truth for Dewey and immediate experience for Whitehead both stand as limiting factors when acknowledging the inherent indeterminacy of language and the need for imaginative and speculative openness in the face of that fact. For discussion of Whitehead's view of the imagination as a "deconstructive" function, see DFL, 179-81. And for more general comparisons of Whitehead and Derrida, see Luis G. Perraja 1999 and Catherine Keller 2002. For Stocker, 30, elements of Derrida's thought can be identified as "process philosophy" in a sense which Stocker does not develop at that point in his text.

exceedingly simple-minded...." He then adds that this simple-minded individual will be in this cognitive state "if he imagines that written words can do anything more than remind one who knows that which the writing is about" (275d). Writing–"external marks" (275a)–is a transcription in words that, as marks, is not subject to change. Spoken discourse is subject to change. But for "one who knows," no difference obtains between *saying* what is known and *writing* what is known. Speaking has an edge over writing in that the former as inherently in process can continually move toward evoking the truth whereas the latter due to its static quality cannot–it keeps saying the same thing forever in that its external marks, once set in print (or wax), always remain the same.

If, however, the one who knows is in touch with the highest level of knowledge and, with Plato, the highest object of knowledge, then there is nothing in the brief comments about writing in the *Phaedrus* which precludes the "one who knows" from voicing that awareness in the static character of print. In this case, the content of writing would become identical with the knowledge of reality acquired by the one who knows. Writing is "supplemental" for the Platonist if the Platonist remains *in process* toward knowing the truth about reality. The reason is that the progress made in this quest cannot extend beyond what printed language has asserted. A discussion, e.g., in dialogue form, engages in process by its very nature and is the preferred route to knowledge. But it is so only because spoken language is more malleable, more immediately open to revision and development. The Platonist intends to move *from* the vehicle of language itself *to* the realm of reality which language is *about*, i.e., its referential dimensions. If the cognitive apprehension of reality could be expressed in language, then for the Platonist it would not matter whether that language was spoken or written. What matters would be the degree of correlation between whatever that language "said" and the structures pertaining to its objects.

Derrida frequently refers to marks, i.e., the orthographic appearance of letters as they constitute words. But in the early essay on the poetry of Edmond Jabès, we read: "Death strolls between letters" (WD, 71). Derrida refers to what appears *between* letters in written words, a remarkable instance of philosophical imagination, and in two senses. First from a strictly empirical perspective, what appears are *gaps*–spaces which separate individual letters from one another. Noticing these gaps at all, much less as philosophically interesting, is certainly incisive. But Derrida has gone far beyond that level of observation by asserting, second, that "death strolls" between letters, for this claim imposes an interpretive schema on that empirical fact. For Derrida, the gaps between letters of written words *have meaning*. Thus the relevant point is that Derrida is looking at the brute data of writing and then *eliciting theoretical consequences from that data.*

However, if a writer says that gaps in print mean one thing–e.g., that they are signs of X, imagistically conveyed but signs nonetheless–then another writer with comparable sensibilities can assert that the same gaps are signs of Y, etc. The question then becomes whether or not it is feasible to interpret these speculative assignments in such a way as to resolve which seems most plausible and revelatory. This inquiry is what metaphysicians do when they question each other and dispute respective accounts of reality. Credit Derrida with the vision which allows a thinker to look at phenomena, gaps between letters, and discern that these phenomena are relevant and important. But once his account is in print (and in that regard, with Plato, saying the same thing forever), it becomes available for inspection and critical review. The account then is open for others to rethink, revise, extend or compress, or indeed to reject altogether as, in Platonic parlance, a "windegg" (*Theaetetus*, 210b).

The consequences from Derrida's reading of such orthography may have epochal sweep. A few pages after confronting the image of death strolling between the letters of words, Derrida writes: "The Being that is announced within the illegible is beyond these categories [i.e., "rationalism" and "irrationalism"], beyond, as it writes itself, its own name" (WD, 77). The "illegible" is what exists prior to everything "written" (by the hand of God, i.e., the totality of creation) including all gaps between all letters of such writing. If we investigate with, presumably, the appropriate deconstructive techniques (deconstruction is not mentioned directly in this essay) what has been written in books of metaphysics wrought by human creators, then what has been "hidden" in Being might be revealed. But at this level of reality–a word used here only as a place name–the discoverable dimensions of being transcend whatever word (or words) may be written to name it. And at this juncture, Derrida comes very close to saying exactly what Socrates proclaims in the *Republic* about the ineffability of the Good as such (509c). Words will necessarily fail us in the attempt to capture the full reality of the referent of those words.

In the early essay "Force and Signification" (1963), Derrida asserted that writing

> **Passage 7F**...is the outlet as the descent of meaning outside itself within itself: metaphor-for-others-aimed-at-others-here-and-now, metaphor as the possibility of others here-and-now, metaphor as metaphysics in which Being must hide itself if the other is to appear. Excavation within the other toward the other in which the same seeks its vein and the true gold of its phenomenon. Submission in which the same can always lose (itself) (WD, 29).

Metaphysics is metaphor, that is, all metaphysical language stands for something other than what it seems to represent. "Being," the proper subject of metaphysics (since Aristotle), must "hide itself if the other is to appear." The question is–where does Derrida derive this sense of "hiding"? The language of metaphysics makes assertions about Being; it has done so from the Pre-Socratics through Husserl and Heidegger and beyond. It has made these assertions *in writing*. How then is "the other" currently "hidden" in Being? When we "excavate" writing per Derrida's image, do we dig into writing as "marks" (Derrida's term) or as "external marks" (*Phaedrus*, 275a)? If the arena of interest is limited solely to marks, then no difference obtains between Platonic and Derridean approaches to writing, assuming that marks on print pages and wax tablets are indistinguishable from one another *qua* marks. Therefore, if writing is importantly revelatory, then either we must draw additional implications from the fact of writing itself, that is, as a purely empirical phenomenon or we must shift attention to what the writing is *about*, what it "says" about its subject, whatever that subject may be. And for us, that subject is metaphysics.

Writing is a palpable fact. Plato and Derrida both agree on this point. The question is what writing *means*. Even if we view writing as a "supplement" to spoken discourse, it would still be a question as to the most appropriate way to situate the fact of writing with respect to what is other than writing. Indeed, if we compare Plato and Derrida on writing, they seem to share the same basic position. To determine the significance of writing, it is necessary to envision its factuality, static marks on a surface, in relation to an account of what is metaphysically most fundamental. The content of this account differs–for Plato it is the Forms and the Good; for Derrida it is what can be divined from the totality of western metaphysics insofar as "Being" yields whatever is accessible from what has remained "hidden" from the interrogative vision of history's most percipient and rigorous metaphysicians. Presumably this account, if it reaches written discourse, will be recognizable in that regard. In other words, the exponent of Derrida's incipient version of "fundamental ontology" will author a text acknowledged as just that, an account of what is most fundamental to "Being" (we recall Derrida's stern strictures on the self-concealing character of whatever word or words come into view as ultimate ontological names).

The conclusion to this interpretation has considerable force. What will decide the staying power of all that Derrida said and wrote on the fact of writing will be whatever is subsequently said about Being or metaphysical matters at a level of high generality. Derrida on writing constitutes an extended exercise in preparation, a vestibule, a holding pattern spanning an entire epoch of sustained reflection from the pre-Socratics, Plato and Aristotle, through Heidegger and into the present arena of non-Derredean metaphysics and beyond. What the future holds in this regard, whether monstrous or mundane, remains unknown.

7.8 Iterability: Singularity and Self-identity

The line of argument in this chapter has shown that iterability depends for theoretical adequacy on metaphysical considerations. Furthermore, when Derrida reasons to a conclusion in what might appear to be a purely linguistic arena of philosophical concern, it is prudent if not essential to question whether assumptions of a metaphysical nature have, so to speak, made their presence felt in the account. The status of singularity and self-identity as they emerge in Derrida's treatment of iterability dramatically attest to the need for such questioning. The following three passages taken as a set illustrate essential features in Derrida's account of iterability; I shall argue that this account is conceptually muddled, if not simply incorrect, and that apart from these internal issues, the account has a series of untenable consequences (stated in the following section of this chapter).

> **Passage 7G** Is it not evident that no signifier, whatever its substance and form, has a "unique and singular reality?" A signifier is from the very beginning the possibility of its own repetition, of its own image or resemblance. It is the condition of its ideality, what identifies it as signifier, and makes it function as such, relating it to a signified which, for the same reason, could never be a "unique and singular reality" (OG, 91).

7.8.1 Repetition and Singularity

We must interpret what Derrida means by claiming that a "signifier is from the very beginning the possibility of its own repetition, of its own image or resemblance." The conclusion of the passage is equally crucial since Derrida infers, "for the same reason," that a signified is not a "unique and singular reality." Thus Derrida denies that uniqueness and singularity pertain to both signifier and signified. Both signifier and signified–presumably *any* signifier and signified–are necessarily plural. It follows that predicating self-identity to any signifier and signified is a fundamental mistake.

Two preliminary points concerning the strategy exhibited by **Passage 7G.** First the passage does not contain an argument–Derrida makes an assertion about the status of a signifier with respect to repetition but he does not support this assertion with any reason or argumentation. The rhetorical question which opens the passage will perhaps dispose the reader to be receptive to the conclusion, i.e., the denial that a signifier has a "unique and singular reality," but a rhetorical question by itself is not an argument.[89] Second, in the context of the passage, is the signified, relative to the signifier, a concept or a referent? Is the signifier functioning with respect to something in the mind or something extra-mental that has somehow led to language as well as to a concept intimately related to that language? We assume initially that signified refers directly to a concept. However, concepts are themselves relational entities–that is, we do not have a concept without that concept being *of something*, i.e., something other than itself. This point concerning the structure of a concept is crucial, as we shall see. We also assume that a concept has its own kind of reality, thereby justifying Derrida's use of the name "signified" to represent that reality in this inquiry.

---

89 See WD, 173–the unique "eludes discourse and will always elude it." However, Derrida appeals to the notions of uniqueness and singularity when it suits his purposes–cf. WD, 254. In addition, note as a comment on philosophical technique that Derrida frequently uses massed rhetorical questions as a device to instill persuasion in his readers. See, e.g., MP, 63, 325; Derrida 2001, 112. However, rhetorical questions arranged sequentially are not arguments.

Derrida refers to a signifier as necessarily possessing "...from the very beginning the possibility of its own repetition, of its own image or resemblance...." The qualification "from the very beginning" apparently refers to the initial coinage and use of a word as a signifier (when words originate is seldom known for certain–rare exceptions being, e.g., Milton's oft-cited coinage of "pandemonium" for literary purposes–but the exact date and circumstances are irrelevant as long as a word has a determinable if not *de facto* determinate history). At this moment of origination in the word's history, i.e., as soon as the word in fact becomes a word in the sense Derrida is currently defining, it is "the possibility of its own repetition."

Derrida writes "image or resemblance." Is this a disjunctive identity, i.e., two distinct ways of referring to the same notion or is Derrida distinguishing between signifier-as-image and signifier-as-resemblance? The context does not offer sufficient background to decide. If the latter, then the repetition under scrutiny can occur in two ways–either a word repeated is "image" *of* or "resemblance" *to* a given word. The question then becomes what Derrida has in mind using "resemblance" (e.g., is the repetition not of the same word but a cognate similar in spelling or sense, or both?). If the former, then "image" and "resemblance" are two different ways to refer to the fact that the repeated instance is exactly the same as the original word. We will assume the latter reading with the note that the burden is on the advocate of the exact wording of Derrida's text to clarify this disjunction.

These preliminary considerations are important but the following discussion has even greater import in its metaphysical implications. Derrida denied that ontology had anything to say regarding interpretation of the notion of a supplement; we have argued that this denial is mistaken. But ramifications from discussion to follow are even more significant as far as disclosing Derrida's unexamined metaphysical elements and assumptions.

Here is an exploration of some aspects of the problem; others will be detailed in due course. We understand "image" as it is used in the phrase "mirror image," that is, indicating a one-to-one correspondence of all features of the given word in the "imaged" word. If so, then the second instance of a given word is a "repetition" as an image of the first instance of that word only if the word itself is in fact *the same in both instances*. Thus a given instance of a word, call it $word_1$, exists *as a signifying linguistic element* only in terms of the possibility that there can be $word_2$, that is, a second instance of the *same* word (we will return to this use of sameness shortly). For if the second instance of the word is indeed different from the first instance of the word in any respect whatsoever, the relation between the two instances would not justify claiming that the second instance is an "image" of the first instance. Therefore, in order to assert that iterability includes the possibility of repetition of an image of a signifier, Derrida must assume that the same word (or "mark") as signifier appears in both instances.

For Derrida, the possibility of repetition of a signifier is necessary to the structure of the signifier and therefore negates the existence of a signifier construed as "unique and singular." $Word_2$ is an "image" of $word_1$ but this relation of imaging the word in transition from the first to the second instances is sufficient to destroy the surmise that the first instance is itself "unique and singular." For example, if I assert "this tree is tall," the word "tree" as a signifier cannot be unique and singular because this instance of the word "$tree_1$" also and necessarily includes the possibility of its own repetition, i.e., "$tree_2$." It follows therefore that the first and indeed any given instance of a word as signifier is always and necessarily plural in this specified sense; as a result, Derrida insists, the first (and any given) instance of a word as signifier cannot be a "unique and singular reality."

But in lieu of an argument supporting Derrida's conclusion, it seems fair to ask why the second instance of the same signifying word, the instance as repeated, effectively eliminates the possibility that the first instance of the same word is "unique and singular"? Consider the fact, argued for throughout this essay, that words do not occur in a vacuum–thus the first instance of a word appears within a context characterized by a set of conditions broadly metaphysical in character; the second instance of the same word appears within a different context, with its own set of metaphysical conditions. The two contexts must differ from one another in at least one respect because if they did not, they would arguably (invoking Leibniz and the identity of indiscernibles)[90] be not two distinguishable contexts but rather a single context. Even if the two contexts were exact duplicates of one another in terms of elements and spatial arrangements, they would still differ with respect to time, i.e., the second context would come after the first context (the same conclusion holds if the contexts were identical in elements and simultaneous, i.e., they would be spatially distinct from one another). Given this framework of contextual elements, the first instance of the word is indivisibly whole and complete entirely by itself–relative to its existence within the first context–quite apart from the occurrence of the second instance of the word within the second context, this instance also complete and indivisibly whole. I suggest that the above analysis of the relation between iterability and singularity is entirely plausible; if not, then we await an argument from Derrida's advocates to support his contention that iterability characterizes a signifier "at the very beginning" and thereby obviates signifier (and, by extension, signified) singularity and self-identity.

7.8.2 Sign and the Structure of Repetition

The above account from *Of Grammatology* states a position which remains problematic in part because it is not argued. By contrast, *Speech and Phenomena* develops at least

---

90 The classic discussion of the identity of indiscernibles and related contexts appears in the *Mon adology*, Section 9.

an approximate argument to support the generation of components constituting iterability in Derrida's aboriginal sense. A textual aside: both works were published in 1967; thus neither the substantive interplay between the two accounts nor their juxtaposition for interpretive purposes at this juncture in this essay should occasion surprise.

> **Passage 7H** A sign is never an event, if by event we mean an irreplaceable and irreversible empirical particular. A sign which would take place but 'once' would not be a sign; a purely idiomatic sign would not be a sign. A signifier (in general) must be formally recognizable in spite of, and through, the diversity of empirical characteristics which may modify it. It must remain the same, and be able to be repeated as such, despite and across the deformations which the empirical event necessarily makes it undergo (SP, 50).

Here sign and signifier appear referentially coextensive–Derrida uses the latter term (or "mark") when he expands the characteristics of the former, but he is not referring to two distinct elements within a single theoretical account. Also, "formally" may pick up the unexplained reference to "substance and form" in **Passage 7G**. Derrida is working within a broad semiological context where substance and form might refer to tangible entities with decidedly empirical characteristics (e.g., an octagonal, metallic STOP sign). In a way, language, whether spoken or written, also exhibits its own "substance and form" as sounds, orthographical conventions, and rules governing syntactical and semantical elements. For our purposes, we will focus on sign as it may appear in language.

The argumentative dimension of the passage begins when Derrida asserts that a sign could not occur just once, from which it may be inferred as maintained in **Passage 7G** that a sign (or signifier) in order to exist "already" contains the possibility of its own repetition, its iterability, whenever that sign first appears. Without this essential possibility, the "substance and form" in question would not be a sign. The references to "empirical characteristics" and "empirical event" presumably refer to "lived" circumstances surrounding an actually uttered (or written) signifier. In other words, Derrida does not intend to evoke a word seen in a dictionary, in isolation and apart from its use in sentences, etc. The acknowledged presence of empirical phenomena will become crucial.

The second passage to be discussed occurs two pages after **Passage 7H** and extends the analysis of iterability, including in its scope an important appeal to possibility.[91]

---

91 Just prior to **Passage 7I** (the passage to be discussed at length), Derrida asserts that "the presence-of-the-present is derived from repetition and not the reverse," adding that "this is against Husserl's express intention...." In the first paragraph immediately after **Passage 7I**, Derrida contends: "Now (and here again the commentary must take its bearing from the interpretation) this determination of being as ideality is paradoxically one with the determination of being as presence" (SP, 52, 53). These comments bookend **Passage 7I** and, I submit, they sanction reading and interpreting this passage as a statement

Passage 7I The concept of ideality naturally has to be at the center of such a question. According to Husserl, the structure of speech can only be described in terms of ideality. There is the ideality of the sensible form (for example, the word), which must remain *the same* and can do so only as an ideality. There is, moreover, the ideality of the signified (of the *Bedeutung*) or intended sense, which is not to be confused with the act of intending or with the object, for the latter two need not necessarily be ideal. Finally, in certain cases there is the ideality of the object itself, which then assures the ideal transparency and perfect univocity of language; this is what happens in the exact sciences.[2] [Derrida's end note] But this ideality, which is but another name for the permanence of the same and the possibility of its repetition, *does not exist* in the world, and it does not come from another world; it depends entirely upon acts of repetition. It is constituted by this possibility. Its 'being' is proportionate to the power of repetition; absolute ideality is the correlate of a possibility of indefinite repetition (SP, 52–italics in text).

Derrida asserts that the ideality in question–"the permanence of the same and the possibility of its repetition"–simply "does not exist in the world." This assertion receives no supporting justification in this passage. But even if this denial were so, it is not clear how it follows from the fact alone that (a) the sought-for ideality "depends entirely upon acts of repetition," and that (b) this ideality "is constituted by this possibility." Derrida's position amounts to a pair of claims, the first a straightforward denial, the second an equally straightforward assertion with two distinct elements. It is appropriate therefore to examine the second claim, both elements separately, to see how it might be supported by argument and, if this reproduced reasoning correctly captures Derrida's end, to evaluate the educed thought in order to determine its cogency. As emergent from a reading of Husserl, Derrida's denial that a certain conception of ideality does not exist in the world depends to a considerable degree on the implied argumentative structure of the position which he proposes in its stead.

A repetition is a certain sort of sequence–all repetitions are sequences but not all sequences are repetitions (e.g., the sequence of natural numbers). If something is repeated, then a thing happens in such a way that it falls into or along a certain sequence. Consider A (e.g., the proposition "this tree is tall."). To repeat A means to assert A again–i.e., "This tree is tall." We now have this situation (i.e., sequence):

1. "This tree is tall."

II. "This tree is tall."

As instances of propositions, I and II are exact duplicates of one another. But if so, then why not say I and II are *identical*? What can we cite in the sequence of I and II which would justify saying that II *repeats* I?

The assumption is that repetition in the desired sense is not equivalent to identity. How then can we discern that repetition (and not identity) has occurred in a given

---

of the argumentation Derrida provides not only as a deconstructive response to Husserl's texts but also as a justification for his own theoretical position on the primordial character of iterability.

sequence? Only if something can be inserted or insinuated *between* the two assertions. Without such an insertion, it is difficult to determine whether identity or repetition should be posited as a category explaining what happens in a stated sequence.

I suggest that the distinction between identity and repetition lies in an articulated conjunction of two distinct characteristics–"otherness" and sameness. The first characteristic is a presupposed "otherness" functioning as an agent of separation between the initial entity (proposition, etc.) and the "repeated" entity (proposition, etc.). Without the separation established by this otherness, repetition collapses into identity.

I and II as individual propositions are identical to one another. But I and II are also numerically distinct from one another. To claim the two propositions are numerically distinct might appear to presuppose that there are two existing trees, the objects of these two propositions ("objects" in this context is vague but it is used that way for a reason, as we shall see). But even if only one tree existed, propositions I and II could still be numerically distinct on condition that they were referring to one and the same tree but were doing so not, say, at the same time. (We will return to referentiality.) I submit that this distinctness or separation following from the fact that the two propositions are numerically other than one another is one element in the structure of what sanctions the appeal to repetition, not identity, as the correct relation between the sequence constituted by I and II.

Iterability remains incomplete, however, if all that analysis reveals about this concept is the fact that otherness must obtain as one of its structural components. There must also be a sequence in which its constituent elements are marked by identity. What remains identical in the linguistic event outlined above?

Iterability presupposes determinable measure of identity in that an element (or elements) in or related to the signifying element must remain the same throughout each and every instance of iteration; otherwise the aspect of "being repeated" does not occur. It might appear plausible to contend that what in the sample sequence is repeated is merely the same four-word formula. Thus if we read I and II above (or hear I and II spoken by someone), then the sets of words, although numerically distinct, are identical in content and order. This kind of identity is therefore sufficient to conclude that repetition has indeed occurred. We note the purely formal character of this identity, depending entirely on orthography. As long as the (numerically distinct) propositions were, for example, in print in exactly the same way, it is immaterial to the relevant point whether or not the propositions were in a language which the reader understood. The reader could *see* the repetition taking place on the page.

However, objections might be raised against this account of the structure of iterability. First, a signifier is a sign and thus is *of* something, presumably something other than itself. Thus words constituting the propositions I and II cannot be taken as mere marks. The identity in question must mean, minimally, that a connection between word and something other than word exists for both propositions. If, for example, the word has definite referentiality in any one instance of a proposition, then it must have the same kind of referentiality in another instance of that proposition, otherwise iterability lacks repetition in a complete sense and thus ceases to be iterability since not everything in the sequence from first signifier to second signifier would have been repeated.

The second objection derives from a close reading of **Passage 7I.** Derrida passes in review a number of apparent distinctions regarding the word "ideality" as the concept is found in Husserl. There is the ideality of the *word*, of the *meaning* engendered by the word, and finally "in certain cases" there is the "ideality of the object itself." For these cases, the ideality of this kind of object "assures ideal transparency and perfect univocity of language," linguistic events which occur presumably only in the "exact sciences."

This section of the passage is remarkable for several reasons. Derrida asserts that one kind of "object" exists so as to produce "univocity" in language, which I interpret as admission that mathematical entities such as triangles or any geometrical figures generate their own kind of uniqueness simply by being what they are.Thus the phrase "right triangle" as the signifier of a given mathematical entity is unique and, as a result, is not subject to the necessity of possible repetition in order to be meaningful–there need not be a plurality of instances of the phrase "right triangle" nor in parallel fashion a plurality of individual right triangles (e.g., with legs of differing lengths) in order for the signifier "right triangle" to function fully and univocally. The reason why iterability need not obtain in this case is precisely because of the distinctive nature of the "object" of the signifying phrase. Signifiers involved with mathematical entities are unique because mathematical entities are unique.

Just as important at this point is the explicit introduction of an "object" which exists by itself, apart from its appearance within the reality of language, but an object which can indeed be *referred* to by language. We have noted that Derrida assumes reference in his theoretical approach to language. But here the referential function is such that it juxtaposes the signifying phrase with the object in a way which obviates the need for iterability. In these cases, signifiers evoke "ideal transparency and perfect univocity," a claim which is exceptional for language since it establishes the actualization of signifiers existing in direct contact with at least one object other than the signifying phrase itself. And since in this case, i.e., the exact science of mathematics (in particular geometry), the activity of discourse produces meaning–e.g., demonstrating theorems–we may conclude that at least in this section of **Passage 7I,** Derrida's context of analysis presupposes language, referentiality, meaning and perhaps even truth.

The question arises whether such objects occur more frequently than in the "special cases" Derrida expressly admits. Recall that in **Passage 7H,** Derrida twice introduced the empirical factor, emphasizing its reality with regard to possible effects on words given the necessity that words continue to maintain their existence intact, able to withstand physical environments in order to become possible elements in iterability. The point to be emphasized now is that the empirical factor is a constant as far as language is concerned. Signifying, signified, iterability, definition, communication, etc.–all these linguistic functions happen, and necessarily happen, within spatio-temporal settings which also include nature and artefacts, frequently intertwined in complex modalities. If this background is granted as a foundational component in the lived reality of language, the sequence of sample propositions yields very different results in terms of iterability than those advanced by Derrida.

We are walking through a forest populated with trees of various heights. This is the setting for the utterance of propositions I and II above. In this setting, the context–or, as Derrida put it, the "empirical event" or the "empirical characteristics"–may encroach on the identity of the words used in language, but this same setting also inspires or provokes or energizes the event of discourse in the first place. Otherwise put, the "modifications" or "deformations" which the empirical setting may produce in words presuppose that those words have already been uttered for some purpose, whether noble or pedestrian.[92]

Propositions I and II are spoken. Derrida appealed to "object" as an appropriate name for a mathematical entity. I suggest that trees in a forest are objects of parallel significance to the extent that their existence as self-contained and self-subsistent entities is capable of instigating language about those entities. But now we may redirect analysis of iterability. For if we approach critically Derrida's position on objects of mathematical inquiry, we may distinguish between a signifier insofar as it (a) functions referentially and (b) contributes to the generation of meaning. Therefore, what is subject to repetition is *not* the *signifier-as-referential* but the *signifier-as-carrier of meaning*. With this distinction in place, the configuration of the pair of propositions I and II above preserves the extra-mental and extra-linguistic integrity of the existing trees and embraces the necessity of iterability but within a dimension of import which remains separate from the existence of the trees (i.e., objects) as such.

Two crucial inferences follow: first, this distinction between referential and semantic functions of the signifier implies that the object of a given signification does indeed, contrary to Derrida's insistence otherwise, exist as a "unique and singular reality." The signifying word or phrase denotes an object accessible as immediately given within the field of experience animated by perception, just as a signifying word or phrase denotes a mathematical entity as object (its existence, interestingly, not established through perception but by some other function of understanding). Furthermore, if A symbolically represents what is and has been subject to iterability, then A must denote a "unique and singular reality" in some sense of uniqueness and singularity. For if A does not exercise such metaphysically precise representation, then no evidence exists that it is A and not something other than A (therefore perhaps completely different from A), which is and has become subject to iterability.

Let A stand for the tall tree in view when Proposition I is uttered. If A is not unique and singular but rather "already the possibility of its own repetition," then either A is not *this* tree but a repeated doublet of this tree or if A repeated is

---

92 Recall Dewey's general account of meaning which includes the necessary presence of extra-linguistic objects as sources of potential and actual referentiality. Cf. Ch5n69.

equivalent to A = A, then the identity of indiscernibles implies that A has never been repeated since A remains equal to A and in no sense "other than" itself. The self-identity of a tree becomes indistinguishable *qua* self-identity from the self-identity of a right triangle. There are many tall trees; there is only one right triangle (as object of mathematical investigation). But the self-identity of *this* tall tree belongs *only* to this particular entity, just as the self-identity of right triangle belongs only to this particular mathematical entity.

The second implication asserts that Derrida has reversed the order of priority–the possibility of repetition depends on the prior existence of identity as a kind of sameness, not the other way around as Derrida maintains. The process of repetition cannot produce, as Derrida claims, the ideality posited as "...but another name for the permanence of the same and the possibility of its repetition.... (**Passage 7I**). The above argument has shown that the "permanence of the same" is not produced merely by the possibility of repeating a signifying phrase. All that is produced by this process is a sequence of repeated words or phrases with nothing for each member of this sequence to refer to as "the same" or (as named here) identical.

Iterability continues to exist on this proposed account. But the possibility of repetition refers to only *one* dimension of a signifier–its referential function, i.e., the fundamental link between the phrase as such, as a set of well-formed words, and an extra-linguistic entity (or referent). Thus Proposition I refers to *one* tall tree, Proposition II–which in terms of linguistic form repeats Proposition I verbatim–refers to *another* tall tree.[93] We summarize this alternate position on iterability: language has the capacity to be repeated and thus to be accurately described as iterable and therefore producing signification, however not, as Derrida maintains, because of the possibility of isolating a given signifier and then claiming that it has "already" repeated itself but because signifying phrases can be connected to entities in such a way as to *mean* the *same* thing in conjunction with the fact that these phrases *refer* to *distinct* entities.For Derrida, sameness depends on a function of language; the counter-position argued here is that language can function in this way *only if* relevant instances of extra-linguistic sameness already exist. Derrida's position on iterability may be seen as just one more application of the principle "nothing exists outside the text." But the argument introduced at this point has been that only if something (i.e., entities) exists outside the text is it possible for language to be iterable in a theoretically coherent manner.

93 It is possible that two propositions could refer to the same tree (i.e., the identical entity) but not at the same time.

7.8.3 Ideality

What then of ideality? Derrida insists that the ideality Husserl endorses does not exist "in the world" but emerges from repetition of a signifier. If signifiers carry the referential function argued so far, then this state of affairs presupposes pluralities of referents either at hand or capable of being experienced–a language user would not say "this tree is tall" in a context which lacked trees altogether or included only a row of bonsai.

The reversed priority argued above elevates identity predicated of distinct trees so that the sameness in question must be specified with respect to "the world;" otherwise Derrida's denial of the place of ideality retains a measure of plausibility. But from the standpoint of classical metaphysics, this sameness can be readily accommodated with, for example, a notion of essence, i.e., the sameness in the tree. What allows a language user to say "This tree is tall"–and to say *in truth* that this proposition is the case if in fact the tree *is* tall–is a determinable nature defining each tree. If we agree that, e.g., California redwood trees are tall, then distinct and individual propositions can be asserted about the tallness of such trees because each tree has "in" it a certain nature.

At this point, we stand before a variety of metaphysical options–the nature just appealed to could be a form existing by itself or a form which is universal but exists only in particulars. Both explanations represent a perfectly feasible type of "ideality." Whether this ideality exists "in the world" as the Aristotelian maintains or "outside the world" as the Platonist holds is the classic issue. But both instances of ideality are integral features of the philosophical imagination and, as such, dwell accessibly to the student interested in developing such ideality. In this respect, the ideality in question is hardly something that just falls from a theoretical heaven; it is the product of deliberate and extended reflection on facets of experience and the ways in which they reveal the structures of the entities we see around us. Although this account with its unabashed retreat into Platonic and Aristotelian metaphysics (and in spite of all the warnings about being as presence!) may appear to be a bloodlessly reactionary move, its intent is only to set up a competing–that is, possible–account to challenge the intelligibility of Derrida's stated position on iterability. In Chapter 10, the ramifications of this challenge are reoriented in a direction which preserves Derrida's motivation but within a more traditionally formulated slate of categories.

7.9 Iterability: The Deconstructive Account vs. the Metaphysical Account

Language is an inherently contextualized phenomenon; many elements are intimately related to many other elements. Philosophical investigation must sort out and connect these elements as accurately and perceptively as possible. The basic argument in this chapter is that Derrida's account of iterability denatures and ultimately dissolves the circumstances in which the type of repetition essential to language occurs. The question becomes whether a theory of iterability can be credibly advanced without considering the setting in which iterability occurs when it emerges as an essential feature from a theoretical examination of given propositions.

It will be instructive to juxtapose this alternative account of iterability–the "metaphysical account"–with Derrida on iterability–the "deconstructive account" along with coordinate implications drawn from that position. This juxtaposition yields a framework of comparison so the reader can appreciate what may be at stake given that differences between the two accounts pertain to a variety of philosophical concerns.

7.9.1 Language and Immediacy

There is frequently a gap between exigencies of practical concerns proper to the lifeworld and spaciousness of theory as it works to encompass whatever spectra of interests and data define its scope and purposes. If Derrida's theoretical position regarding meaning, signification and ultimately referentiality itself were to stand as a foundational element in the ongoing realization of actual occurrences of "lived" language, then it seems difficult to deny that an essential, built-in skepticism animates every linguistic event in the lifeworld. No one would or should ever be certain about whether any intended meaning has ever successfully crossed the divide between the consciousness of Speaker A and the consciousness of Auditor B because neither A nor B can respond cognitively to the world in such a way as to command the means required to enable such transmission.

Any given signification is and must be iterable. This possibility is real and must be understood as necessary to the structure of that signification. But the question that arises in a given context with a given proposition or utterance is whether this possibility is significant for determining the meaning and referential dimensions of *this proposition uttered in this spatio-temporal context.* Consider the example from the Forest Scene, "That tree is tall." To what extent is it necessary to consider this proposition as necessarily something other than itself because the proposition as a unity must be iterable? Is this sample proposition fully meaningful in the context in which that proposition originates and is uttered?

Derrida does not allow autonomous unity to apply to words grouped into propositions; this unity becomes subservient to the diverse, scattered range of signification these words may exhibit if contextual limits are removed and conceptual structures are dissolved. By contrast, there is a foundational movement in the metaphysical account of iterability in that a proposition enjoys autonomy by an established connection to a stable metaphysical element–e.g., a nature or essence in the case of the sample propositions concerning a property of trees. There is no component of potentiality or possibility, or otherness theoretically operational in that a proposition becomes significant only because it can exist and function in relation to other existing propositions. The metaphysical account acknowledges necessary linkage joining distinct propositions but this linkage reflects connections established between and among metaphysical elements which lend themselves to individual instantiation at the level of propositional immediacy. Signification and meaning, ultimately referentiality itself, reflect the structure of reality construed as in many respects stable; these linguistic functions are not generated by processes involving the application of particular instances of these functions, as the deconstructive approach maintains.

7.9.2 The Conceptual Realm

Consider the relation maintained by deconstruction between the signifier (a) as referential and (b) as vehicle for the generation of the signified. If a change in context, e.g., alters the signifier so that it does not and cannot have a "unique and singular reality" from one context to the next, then Derrida maintains, "for the same reason," that the same conclusion concerning constant variability holds for the signified.

If we take a concept to be the uniform bearer of significance–e.g., the concept of tree–then Derrida's position implies that a concept must be capable of being repeated and in this process it voids its status as a "unique and singular reality." Thus if the signified is equivalent to the concept, then the concept becomes impossible since signifier evoking concept cannot be transposed beyond its putative existence in any given set of originating circumstances in order to apply with equivalent conceptual relevance to another set of circumstances. How then are concepts possible at all? Furthermore, a fundamental disconnect arises between the tree itself–if we assume that the natural entity exists as a "unique and singular reality" even if the signifier for the tree does not exist in this way–and the concept of a tree. For if the concept of a tree cannot itself be unique and singular, then no guarantee exists that claims about trees at the conceptual level are either limited just to trees (since the concept might be broader than the nature of trees) or have exhausted what is available about trees at the level of a concept (since the concept might not be sufficiently inclusive to capture the nature of a tree as a self-subsistent entity).

We can pose the problem without attempting to resolve it–what is the relation between concept and its object; for example, does a concept refer to an object as a member of a set or to a type of object? There is, however, an important sense in which this question of conceptual scope cannot even be raised since if Derrida's stated implication holds, then the very existence of concepts is at best tenuous. For how a concept can exist at all is not obvious if it follows, as **Passage 7G** maintains, that the signified, which we are now interpreting as a synonym for concept, cannot be unique and singular. It seems necessary to conclude then that the stated theory of signification entails the impossibility of producing concepts whether for practical purposes in the lifeworld or for purely philosophical considerations at the level of theory.

In addition to these structural problems, note also the self-referential implications for the presentation of deconstruction as a series of programmatic gestures or aggregations. Either Derrida's concepts are impossible to formulate in the first place since all concepts are impossible to formulate, or if deconstruction as an ensemble of techniques is to retain its formal autonomy then the language and conceptual configurations Derrida develops must be kept apart, a *sui generis* philosophizing impervious to any implications derivable from the potential application of deconstruction aimed at its own formulation.

The metaphysical account approaches the conceptual order by working with a pliant sense of unity. Unity in the desired sense is analogous to a characteristic common to particular extra-linguistic instances of experience (e.g., the trees in the Forest Scene of Chapter 4) which engender the use of language. Once adapted to the cognitive realm, this functionality grounds the possibility of establishing carriers of significance–concepts–as an essential component of knowledge. Kantian overtones become visible insofar as a principle of unity underlies establishing the possibility of a concept. The structural connection between the notion of a concept and the notion of unity seems undeniable. But however the integration of unity with the "makings" of a concept be resolved, the fact of difference between objects perceived and the realm in which objects become the matter of cognition is a given for the metaphysical account. In this sector of the complex range of iterability, the metaphysical account must investigate at least two areas–the relation between concepts and words, and the formation of concepts with respect to their structure and the modes in which they refer to an object or objects outside themselves.

7.9.3 Particularity

Derrida's position maintains that at the level of concepts, no transcendental signified can be generated in the relation between signifier and signified when the latter is a concept of, e.g., a tree. And in **Passage 7G**, Derrida asserts that the same denial of status as "unique and singular" holds for signified by virtue of identical reasoning applied to signifier. But if "signified" also refers to an extra-linguistic entity (e.g., a tree existing in nature in relation to the signifier "tree"), then it follows that the tree itself cannot be construed as "unique and singular" because we are told that what holds for the possibility of repetition for the signifier "tree" also holds necessarily for what is signified, the tree in nature.[94] This position sanctions the unusual inference

94 Derrida forthrightly states this implication concerning the inherently indeterminate status of particulars as a phase in his reading of Rousseau's reliance on supplementarity: "The enjoyment of the *thing itself* is thus undermined, in its act and in its essence.... One cannot therefore say that it has an essence or an act (*eidos, ousia, energeia*, etc.). Something promises itself as it escapes, gives itself as it moves away, and strictly speaking it cannot ever be called presence" (OG, 154–italics in

that there could be a given tree if and only if it were possible that more than one such object of the same type could exist in nature. Derrida's claim thus rests on assuming that particularity presupposes plurality–a substantive metaphysical stance. But why must this conclusion follow? Why should the conditions defining the existence of a signifier if the signifier is understood to be purely linguistic also apply to the reality of what is signified by that linguistic vehicle?

The deconstructive position presupposes that parallel reasoning obtains between two very different orders of reality–language and objects referred to by language. Furthermore, even if the deconstructive position does not *de facto* deny particularity, we may fairly insist that this position rests on assuming that particularity plays no active role in philosophical considerations about language. This inference is neatly consistent with the implication drawn in Part I regarding the impossibility of Derrida advancing a substantive metaphysics dealing with the reality and status of particular entities. For the deconstructionist, sheer repetition of signifiers–i.e., language about particulars–does all the "work" as far as situating and integrating the fabric of general terms is concerned.

On the other hand, the metaphysical account posits the independence of the self-identity of extra-linguistic entities from considerations pertaining to the self-identity–or, if Derrida is correct, lack of self-identity–of signifiers in the process of representing those entities as signified. Thus the self-identity of an extra-linguistic entity *qua* self-identity is not reducible to and in fact remains completely apart from all linguistic considerations. The metaphysical account of iterability respects the independent reality of particulars. Furthermore, signification is not a consequence of the possibility of repetition of linguistic formulas alone as the deconstructionist account maintains; rather, signification becomes a consequence of the perceptual experience of entities in the extra-linguistic world and the consequent concern to voice that experience. As Socrates puts it in the *Phaedo*, we predicate or "say of" particulars that they possess certain properties (*Phaedo* 102b). But iterability is a function of the multitude of particulars and the capacity of language to evoke particular instances of that multitude as well as individual properties adhering to those instances, not the other way around.

Derrida correctly insists on the necessity for iterability but his conclusion concerning the metaphysical status of a given signifier must be carefully qualified. The first instance of a signifying mark displays characteristics which justify, indeed necessitate, that this instance is in point of metaphysical fact "unique and singular." And for each subsequent instance, the signifying mark can also be read as "unique and singular" with respect to its participation in a metaphysically complex context.

---

text). As a result, none of the classic Greek metaphysical concepts identified can apply to the "inner," metaphysical nature of things, a conclusion paralleling the radically skeptical position reached by other means in Derrida's approach to iterability and its extra-linguistic objects.

7.9.4 Self-identity

One of the most important consequences of the deconstructive account of iterability is the apparent denial of the relevance if not the very existence of self-identity. If neither signifiers nor signifieds exist as realities which are "unique and singular," then it is not obvious whether the very notion of self-identity abstracted from concrete realities which provide the "faces" of signifiers and signifieds is anything more than a name lingering as metaphysical residue from the effects of being as presence.

According to the contrasting metaphysical account, self-identity gives particulars access to the domain of logical principles. As predicated of a particular, self-identity is purely formal and can be said of any entity. However, this conferral establishes sufficient stability for this variable (as in A = A) such that it and any entity are deployable according to all pertinent logical concepts and principles. Furthermore, self-identity as a notion capable of isolation and abstraction from its predicational use has the following features: it exists as one and the same mark; as one and the same mark connected to a certain basic meaning; as one and the same mark with the same basic meaning within a context; as mark retaining this meaning regardless of the pressure and interaction between the mark and everything that exists in the context within which the repeated instance of the mark occurs.

If self-identity as a concept fundamental to logic is questioned–indeed, rejected–then it is fair to conclude that the status of "logical" concepts in general has also been thrown into deconstructive indeterminacy. The word "logical" is in quotes at this juncture to emphasize the formality of a notion such as self-identity. But for the advocate of deconstruction, an appeal to "formality" as it putatively applies to a range of abstract, ordering concepts is only as pertinent to inquiry as the extent to which self-identity and all such notions can be analyzed and preserved in the face of their long-standing metaphysical heritage.

In an investigation deployed at this level of high generality, an essential feature of inquiry is the enumeration and analysis of the *implications* which ensue if any much less all such concepts and principles are thrown into question. To suggest that these notions can be questioned in order to appreciate and, if possible, to discern their underlying character with respect to their extended habitation within the history of metaphysics is one thing, but then to hold these notions in abeyance as restrictive if not distorting is something else entirely. If no one entity is ever "single and unique" because self-identity is a residue of limited metaphysical thinking, the consequences of this lack of unity and stability must be determined and assessed–including the effects derivable from the rejection of the notion of self-identity as such. We will return to this fundamental question in Chapter 9.

7.10 Iterability: A Transcendental Snapshot

Derrida's notion of iterability is pivotal to his theory of signification, but its presuppositions and implications are also essential elements in formulating the strategies proper to deconstruction as a mode of textual interpretation. As we have shown, iterability is complex and qualitatively diverse in its metaphysical structure. The following is a description, sketched in sharply outlined tones, of the structure of iterability given the passages analyzed in conjunction with the interpretive and speculative arguments developed in this chapter:

Iterability depends on the notion of possibility but not all dimensions of possibility are relevant to account for this dependency. Thus the possibility of what might happen in the future to the import of a word or signifying phrase is not necessarily relevant to what has happened and is happening in the present to those linguistic elements.

Iterability depends on a notion of possibility which incorporates absence as a kind of negation. This absence ultimately encompasses a totality field characterized with as much potential for metaphysical development as is allowed short of the generation of entities, events and states of affairs that, whether blatantly or after close scrutiny, are depicted in ways which threaten to violate the principle of contradiction.

The possible avenues the evolution of language as a structured phenomenon will take in the future cannot be known in the present. Therefore, the range of possibility includes future temporality but only if we hold the future in our imaginations in certain ways. We must imagine the future and also imagine that events in the future will resemble events in the past leading into the present even if the iterated or repeated instance occurs just once in the future or, more likely, a finite but fairly high number of instances. As already noted, we must also imagine that any changes that occur in the future will not materially affect anything of substance in articulating the components of iterability as presently characterized, given that its cogency in a theoretical context depends on a projection of elements from past and present into the future. Whatever potential changes fall under the purview of this requisite imaginative projection will include (a) entities existing so that they are capable of being experienced as well as (b) our experience of those entities transpiring in modes accessible to language and finally, that anything which changes in any way within either (a) or (b) will not affect the stated structure of iterability.

It is necessary to identify a minimal set of contextual conditions underlying instances of possible iterability in order to determine whether these instances indeed display relevant dimensions of iterability. In other words, iterability is necessarily (a) contextual and (b) characterized by ineradicable aspects of space-time and materiality. In addition, iterability presupposes sameness in that some element (or elements) in the signifying element must remain the same throughout each and every instance of iteration, otherwise "being repeated" does not occur. These elements must also display a formal sense of self-identity in order to preserve the difference factor required for the possibility of iterability as a repetition within a given sequence of linguistic elements.

7.11 Summary and Prospect

Derrida notes that John Searle agreed with positing iterability as a necessary component in a philosophically viable account of language (cf. **Passage 7C**). Thus the necessity of iterability as a component of a theoretical treatment of signification is not the issue. The problem concerns the ways in which Derrida formulated, defended and developed iterability in tandem with the assumptions and implications connected with this approach.

Derrida made an extensive number of implicit metaphysical commitments including, especially for establishing iterability, the use of necessity and possibility as key modal operators throughout his formulation of all the elements of deconstruction examined in Part II of this essay. Furthermore, examination indicates that the notions of signification, context and negation–each of which includes its own set of metaphysical elements–are pivotal to iterability, a cardinal feature defining the theoretical underpinnings of deconstruction. Therefore, the extent to which obscurities and inconsistencies inflect the notions constituting iterability is the extent to which iterability, taken as a essential component of deconstruction which must bear the traditional logical character of internal coherence to guarantee its cogency, is also subject to an equivalent measure of scrutiny. In order to give point to the array of problems inherent in Derrida's account, the present chapter included a counter-position describing the structure of iterability from a more traditional perspective as well as a comparative summary of these two approaches with emphasis on their differences and the implications derived from those differences.

The criticisms introduced in this chapter revolve around notions of high generality–negation, difference, unity, possibility–all of which have vigorously animated the history of metaphysics. In fact, what might be termed the "transcendental tempo" marking Derrida's reflections on language has reached its highest point of sustained application and its greatest intensity of import in the formulations of and derivations from the notion of *possibility*. The movement of responses connecting the critical comments in the four chapters of Part II has been a progression toward possibility; indeed of the set of generalities and abstractions animating Part II, the concept of possibility–especially the statement and implications derivable from the multi-form complexity of that concept–has proven to be arguably the most problematic for the intelligibility of Derrida's notion of iterability.

A seminal phase of the argument in Part I was that being as presence represents an approach to totality. Derrida's position on iterability also extends to totality, even if only implicitly. Furthermore, totality and possibility are (or can be, depending on conceptual boundaries) closely linked. The concern raised in this chapter is that the position taken on iterability fails to distinguish among different types of totality–e.g., what exists as possible in the sense that it is imaginatively real, a projection into the future required by Derrida's position on the necessity of repetition. And yet this projection remains entirely in the realm of supposition, in contrast to the arena characterized by perceptual experience which enjoys the status of yielding publicly verifiable facts and also is of dominant immediate interest in the lifeworld. We also must observe that the unity of a factual totality is not and need not be identical to the unity of an imagined world–indeed, since the number of such worlds lacks an obvious limit, imagined *worlds*. Various differences mark the juxtaposition of these contexts as they play into the general notion of possibility, and yet Derrida ignores these differences in the formulation of iterability as well as in the inferences he draws based on this formulation and applied to language in general.

Derrida's writings on iterability conflate different senses of possibility (presupposing the actualization of totality as a necessary condition to support the cogency of his position). My contention is that Derrida is appealing to a kind of possibility with content at least as broad as that envisaged by Hume in terms of what could actually be the case. The main difference between Derrida and Hume in this context is that for Hume, the experience of a given fire and a given proposition about that experience–e.g., "This fire is hot"–is decisive and controlling *while this experience currently circumscribes the context of the person uttering this proposition*. But for Derrida, the empirical givenness of the fire, while not denied, is not the terminus or ground for the ultimate significance of the proposition "This fire is hot." For Derrida, all imagined possible states of affairs are *necessarily* attached via iterability to the import of *the single proposition* "This fire is hot" when this proposition is uttered by a certain person in a certain context. As a result, the philosopher must take seriously that many possible states of affairs could eventually transpire relative to what has been asserted in the present in order to justify theoretically the meaning of this proposition when uttered as a purely linguistic component in *this* complex state of affairs–when *this* individual is near *this* fire as it burns at *this* place and at *this* time, all circumstances culminating in an individual's decision to speak. Derrida has invalidated the distinction (argued for in this chapter) between two types of material possibility, thereby seriously undermining the coherence of his account of iterability.

We will analyze ramifications of this line of interpretation and its conclusions in Chapter 8. But regardless of what remains to be said in this essay, it seems fair to contend that much additional analysis seems warranted in order to clarify Derrida's texts involving possibility and how this concept pertains to iterability and its theoretical justification with respect to deconstructive concerns Derrida wanted to foster. This analysis will emphasize metaphysical matters of foundational importance and ancient lineage. For in order to generate argumentative associations intended to dismantle the traditional concepts forming the framework for a philosophical understanding of language, these matters (a) must be made consistent with the strategies of deconstruction and (b) must also be reinterpreted and redescribed in terminology which remains as far distant from the discourse of the history of metaphysics–and the myopic character imposed by being as presence–as intelligible discourse can be.

The texts examined in this chapter and throughout Part II promise substantive, positive doctrine regarding the "deep" structure of language as deconstructive analysis discloses that structure. But the most striking and arguably incongruous phrase in the passages analyzed in this chapter is Derrida's assertion that the considerations specifiable in his reply to Searle, especially with regard to the pivotal matter of iterability, constitute a "structural law" (**Passage 7C**). What conditions must be in place in order to justify a theoretical appeal to a "structural law," *any* appeal to such a designated law but especially when the context of inquiry concerns a philosophical analysis of language? This question involves three distinct but related issues within the framework of Derrida's thought: Does the "structural" phase of this law rely for its integrity–indeed, its very existence–on any basic concepts of classical metaphysical and its terminology? Second, whatever may be the notional elements of the law in question, how do such applied instances of generalities avoid the inherently limiting if not stultifying aura of inclusion within the ambit of being as presence? And third, how can this appeal–any appeal–to a "structural law" be logically compatible with deconstruction's leveling effect on any, presumably all general principles?

To maintain even the feasibility of a structural law, Derrida must "shake" not only the stated content of the history of metaphysics but also the derived traits of being as presence until some elements achieve the requisite stability to justify positing anything "law-like," especially when the context is as fluid and expansive as that of human discourse. The elucidation and justification of iterability by itself as well as the assertion of a structural law as a pivotal moment in the defense mounted against John Searle's critique are commensurately dependent on highly abstract conditions. The question again: how do these positions, ultimately derived from cardinal concepts in the history of metaphysics, cohere with the inherent repressiveness of being as presence? And in a related vein, how do these positions themselves remain impervious to the critical treatments brandished by deconstruction and aimed at instances of being as presence? Part III of this essay addresses these and equally fundamental questions.

Part III: **Presence, Language, Metaphysics**

8 The Foundation of Deconstruction: Generalities at Play

In Part II, we critically evaluated a series of fundamental elements in Derrida's philosophical analysis of language. Questions were raised for each element, questions relying for impetus and argumentative direction on the viability of certain cardinal metaphysical notions. The issues involved might be called *regional* criticisms since their effect derives from a series of delimited metaphysical concerns. But another dimension pertinent to Derrida's deconstructive response to linguistic concepts emerges if we stand back from the immediacy of Derrida's analysis and pay attention to backdrop considerations which play into the substance of his more explicitly focused critiques. These considerations, duly developed, have important consequences for the basic intelligibility of being as presence as perhaps the fundamental element in Derrida's thinking, at least with respect to lessons learned and applied from the history of philosophy. Furthermore, as we shall see later in this chapter as well as in Chapter 9, the development of these questions raises commensurately important concerns for the cogency of deconstruction as a philosophical methodology.

8.1 Dimensions of Totality

If being as presence has permeated the history of metaphysics to the extent that Derrida claims, then it would be not only consistent but also prudent to investigate whether Derrida contoured his critique of standard linguistic concepts so that the deconstructive motion active in Derrida's thought itself replicates or even merely incorporates aspects or phases of being as presence. The assumption grounding this inquiry is that Derrida himself continued to be a product of the history of metaphysics even as he was in the process of formulating techniques to confront critically that history.[95] As a result, the investigation pursued in this chapter will proceed along guidelines set by being as presence itself, that is, along conceptual parameters marked by extreme generality and comprehensive scope.

In order to appreciate and understand the extent to which Derrida must activate metaphysical considerations to establish his appraisal of language, consider, perhaps as a fundamental challenge to the committed advocate of deconstruction, that Derrida's critique of linguistic concepts depends on the implied existence of a totality characterized by at least three determinate dimensions. These dimensions–Temporality, Context, Meaning–are not merely nascent representations of abstract possibilities; we will

95 See sources cited in Ch1n2 for Derrida observing the historical and philosophical necessity of remaining within metaphysics even while intending to move beyond its limits.

show that each dimension must be fully actualized in order to justify the conclusions Derrida explicitly draws. Furthermore, these conditions taken as a set also mirror the stipulated character of being as presence, assuming we grant that equivalent totality-considerations are necessary as preconditions for both investigations. This mirroring function has especially significant consequences, as we shall see.

The modality of possibility, i.e., its lack of theoretical limitation, provided the impetus and foundation for the criticisms raised in Chapter 7 against Derrida's concept of iterability. However, additional dimensions of possibility undergird Derrida's formulation of deconstruction as a programmatic approach to interpreting language especially with respect to the various guises and functions of language in philosophical and literary texts. Derrida's position requires that these dimensions, once identified and their import fully appreciated, move from what *could be* the case to what *must be* the case. The initial dimension of this totality revolves around a relation between possibility and time.

TEMPORALITY: All dimensions of time as traditionally demarcated coalesce in one instant or as one duration of determinate bounds insofar as these designations ground the temporality of a given context.

It is readily possible to imagine the linguistic sequence "Please bring me a glass of water" occurring in various contexts, including not only widely diverse variations on a given natural and technological setting but also similarly distinct variants in psychological states of the speaker and person so addressed. Each imagined state of affairs would result in variance of reference and meaning as emerging from occurrences of an identical linguistic sequence within their respective contextual environments. But in order for this set of complex imaginary possibilities to come to actualization, it would be necessary to collapse any temporal differentiation relative to discretely distinct possibilities within the set.

The event comprising the request for a glass of water happens during a certain segment of time. The same event can, however, be imagined as happening at some point in the past–same linguistic sequence, different, perhaps very different contextual surroundings–result: different meaning. Or the same event can be imagined as happening at some point (or points) in the future–same linguistic sequence, different, perhaps even more different contextual surroundings–result: again, very different meanings. But to have these different possibilities impinge on the meaning of the linguistic sequence announcing the request at *this* point in time presupposes the superimposition of *all past* time and *all future* time on *this time*. Thus if different instances of the same word appear in different times, then Derrida's position assumes that *all of time* relative to the existence and use of that word, both past and future, coalesces at a given instant or, alternatively, within a continuous and undivided duration. This temporal confluence is a necessary and decidedly metaphysical condition for the claim that no instance of a word in a given context can be assigned a definite meaning and a definite instance of what Derrida has termed transcendental signification.

Derrida's theoretical pronouncements when he develops foundations of deconstruction implicitly hinge on a notion of time in which limits cannot in principle be introduced. But introducing temporal limits is surely a necessary condition not only for the possibility of distinguishing between and among events but also with regard to situating the human capacity for participating in these events. If a speaker could be consciously aware of or even just actively believed in such an all-engulfing temporal universality, the upshot would be complete paralysis of discourse–it would be impossible for the speaker to focus on a single intended meaning congruent with and arising from the circumstances facing the speaker at *this moment* in time, and in the company of these individuals and entities at *this moment* in time.

CONTEXT: All (metaphysical) features of all (possible) contexts are superimposed on the specific content of a given context.

Furthermore, linguistic sequences do not occur in a setting characterized only by time. Time is one feature underlying the utterance of a linguistic sequence but it is one of a set of factors. Another is space. Yet another might be called materiality, a general term including everything composed of any kind of physical matter. These features coalesce into contexts, spatio-temporal settings which include material things (things defined in a very wide sense).

Recognizing these additional factors is vital since their existence with respect to a given linguistic sequence has consequences for Derrida's approach to language, in particular the principled rejection of fixed meanings as well as determinate and stable signified elements. Thus another necessary condition to sanction deconstruction's denial of determinate meanings is not only that all time must coalesce into an instant or a single indivisibly continuous duration but also that all possible contexts relative to the occurrence of a given linguistic sequence must be, in fact, superimposed on the specific boundaries of the context within which the given linguistic sequence is uttered.

Consider again the request for a glass of water. Another context in which that same linguistic sequence could occur is a stage play where fulfillment or rejection of the request is immaterial because of, say, an impending turn in the dramatic action. The performers are fully aware that they are only acting with respect to this request whereas two persons in the extra-theatrical lifeworld know just as well that such a request is not a linguistic exercise guided by the intent to pretend a course of action for a sitting audience in a building shaped and constructed for precisely that purpose. In sum, one factor situating awareness of the difference is the undeniable facticity of the surroundings. Thus to contend as Derrida's position does that (a) the staged instance of the linguistic sequence is just as real in its own way as (b) an actual occurrence of that sequence and therefore that no privileged meaning for that sequence can be determined presupposes that (c) the context underlying and informing (a) above *must* be incorporated into the contextual setting of (b) so that the distinctive meaning of the linguistic sequence in (b) is necessarily compromised. But such incorporation requires that (a) and (b)–separate contexts defined by unique spatio-temporal considerations as well as diverse sectors of materiality–become indistinguishable from one another.

It is possible, i.e., imaginatively possible, to move from identification and description of the features of a given context to complex and highly diversified variations derived from the set of original elements. As imagination becomes more and more energetic in engendering expansive contextual scenarios, the boundaries of the original context are extended into the outermost reaches of imaginative possibility. But, and this point is crucial, such extension exists *only in the imagination*. To engraft a purely imagined extension onto a context exhibiting its reality in a perceived spatio-temporal setting occupied by extralinguistic entities and permeated by agents occupied with personal and interpersonal concerns is to insert an entirely distinct realm–one existing *only* in the imagination–within a dimension of reality which, as such, is and can remain divorced from the vagaries of imagined scenes in imagined worlds.

This foundational dimension of reality is the lifeworld. When an imagined insertion is pursued and embraced for theoretical reasons, the investigator has adopted a form of inquiry with a structural impetus so comprehensive and precise as to overwhelm the demands of immediacy. For immediacy is an integral element in the fabric of the lifeworld as actually lived by its inhabitants. We observe then that Derrida's theoretical concerns overpower his appreciation for the requisite and immediately recognizable here and now, the radical character of the moment, when decisions are made, whether right, wrong or morally neutral, and life becomes joyous, tragic, or just continues in humdrum fashion.

The individual speakers involved in (a) and (b) above can discern the difference between their respective contexts; to deny such discernment on their part beggars belief. The claim here, against Derrida, is that his concerted skeptical conclusions concerning distinct meanings and referential content presupposes metaphysical considerations which, if implemented, would necessarily collapse apparently undeniable features in the lifeworld–e.g., that asking for a drink of water after running a race occurs in a location other than that where a character in a play might make precisely the same request without any intention that the request be satisfied (given, e.g., certain circumstances in the plot).

The point is not that a given linguistic sequence might be uttered in very different contexts thereby resulting in very different meanings for that same sequence; the denial is that (a) the individuals involved in these contexts are always and necessarily subject to a kind of deconstructive indeterminacy as to what these meanings might be and (b), more fundamentally, that these contexts, quite apart from the degrees of awareness of the individuals residing within them, exist without exhibiting their own spatio-temporal fabrics which as such cannot be dismissed because of an interest in posing a theoretical conclusion aimed at establishing a fundamentally skeptical linguistic indeterminacy.

Derrida's treatment of language is defined by a relevant parallel the effective range of which spreads over two distinct but related dimensions. An essential factor in his denial of a transcendental signified is the fact that the context asserted to underlie any given linguistic event can be restricted only by an arbitrary and temporary gesture. Now the notion of context when applied to a linguistic event includes the spatio-temporal circumstances surrounding the words resident in that event. As we have often noted, the speaker of language does not do so in a metaphysical vacuum; nor do words on a page of a literary text function irrespective of a set of circumstances metaphysically contouring the import of those words. It follows that the denial of the relevance of context for a linguistic event must incorporate this set of given metaphysical circumstances and then, necessarily, extend the range of this incorporation until it is not possible to delimit a context which can be taken as stable and definite, the underlying ground for fixed reference and meaning. But in order to secure as a theoretical fact that any such delimitation of a context is not feasible, this extension must span the limits of reality. For if the spatio-temporal factors defining a context were not so extended as to encompass totality, then it becomes theoretically possible to indicate a set of factors which could for a given linguistic event remain fixed and stable–not just "negotiable," as Derrida admits–and thereby serve as the metaphysical foundation for a context which was itself fixed and stable. However, Derrida insists that no such context can be so identified.

This conclusion may be restated as follows: a context must include someone using language, whether spoken, written or thought, and a set of spatio-temporal conditions which circumscribe the production of that language. These conditions will also include whatever material or tangible entities populate the space-time contoured by these conditions. But since, according to Derrida, it is never possible to establish a given context by stabilizing that context into something resembling a unity, the only way to ensure that this possibility can never in fact be so actualized is to extend the range of all metaphysical components of any immediately given context–or, per Derrida, what appears to be an immediately given context–*to the outermost limits of whatever could function as an element with respect to such metaphysical components*. Thus, any context is and must be effectively bounded by totality in order fully to justify the claim that no single context can ever be structurally limited as a factor in determining what could be asserted as "the" meaning of a given linguistic utterance which occurred within the boundaries of that context.

Notice what follows: totality must be to some degree present to the immediacy of a given context in order to guarantee that the limits of that context are, in fact, only apparent. Each context must therefore effectively exist *in the presence of* totality, so to speak. For without the limits of totality being present in this way to each and every apparently distinguishable context, it would be possible, as just noted, to situate a context so that its limits were indeed determinate. And again, with determinate limits in place, it becomes theoretically possible to accomplish what for Derrida cannot in principle ever be accomplished–determining a fixed and stable meaning of a given utterance or written sentence.[96]

96 At PS, 144, Derrida claims the necessity to "negotiate" in order to reach an approximate context in the usual sense. By contrast, however, Derrida appeals to the "stable" contexts established in his own work in a variety of sources (cf. Ch5n67): LI, 79, 81, 157; 1992a, 64.

MEANING: All future meanings of a given linguistic sequence insofar as these meanings are rendered possible by alterations in future contexts in which that sequence could appear are superimposed upon the initially determinate meaning of that sequence as that instance of the sequence occurs in any one given context.

The third presupposition can be taken as implied by the analyses of Temporality and Context but its importance deserves special emphasis. The possibility of multiple meanings for a given linguistic sequence can be readily imagined but analysis of this possibility reveals that it is itself a residual effect drawn from the confluence of all contexts in each of which a given linguistic sequence or sequences occurs as well as the collapse of any sustainable divisions or separations in the movement of time. Only if all possible meanings of a given word are present simultaneously is it possible for Derrida to assert that, in point of deconstructive fact, no instance of a word or ordered group of words in a given context can be assigned a definite meaning. Such meanings are little more than temporary resting points amidst the ongoing flux of perpetual indeterminacy which grounds Derrida's implicitly situated metaphysics deployed as an affective foundation to justify the possibility of deconstructive conclusions.

8.2 Convergent Totalities: An Isomorphic Play

In Part I, we established various senses in which being as presence functions with respect to totality understood as a metaphysical ultimate. But it is now apparent that an implicit appeal to totality also underlies Derrida's more restricted critique of individual concepts pertaining to language.The triple dimensionality just introduced is a diverse partitioning of a single totality. This totality spans as well as underlies the narrative accounts detailing the aggregations of theory which, for Derrida, constitute the interpretive schematic known as deconstruction.

The absence of theoretical stability with respect to referentiality and meaning in all linguistic dimensions is a foundational consequence for deconstruction. This absence has been described in the accounts of pivotal components of deconstruction presented in Part II of this essay. However, the justification of this consequence depends on accepting givens of decided metaphysical function including the undifferentiated unity of time in tandem with the contextual dimension–that is, the possibility of a settled demarcation of events and spatio-temporal limits. As traditionally understood, these limits control the import of discourse of any kind according to boundaries established by singular contextual demarcations. Of course, for Derrida there can be no such demarcation at least in terms of providing theoretical certainty or precision. This denial is the ultimate ground establishing why deconstruction rests on a unity of all possible contexts, each of which encroaching on all the others, a continually overlapping process which does not cease until it is all-englobing and comprehensive in every possible sense and dimension of reality. The spatio-temporality defined by limits of a given context seemingly derived from the ineradicable difference between the durational present of that context and its respective past and future in inextricable conjunction with regions of space expands until its unlimitedness parallels that of any and all unbounded contexts. Indeed, the use of the plural, "contexts," is for Derrida's purposes misleading since the explicit plurality denoted by the plural form of the word exists *as* a plurality, as a multiplicity of distinguishable contexts only for the sake of convenience on the part of speaker or writer.

We have seen Derrida admit that metaphysics as a precondition for all such thoughtful endeavors intending deconstructive results cannot be done away with completely (**Passage 4F**). However, it may fairly be asked: to what extent does reliance on metaphysical considerations affect the formulation of deconstructive principles and strategies as well as the subsequent application of those principles and strategies to given philosophical and literary texts?

The answer is straightforward: *the same kind of totality framework contours* (a) *being as presence with respect to the history of metaphysics and* (b) *the structural foundations allowing Derrida to conclude that meaning, reference, the transcendental signified and context are in principle impossible to determine with any degree of certitude or inherent limitation.* Thus a delicately poised isomorphism exists between the unity underlying the conjunction of the conceptual configurations inherent in Derrida's notion of being as presence and the unity underlying the set of conditions which must obtain in order that Derrida's deconstructively motivated writings on language, both their ideational structure as well as their emergence when applied to texts, can accomplish philosophically what he claims they accomplish.

The resulting isomorphic unity is an environmental totality inflected with diversifications and oscillations between and among elements which fall under the circumscribed scope of the respective individual totalities. However, this congruency does not debar a pertinent difference between the two global perspectives. Although totality is dominant in both phases of Derrida's thought, the unity grounding being as presence basically arises from textual considerations in that its invariant effect originates from the history of metaphysics, from *written accounts* which purport to disclose how being should be understood. Derrida derived the notion of being as presence by reading various texts appearing in the history of metaphysics, reflecting on that material and then drawing conclusions about the assembly of texts so read.

To this series of textually based metaphysical systems, Derrida added the distinction between sameness and identity, then formed the concept of *différance* as the animating force controlling the history of metaphysics and limiting its sole dimension of sameness to being as presence. By contrast, the foundational unity for the deconstructive approach to language is *itself fully metaphysical* insofar as it relies on an omnipresent temporality and a commensurately present set of contexts lacking an effective limit. If this italicized phrase is too hard-edged and historically conditioned, replace it with the following, more nuanced but referentially equivalent description:"configured in ways employing terms and directions of thought either indistinguishable from or fully compatible with much traditional metaphysical thinking."

In this respect, Derrida has contributed addenda to the history of metaphysics even as he proclaims, while developing deconstruction, that it constitutes an assemblage of attitudes and strategies that is independent of or at least to a degree separate from that history. Of course, deconstruction is not theory for the sake of theory. Deconstruction names an approach to a critique of traditional concepts animating the structure of language and a series of techniques for reading texts, whether in language typified as philosophical or literary. To the extent that the latter dimension of deconstruction–its application to given texts–admits of formulation in something like a theoretical setting,[97] that setting reflects the general and abstract factors which control and direct the critique of language proper.

It is, however, important in the present context to be fully aware of the fact that the isomorphism just described represents only a threshold perspective concerning the extent to which deconstruction depends on metaphysical considerations. Once a more detailed description of the scope of this dependency comes into view, crucial consequences follow regarding the cogency of deconstruction as a conjoint set of theoretical gestures with common interpretive goals.

In the next section, I will illustrate the extent to which three concepts fundamental to western metaphysics are endemic, seemingly of necessity, to Derrida's formulation of deconstruction's basic notions and strategies. Once these accounts have been developed, I will then explore potential ramifications regarding whether or not or at least the extent to which deconstruction as formulated by Derrida's texts remains a viable philosophical procedure. These ramifications arise, again, on condition that deconstruction stands as a unified theory broached to accomplish determinable ends.

8.3 "Formal" Entities and the Texts of Deconstruction

A point of interpretive strategy: since Derrida uses "form" and cognate notions often and in various contexts, we briefly indicate how "formal" will inflect the interpretation of Derrida's texts argued at this point. The appeal is to the frequent appearance of this term throughout the history of metaphysics and to emphasize the malleability of predication and coordinately abstract status of common referents of "form" and "formal." Subsequent analysis in this essay concentrates on a series of such terms. Thus *unity*, *sameness* and *difference*–the same set of fundamental notions that grounded Chapter 3 and its analysis concerning the structure of being as presence–

97 Derrida denied that deconstruction was a method and that it was possible to codify its practices into a set protocol. See: Kearney, 155-6; LI, 141; N, 101-2, 103; Derrida 1987, 262. Despite Derrida's in sistence in this regard, a number of his advocates try to do exactly that in greater or lesser degrees of specificity: See Wood, 3; Caputo 1997, 77; Hill, 117-8, Naas 2003, 161-2; Harvey, 25, Gasché 1989, 3,4.

are typically described as "formal" in the sense that they apply indiscriminately and without any theoretical residue to many distinct things and types of things, and that they do so in ways which convey significance when these realities are viewed at a certain level of abstraction.

That formal factors derived from these terms–unity, sameness, difference–explicitly propel as well as implicitly contour pivotal junctures in Derrida's texts regarding both the theory and practice of deconstruction will become evident if the reader of Derrida observes the rhythm and detail of Derrida's argumentation. But noticing that such terms animate Derrida's work does not by itself discursively identify their exact status and function. By "status," we mean a determination specifying *whether* students of Derrida should understand the import of these terms, employed variously by Derrida, according to second-order discriminations such as degree of formality, nominalism, or some other alternative as yet unspoken in philosophical analysis and reflection. The contention then is that *how* the student of Derrida and deconstruction decides to interpret the status of these terms will have implications of great importance.

We now describe some of the most manifest ways in which specific metaphysical elements dictate the flow of Derrida's version of deconstruction; we will also sketch preliminary lines of commentary and criticism based on these accounts. The intertwining metaphysical themes at work in this broadly comprehensive context will be more fully and precisely appreciated and understood if we approach their analysis gradually. Although each of the metaphysical elements identified in this chapter has a variety of diversifications within the long and variegated history of metaphysics, the accounts ascribed to these elements are sufficiently developed to stand as representative of that tradition.

After articulating a number of settings in which unity, sameness and difference function integrally in advancing the ends of deconstruction, we will pose this question: if unity, sameness and difference–all three as a conjoint set and each taken as such–are indeed necessary to the texts required in order to formulate and apply deconstruction, should this trio of fundamental realities and their conceptual indicators be abstracted from the letter of Derrida's writing, considered as isolated and independent elements in their own right, *and then also be subjected to deconstructive analysis*?

8.3.1 Unity

The Aristotelian maxim that unity (like being) can be said in many ways is a useful entry into the relevance of what might appear initially to be a concept too vague to be relevant.[98] In Chapter 3, we noted the unity exhibited by the history of metaphysics in the sense that it had a beginning, an "organic" duration and a termination of the sort

98 At *Metaphysics* 1040^{b}16-20, Aristotle asserts that unity, like being, can be said in many ways.

Derrida describes as "closure." If we approach deconstruction as a set of approximately stable interpretive techniques, then unity also applies to the history of metaphysics in the sense that this history encompasses a collection of texts with a determinate number of members. There is only *one* history–one series of texts, constituting western metaphysics.[99] The fact that some works may or may not be apocryphal (e.g., several dialogues attributed to Plato) does not tell against this strict sense of unity. A given work will or will not belong properly to the history of metaphysics; whether a given work is or is not apocryphal bears on epistemological concerns relative to what is known about the work, not on the status of the work as such. That decisions may have to be made concerning whether or not to include a given work within that series of texts does not affect the unique unity of that set insofar as it represents, when finalized, the sum total of western metaphysical thought. Thus if the history of metaphysics is in any respect broader or more inclusive than that collection of texts circumscribed according to Derrida by being as presence, then the closure of metaphysics, a feature of the organic history of metaphysics which is crucial for Derrida's overall position, may not in fact be the case.

This sense of unity is important to bring into the open because it establishes a limit to any inquiry which attempts to respond to "the history of metaphysics." If a work with philosophical intent is written to summarize and speculate *about* "the history of metaphysics," that work must be in significant respects *other than* that history. An interpretive distance must exist between "the history of metaphysics" as a finished and complete set of texts and whatever thesis a work about that history maintains in conjunction with the articulated justification of that thesis.

This distance presupposes a second sense of unity, i.e., deconstruction itself as "other than" this unity. The textual constitution of deconstruction must be understood as a programmatic unity of integrated elements which transcends all differences exemplified by the various systems constituting the history of metaphysics. This medley of interpretive techniques remains a unity; it is also stable, indeed it is invariant in that its regulative strictures–e.g., the question of appropriate strategy with respect to opposition if one opposite has been privileged throughout history over its counterpart–always remains the same regardless of the content and divergent character of a given metaphysical system. This unity thus establishes a perimeter of limits regarding the extent of deconstruction's effects on its subject.

---

99 Recall Derrida explicitly saying that there is "only" a single history of metaphysics. See Ch. 1, **Passage 1A.**

8.3.2 Sameness as Self-identity

Unity also has a useful and important heuristic dimension when the context of concern is focused on ways to approach the status of formal realities as they function in Derrida's texts on deconstruction.[100]

The primary distinction between the unity of the history of western metaphysics and the unity of *différance* is one of sheer size and internal complexity, but it is crucial to a critical understanding of both elements in Derrida's thinking that they be approached, initially, *as unities*. *Différance* can be construed according to a traditional structural template as a whole of parts. Thus sameness and identity–parts–cohere with one another in Derrida's analytic account of *différance*–a structural whole.[101] Furthermore, *différance*, with its special sense of sameness, is essential to the theoretical formation of being as presence. The following question emerges: how are (a) the sameness of *différance* and (b) deconstruction–an aggregation of techniques formulated to function in ways other than everything metaphysical collected under the stanchion of being as presence–congruent with one another? An answer to this question, if it can be secured, will go far in showing what Derrida borrowed from the history of metaphysics while in the very process of establishing a methodology of correlated gestures (all, presumably, with the same goal collectively in view) precisely to "deconstruct" and ultimately thereby disclose what has been hidden during that very history.

In his account of *différance*, Derrida distinguished between sameness and identity but without developing the distinction in order to clarify their use, especially in settings marked by high generality and extensive application. But the self-identity pertaining to sameness as an integral component in the articulation of *différance* must be distinguished from identity as a concept which, according to Derrida, is not equivalent to sameness. We saw in Chapter 3 how it is vital to recognize the difference between identity and self-identity as well as the complexities involved in determining the respective properties

---

100 Aristotle provides useful direction at *Metaphysics* 1018^{a}8: "Clearly, therefore, sameness is a unity of the being either of more than one thing or of one thing when it is treated as more than one, i.e., when we say a thing is the same as itself; for we treat it as two." The question becomes whether Derrida's sameness in the context of being as presence operates in this way–when the "doubleness" of sameness is generated only because we treat a single thing "as two"–or whether Derrida intends sameness to differ substantively. A Platonic Form is the same as itself (formally, by the principle of identity) and an Aristotelian form is the same as itself (for the same reason) but it does not follow that a common sameness applies to both types of formal reality. However, this commonality is precisely what Derrida discerns, its scope encompassing the totality of western metaphysics and with all constitutive differences inflecting that totality fully preserved. Part I of this essay demonstrated the challenges involved in coherently eliciting this commonality.

101 Derrida's denial that *différance* is a concept understood in any standard or traditional sense of the term does not obviate the necessity of distinguishing between sameness and identity (i.e., as parts) in order to comprehend how *différance* functions (i.e., as a whole of these parts). This point will be discussed later in Chapter 8.

of each, especially when self-identity is predicated of sameness (and to anything else). Self-identity is *necessarily* predicated of sameness within *différance* in order to guarantee that this sameness, and by extension *différance* itself, functions *in the same way* throughout its application to the entire history of metaphysics.

Derrida contends that *différance* drives being as presence and that *différance* as such does not exist in the same way as any concept resident in the history of metaphysics; therefore, *différance* should not be approached as just another addition to the storehouse of traditional metaphysics' notional entry points. But Derrida's reader has only to observe and itemize all that *différance* does with respect to the formal and systematic structuring of the history of metaphysics to appreciate that Derrida's denial of conceptual status to *differerance* is without explanatory merit. It does not matter how we style "*différance*" in terms of second-order vocabulary, i.e., whether it is named a "concept" or some other sortal designation. It also does not matter if we allow "*différance*" to remain unique in all respects–it is and remains a mark serving a designated function within Derrida's texts, although apparently it is inappropriate if not damaging to the integrity of "*différance*" as a vehicle of epochal import to interrogate "*différance*" as such in order to determine a discernible structure. This procedural inappropriateness may be acknowledged, but the crucial point regarding Derrida's project still stands. What *différance actualizes* within the theoretical confines of deconstruction policy statements is the crucial factor.[102] For if what *différance* produces throughout the history of metaphysics does indeed remain the same, then this commonality testifies to Derrida's implicit acceptance of sameness as, with respect to the formulation and application of deconstruction, an instance of self-identity. If so, then Derrida's sharp distinction between sameness and identity in this context no longer can be maintained since the sameness of *différance* must remain identical in a number of crucial respects in order that *différance*, through sameness, can generate what Derrida claims that it generates.

What evidence has Derrida provided that *différance* as he has formulated its structure and effects should–and, indeed, does–remain resolutely *the same as itself* throughout its engendering function with respect to the history of metaphysics as the subject of deconstructive agency when that same history is replete with differentiation, variation, and dissimilarity as far as its content is concerned? Derrida employs this "formal" approach to sameness, thereby presupposing its self-identity, in at least five distinct senses:

102 For development of this claim, note again the discussion of *différance* in Chapter 4 of DFL.

8.3.2.1 *Systemic Différance*

Différance as systemic is constituted by a set of distinguishable elements–e.g., sameness and difference. But elements of this plurality are related to one another as a unity named *différance*. This unity functions as one pivotal component in the application of deconstruction and, as such, remains the same throughout the history of metaphysics. As time, culture, political realities, circumstance, doctrinal specification vary throughout metaphysics' long history, *différance* remains self-identical with respect to elements of its structure, sameness and difference, the logical priority of these elements relative to one another, and its unvarying application to the variegated writings found in that history.

It does not matter what period in the history of metaphysics or what system during that period comes under scrutiny, the application of the gestures of deconstruction will result in, for example, the denial of both a transcendental signified as well as a stable sense of context. This consequence does not vary throughout the complete interplay between the history of metaphysics and deconstruction; therefore, with respect to substantive conclusions elicited from deconstructive principles, sameness prevails.

8.3.2.2 *Functional*

Sameness is the key element in *différance* and the grounding formal notion for establishing being as presence; therefore, presence stands as the one and only uniform element throughout the history of metaphysics. As such, presence illustrates another dimension of sameness, e.g., it extends over all possible dimensions of space-time insofar as space-time has been integrated within given metaphysical systems. Sameness also encompasses all explicitly asserted features in all the metaphysical systems which constitute the history of metaphysics. Sameness functions in a precisely self-identical way throughout this welter of complexity regarding the generation of a single outcome–being as presence–the dismantling and penetration of which is the unifying goal of deconstruction.

Any and all variations in metaphysical systems regarding the content of the structure of space and time remain subject to analytical entry into those variations provided by common components of deconstructive agency. These components, listed above under the "systemic" heading, always function in the same way throughout the history of metaphysics even if results of their application differ in content.

8.3.2.3 *Experiential*

In this context, the pertinent parallel is between sameness of being as presence and sameness linking perception and imagination–something must be common to those two distinct modes of consciousness in order to justify, in the application of deconstructive strategies, the ready transition from one mode to another mode with equal force in terms of evidentiary effect. Only if the two modes of experience are, in some way not indicated in Derrida's texts, the same will an appeal to both modes be equivalently relevant to the point concerning the indeterminacy of contexts.

The tangency between the elements typically separated by difference (i.e., perception and imagination) will generate a condition similar if not identical to the presence which, for Derrida, has permeated the entire history of western metaphysics. The limits of the perceptual adjoin the limits of the imaginative with an intimacy of such degree that experience and entities existing solely–or seemingly so–in the former realm inevitably and necessarily slide into a parallel level of existence in the latter realm. Furthermore, this experiential sameness remains identical to itself throughout the entire sequence of metaphysical systems. Imagination and perception continually intersect in order to ground the unvarying denial of contextual unity and stability of a given context. This denial obtains regardless of the complexity or simplicity of circumstances surrounding a given context and regardless when in the history of metaphysics or in the domain of common experience an inquiry might be initiated to secure such stability as a necessary condition for grounding the other elements. Derrida rejects this stability.

Sameness must also apply to time as an unstated factor in deconstruction when it is situated in the experiential phase of justifications in the critical analysis of concepts pertaining to language. (The following discussion supplements the account of temporality featured at the outset of this chapter.) When, for example, the unity of a concept is denied because perceived objects apparently delimiting a context can become imagined objects extending and removing those limits, then the same time must underlie and ground both the perceptual and imagined experiences. If the time underlying the experience of perception and the time correlatively grounding the experience of imagination were somehow distinct from one another, then it would not be possible to argue for the lack of limits of a concept in the manner just described since there would be no guarantee that perception and imagination shared a common spatio-temporal environment.

That this is a major metaphysical assumption about time as a resident sameness is demonstrated by attending to the temporality of an imagined scenario. The shifts and disruptions in tempo commonly characterizing an imagined state of affairs suggest strongly that time animating the sphere of imagination may operate very differently from time when the grounding experience is perception in its immediacy. Therefore, Derrida's position also requires ascription of sameness predicated of time; such predication is necessary to concluding the inherent indeterminacy of contexts.[103]

---

103 Derrida maintained early in *Writing and Difference* that the question of a revamped understanding of imagination was a prerequisite for developing "deconstructed" positions in relation to their encrusted counterparts within the history of metaphysics–WD, 4; see also WD, 7. Eva Brann's majestic *The World of the Imagination* may serve as a point of departure for investigations in this area. Brann's fundamental approach to the imagination is in several ways thoroughly traditional. She says in her

8.3.2.4 *Applicability*

Instances of a basic relation typically named "opposition" appear often throughout the history of metaphysics and Derrida's texts present various lists of such opposition as objects of interest and concern. Furthermore, deconstruction as applied frequently confronts these instances and reconfigures their relation in various ways.

Once distinct components of a given instance of opposition are related to one another *as opposites*, the difference between the two components duly established as that of opposition must remain the same throughout a given duration of historical time. Thus if matter and form (one example among many) differ from one another and are also construed as opposites, then this difference must be preserved as a difference even if the particular instances of matter and the particular instances of form vary over given segments of historical time. If this kind of difference would cease to remain the same, e.g., the positions of two philosophers both of whom appealing to matter and form in their metaphysics, then any strategy to subvert or overturn a dominant opposite by way of inference derived from a generalized sense of opposition becomes impossible for the simple reason that no such generalized opposition would endure and thus exhibit a mode of existence capable of becoming subjected to the application of deconstructive strategy.

In general, opposition as a unique referential indicator ceases to exist unless it encompasses a complex web of structural stability. The two components in question (i.e., as opposites) would perhaps be *related to one another* but not necessarily linked with the kind of strict directionality within a common area of reference justifying the claim that the two components are *opposites of one another*. Thus the difference between the two poles of every instance of opposition must remain the same; the relation between the respective weight of each dominant pole (if such dominance indeed obtains) with respect to its coordinate opposite must remain the same and do so throughout the history of metaphysics. If either of these two designations were to become reversed in order of opposition or to have this order fluctuate through history, then deconstruction would be unable to pinpoint a target for purposes of dissembling the dominant opposite (since, simply stated, it might not be the dominant opposite) or even initiate an analysis at all (since the relation of opposition might no longer be in play).

This single conceptual configuration, opposition, must necessarily retain its unity throughout the history of metaphysics as well as throughout the pressure

Preface: "There is an imagination; it is a faculty or a power; specifically it is a faculty for internal representation; these representations are image-like; therefore they share a certain character with external images; in particular, like material images, they represent absent objects as present; they do so by means of resemblance." Brann then adds that although "these articles may seem pretty obvious," each of them "is hotly contested;" furthermore, "the debate is often of great intellectual interest and sometimes of fundamental human significance," 5. Her treatments of consciousness, 213-7, and "the space of imaging," 581-602, are clear and bold in terms of speculative possibility. For additional discussion of the imagination with regard to Derrida's projects, see DFL, 113-6, 121-35, 179-81.

placed on that history by deconstruction in its quest to reorient, or at least to throw into question, the direction of privilege traditionally stamped on the given instance of opposition.

In summary: To the extent that deconstruction aims its instruments of analysis toward assumptions encrusted within a highly diversified array of particular instances of opposition in order to achieve critical and credible results, to that extent deconstruction depends on the unvarying reliability and sameness of opposition construed according to a strictly defined set of abstract conditions. Opposition in this sense is a metaphysical primitive which must be taken as the same throughout all its appearances, regardless of the subject matter deployed oppositionally. It will be observed that this sameness also extends to all substantive conclusions derived from the application of deconstructive strategies.

8.3.2.5 *Modality*

Deconstruction depends on the implicit juxtaposition of a kind of possibility which becomes, given Derrida's approach to being as presence, indistinguishable from totality as actualized. Thus the omnipresence of presence with respect to difference, that is, the effective denial of difference when applying deconstructive strategies procedurally nullifies any distinction between possibility and actuality with respect to any single occurrence of written or spoken language. As a result, the deconstructionist can appeal to *anything* that has, is, or will happen linguistically in order to dislodge the intent to impose limits on any of the traditionally stable concepts in the analysis of language. All sectors of possibility remain the same throughout the application of deconstructive agendas in the sense that anything that has or will occur with respect to linguistic usage is always a legitimate source of evidence to justify the dismantling of an apparently established core of meaning.

8.3.3 Difference

As shown at the outset of this chapter, the advocate of deconstruction must deny that traditional and seemingly unassailable differences obtain within each of three dimensions of this totality–the difference between and among distinct sectors of time, the difference between perception and imagination with respect to the existence of contexts, the difference between referential entities and meanings derived from a given context and related but distinct referential entities and meanings derived from other possible contexts. This variegated interplay of denials results in a kind of Heraclitean indeterminacy, a veritable maelstrom of undifferentiated unities on which in their collective totality Derrida's various criticisms of standard theoretical apprehensions of linguistic concepts depend.

There are, however, various senses in which *difference as such* must stand by itself as a formal requirement in order to establish fundamental features of the deconstructive agenda. It is therefore essential to detach difference from the connection Derrida assigns to it as a feature of *différance* and to recognize that difference exercises a series of crucial roles apart from its function within the theoretically determined confines of *différance*. We discuss three such roles.

8.3.3.1 *Textuality*

Derrida frequently uses "marks" as a preferred term referring to basic elements of language, e.g., to words whether taken individually or as components in sentences and propositions. Thus the differences between and among all the letters comprising an alphabet, the differences between and among all the individual words comprising a given work (and of course a given passage excerpted from that work for deconstructive consideration), and the fact that the totality of, e.g., Plato's texts differ from the totality of Aristotle's texts with respect to the "marks" constituting both sets of readings–all these brute empirical characteristics illustrate, irreducibly so, the fact of difference. Even if the ultimate philosophical import of both sets of texts (using Plato and Aristotle as examples) is ultimately governed by *différance* and subsumption of all given textual differences under sameness, the statement of this import derives from an extensive array of words and propositions requiring as an empirical necessity the fact of difference in order that this language exists according to traditional formats thereby becoming open to deconstructive interest.

The Platonic Socrates says about written words that "they go on telling you just the same thing forever" (*Phaedrus*, 275d–Hackforth's translation). The sameness to which Socrates refers depends in part on the fact that the words in question, as written, will always remain the same orthographic units, "marks" in Derrida's vocabulary. Even if one wanted to contend that exactly the same words when used as elements in texts can admit divergent readings, these readings will derive from the same set of original words; the divergences in question are produced by readers' reactions to those words, not by words as orthographic units themselves. In this respect, all words upon entering into existence within a given text and being accepted as such by scholarship remain the same forever. The factor of difference taken in a purely formal sense is necessary to establish the possibility that language as an articulated phenomenon exists at all, since a passage from a Platonic dialogue (or, obviously, from any text) will always be determined as that passage only through the same set of primitive, empirically defined, textual differences.

8.3.3.2 *Interpretation*

Consider a text, e.g., a Platonic dialogue. It is a hermeneutical truism that one and the same text can generate many divergent readings. But all readings of the same text as divergent presuppose identity of the original text. If that text were to vary over and above its orthographic sameness, then it would be premature to assign difference to distinct readings since such divergent results need not in fact be divergent if each reading was based on what amounted to a different text. This is an obvious fact about the structure of texts but it is worth mentioning given Derrida's frequent appeal to words appearing in textual situations "already" and continually "splitting" or "dividing" themselves in a variety of ways, especially with respect to the possibility of establishing a fixed meaning in a given context.

It would be possible at least theoretically to interpret diverse texts of any two philosophers in such a way that the respective meanings of these texts, once conjoined, would be the same. But such interpretation would have to go far regarding the extensive ranges of traditional meanings ascribed to the different texts in order to transform those meanings into a unified message. The possibility of such compressed if arguably reductive interpretation raises the question whether or not deconstruction will allow a difference between justified and unjustified readings of a given text. If "anything goes" is not a legitimate maxim of interpretation according to canonic deconstruction, then it seems necessary that at least some readings of texts offered in the name of deconstruction must be disallowed.[104] It follows then that there is difference separating a justified from an unjustified deconstructive reading of a given text. How precisely the distinction between justified and unjustified will be drawn in this context is irrelevant; the point is the necessity of contrast between the two possibilities. And it is this contrast which at a fundamental level of formal intelligibility "difference" establishes and preserves. We will encounter in Chapter 9 a striking example of irreducible difference animating a strictly interpretive context; Derrida's participation in this exercise in difference will have far-reaching implications for the ultimate cogency of deconstruction.

8.3.3.3 *Formality*

Derrida claims with aphoristic force that those who have not recognized the priority of *différance* over difference have been "left far behind" (P, 101n13). The point is that in terms of ultimate philosophical priority, *différance* for Derrida is prior to difference, not vice versa. But against this thesis, we note: (a) difference as such must remain self-identical and impregnably so, i.e., difference itself cannot become subject to difference, especially within the confines of *différance*. For if difference were to metamorphose into anything other than itself, then the predication of difference to all distinct elements of a given metaphysical system would not be

104 See LI, 144-5 on the "anything goes" possibility for applied deconstruction. For a discussion of limits with respect to the application of deconstructive gestures and resulting interpretations of texts, see DFL, 165-6.

possible. The result would be the loss of the unique character of that system in terms of (i) the relation of its elements with respect to each other within that system as well as (ii) the difference between that system and all other systems, since none of the other systems would also be able to preserve their internal structure in such a way as to maintain their own individual identity.

It is necessary then that (b) difference function as a logical primitive in order to preserve a distinction between, e.g., the Platonic and Aristotelian views of reality, or any other view put forth in the history of metaphysical speculation. Only when this difference obtains with respect to the content of the history of metaphysics in its entirety can Derrida formulate *différance* in order to generate by this notion a theoretical consequence regarding that history–i.e., being governed in its entirety by being as presence.

As a purely formal principle of regulation, difference as such is necessary in order to distinguish between identity and sameness within the structural limits of *différance*. Derrida can postulate this distinction as applicable to *différance* as a single, unified operator with a specific function in relation to the history of metaphysics only because he could assume that difference as such is free to penetrate a conceptual configuration of his own design and to establish "spacings" between and among its elements. Thus, the very possibility of *différance* explicitly stated as a key notional component in deconstructive strategies depends on difference, not the other way around as Derrida would have it.

8.4 The Clash of Isomorphic Totalities

For Derrida, being as presence epitomizes an epoch in western thought. It may fairly be said, however, that its primary function in the overall context of Derrida's work is more strategically negative than affirmative in that it forces attention on what has been overlooked or suppressed during that epoch. By contrast, deconstruction is mainly affirmative if viewed as an ensemble of gestures for realizing and perhaps even articulating what being as presence has concealed.

If, however, we examine more closely the just-concluded exposition of components involved in the confrontational play linking being as presence with deconstruction, a consequence of considerable weight emerges. The totality dimension driving being as presence rests on the sameness, i.e., the unifying core of *différance*, which justifies predication of the one name, "being as presence." Although Derrida's notion of *différance* explicitly contains an application of difference, that is, between and among distinct metaphysical systems within the entire history of metaphysics, this is a dimension of difference with, given the sweep of history, what might be termed inherently vanishing effect. For as the movement toward the totality of the history of metaphysics–and, for Derrida, ultimate closure–commences and continues, all differences gradually become reduced so that when closure is in fact realized, difference has become attenuated to the point where its function as a source of distinction between and among entities as well as between and among experiences of entities is rendered null with respect to potential insight into metaphysical considerations. Derrida did indeed insert as a structural feature of being as presence the possibility of degrees of presence (a claim not developed). But there is an undeniable sense that the history of metaphysics insofar as it culminates in a closed and apparently exhaustive process of systematic thought just *is* "being as presence." Sameness and the unitary functioning of being as presence triumph over difference.

As soon as Derrida announces that the history of metaphysics must be read as marked by closure, being as presence–the single name for the entire history of metaphysics–will, it seems, affect if not determine the ultimate revelatory capacity of every moment within that history. With the advent and work of Jacques Derrida (1930–2004) and after closure has rounded out the history of metaphysics, the diversified strategies proper to deconstruction have appeared, but not without their statement arising on the basis of serious and sustained foundational activity on the part of its originator. Time in all its dimensions, all apparently limited contexts extended into all possible dimensions, meanings and referents extended in all possible dimensions: we demonstrated at the beginning of this chapter that these three totality-features are and must be in place as a set of necessary conditions in order to ground perspectives on additional metaphysical formations in such a way that Derrida can maintain the conclusions he wishes to draw on seminal topics in his treatment of language–transcendental signified, contexts, negation and iterability.

Furthermore, it is evident upon examination that realities underlying the theoretical configurations characterizing the basic gestures of deconstruction are classic instances of elements found throughout the history of metaphysics–unity, sameness, difference. It follows immediately that if Derrida's deconstructive reading of that history is correct, all these realities are permeated with being as presence. The reason: being as presence pertains necessarily to the entire history of metaphysics and therefore to every constituent element defining that history–including, of course, unity, sameness and difference, however they may be or have been configured when inscribed within a given metaphysical system.

This ontological fact, so to speak, has crucial consequences. For it may be inferred that the totality Derrida closes off when he reads the history of metaphysics and pronounces upon its deep structure with respect to being as presence *is isomorphic with* the totality he implicitly depends on when he discourses analytically and normatively on the structure of language by appealing to dimensions of unmistakable metaphysical import, in particular dimensions characterized by totality-considerations. In other words, in order to establish the possibility of the critique of meaning and signification in his analysis of language, Derrida must have in place–albeit deployed according to varying senses of implicitness–a structure fully metaphysical at its base and in all necessary respects isomorphic with that attributed to the full sway of the history of metaphysics emergent in and through being as presence. However, and the point bears repeating: the concepts standing as the foundation for this structure are, if Derrida is correct and his position is consistent in this regard, thoroughly impregnated with being as presence.

The result is a single isomorphically structured totality marked by the clash of diametrically opposite consequences: (a) one closed system with respect to the articulation of the structure of being throughout the history of western metaphysics interlocking with (b) an assemblage of viable concepts and realities necessarily continuing and open-ended in its applicability–in a word, deconstruction. This assemblage is collectively aimed at analyzing the structure of language as well as texts written in light of that structure and, Derrida assures us, providing precisely the means to disclose what the language of metaphysics has epochally and pervasively concealed.

The fundamental question arising from this clash of opposite destinations concerns the inferences which follow given the inescapable fact that although opposed in the direction of their effects, both destinations–being as presence and deconstruction as a methodology–result from appeals to and derivations from the same basic realities. Why should the array of Derrida's deconstructive criticisms be accepted if the vehicles by which these criticisms are asserted are themselves exactly what Derrida's position is intended to throw into question? To use Derrida's own language, it appears imperative that the foundations of deconstruction should be "shaken" with just as much energy and rigor as those foundations animating the history of metaphysics: the reason–they are, in fact, the same foundations!

8.5 Self-reference: Deconstruction and Interpretive Limits

As we have just shown, unity, sameness and difference are necessary to the very formulation of the elements of deconstruction. If so, then this functionality must be recognized and the elements so named must be extracted and considered on their own, formal factors essential to the utterance and expression of deconstruction as a methodological gambit. But this identification and subsequent isolation establishes unity, sameness and difference as yet another appearance embedded in a seemingly radical if not "monstrous" (Derrida's descriptor) setting of classic instances of abstract concepts and realities. Therefore, given the narratives advancing the gestural content of deconstruction–a configuration of texts spread out over a number of years in the history of Derrida as a writer (not to mention a variety of attempts by others to follow Derrida's lead)–the advocate of deconstruction must fully appreciate how the details of deconstructive arguments exemplify difference, sameness and ultimately the predication of unity. Another consequence follows, with even graver import. The advocate of deconstruction must also and necessarily determine and justify how such abstractions are to be understood by themselves, as isolated phenomena, since these abstractions structure the formation of Derrida's discourse and deconstruction generally, however pliant and variable and resistant to the blandishments of metaphysical concepts its presentation at the level of theory may appear to readers' eyes.

Consider *différance*. If Derrida depends on *différance* as a necessary factor in the existence and effectiveness of deconstruction and if *différance* itself possesses a unified function, then the question becomes whether the requisite formal characteristics of *différance* can be–and should be–abstracted from *différance* and then be made subject to inquiry in their own right. The following argument shows that the answer is in the affirmative. It is necessary that *différance* as described in Derrida's texts *functions in the same way* throughout the epoch of western metaphysics. In a word, *différance* as such, taken as a self-contained theoretical unity, exhibits sameness; indeed, without sameness as one of its formal properties guaranteeing uniformity of function, *différance* would not be *différance*.

It follows that Derrida has imported a certain understanding of sameness with respect to formal characteristics–thus sameness must retain its structure over vast amounts of time encompassing extensive changes in cultural and political modalities. As a result, sameness must exhibit self-identity, it must remain identical to itself. The sameness of *différance* also retains its structure when its objects–systems comprising the history of metaphysics–generate considerable diversity. These implications elicit and accentuate the self-identity of sameness as an integral component of deconstruction. If this claim is granted, the following question seems unavoidable: must all the available devices of deconstruction also and necessarily be directed at evaluating sameness, a formal property of *différance*, given that this sense of sameness–an exemplification of self-identity which, as a formal reality, is capable of extraction from that to which it can be applied–stands as a necessary condition for the very existence of deconstruction in the first place?

The answer is yes. In fact, all these theoretical approaches to and borrowings from unity, sameness and difference–however they might be ultimately formulated in their foundational function for the emergence and stability of deconstruction–*should themselves all become the target of full-fledged deconstructive investigation*. The reason: the foundational status of these three components locates them both individually and as a set squarely within the history of metaphysics. And since it is the primary purpose of deconstruction to break down the hegemony of realities of this sort insofar as they have stood as the dominant features of western metaphysics, there is no apparent reason why deconstruction should not also be aimed at the foundational elements which underlie the very possibility that deconstruction itself can emerge as something of putative theoretical value and as an aggregation of techniques evocable in language.

Important implications follow either way–whether deconstruction does or does not investigate these realities. If deconstruction remains silent in this respect, then it appears impossible to avoid the implication that there are privileged senses of unity, sameness and difference; in its silence when facing these realities, deconstruction has run up against the limits of its own effectiveness. If, by contrast, deconstruction does address the status of its own foundational notions, then the successful disassembling force of deconstructive inquiry will obviate the very possibility that deconstruction as a set of interpretive approaches embodying sameness and difference can even be formulated and expressed in language, much less applied to extant philosophical and literary texts. The very attempt to state anything of a deconstructive cast, whether theoretical or applied, would collapse upon its own self-destructive import.

These implications are crucial in critically addressing the cogency of deconstruction if it is taken, as some on the current scene certainly do, to be a viable approach to pursuing the philosophical enterprise. But given the concerns detailed above, how viable is that approach? At one point, Derrida expressly admitted that deconstruction can be directed against itself, so it is clear that Derrida was aware of this dimension of his own position.[105] What is perhaps not clear is the extent to which an articulated set of self-referential implications bear on the very possibility of deconstruction itself.

8.6 The Integrity of Deconstruction

We have argued that metaphysical considerations pertinent to deconstruction are discernible when deconstruction is advanced as a coherent theoretical posture characterized by a determinate conceptual scheme. If this interpretive approach is feasible, then metaphysics is presupposed in two senses: first, as the source of an implicit series of embedded and interlocking positions and assumptions necessary to justify the deconstructed critical evaluation of traditionally defined and developed linguistic concepts (Chapters 4-7 above); second, as a conceptual field yielding a range of concepts necessary to develop a concatenation of gestures standing as methodology for a mode of reading which intends to unearth the concealed import of philosophical and literary texts (shown earlier in Chapter 8). These revealments serve as prerequisite for establishing a "monstrous" retelling of basic and traditional metaphysical and philosophical principles.

The following are facts about deconstruction as it emerges from Derrida's texts:

a the articulation and discursive rhythm of its gestures depend on general terms;
b these terms are not explicitly identified as functional in this respect;
c these terms are not themselves subjected to deconstructive treatment.

The fact that Derrida's texts leave the status of general terms in a state of unanalyzed givenness opens his overall position to criticism on several fundamental

105 Derrida admits that texts advancing deconstruction as a program of analysis can themselves be treated according to these specifications–see Kearney, 155-6; Derrida 1992, 46, see also in this regard LI, 141. For discussion, see Harvey, 43, 96-7; Jürgen Habermas 1987, 185; Berman, 24; Hobson, 22, 24; Naas 2003, 28, 77; Norris 2012, 82; Daylight, 50, 60. None of these secondary sources pursue to any extent implications of this self-referential contextualization.

levels, with decisive influence on the cogency of deconstruction both as theoretically formulated as well as in its practical application to existing texts.

Indeed, this position generates a pair of inconsistencies identified below as "internal" and "external." It may be observed that such opposition, so bluntly posed, stands as a ripe target for deconstructive discussion and dismemberment–and immediate rejection on grounds of programmatic irrelevancy. The challenge therefore is determining whether deconstruction can explain away this opposition (and attendant implications detailed below) without succumbing to the fallacy of begging the question, for the point of this internal/external rubric is precisely to demonstrate that deconstruction cannot coherently even initiate much less execute a response to such a challenge.

8.6.1 Internal Inconsistency

Deconstruction relies on general terms for both its theoretical formulation and practical application to texts. However, the primary stated purpose of deconstruction as a mode of philosophizing includes disassembling such general terms–the dominant conceptual "givens" that have contoured and animated the history of metaphysics, presumably wherever these terms exist in the canonic texts of metaphysics and however philosophers thought to deploy them. Thus deconstruction is unstable regarding the lack of explicit recognition of its own reliance on general terms; the statement of deconstruction in Derrida's texts lacks an appropriately justified theoretical base and therefore that statement as itself a textual given is incompatible with the stated goals of deconstructive analysis.

Furthermore, once these general terms have been explicitly brought into view, deconstruction becomes internally inconsistent, a much more potent objection than the initial condition of instability just posed. Let deconstruction do its work on unity, sameness, difference. If these terms are in fact analyzed and deconstructively shaken and exploded in accordance with the stated purpose of deconstruction, then deconstruction as explicitly formulated destroys itself even as a mere possibility, as an aggregation of terms potentially applicable to texts (regardless of authorship) and with a determinate significance for a contemporary audience. Deconstruction implodes by virtue of its own self-referential rigor. But if deconstruction indeed breaks down the very realities which allow deconstruction to come into existence as a theoretical stance–if deconstruction is self-referential–then it seems that deconstruction necessarily and virulently sows the seeds of its own destruction, at least if deconstruction is consistent and complete in its application to the subject matter the hegemony of which has ultimately engendered the need for deconstruction in the first place. Every attempt to formulate, that is, to put into words and propositions Derrida's thoughts on how to read texts, disintegrates and vanishes into a string of logical incompatibilities while in the very act of postulating those formulations.

In general then, the extent to which deconstruction as theoretical excursus depends on concepts of high generality–unity, sameness, difference–is the extent to which deconstruction becomes either inconsistent, since its very utterance as a theory of interpretation embodies dimensions of these three realities which individually and collectively must remain identical in structure and therefore impervious to any critical reflection or, at best, unstable and incomplete if its development as a theory does not address the specific character of these generalities.

8.6.2 External Inconsistency

The problems just identified pertain to deconstruction within its own ambit as an aggregation of theoretical pronouncements. But deconstruction so formulated does not appear in Derrida purely in its own right, as a self-contained vehicle for reading and interpreting the "meaning" or content of anything literary appearing within that tradition. Deconstruction in fact emerges from a much broader conceptual context. This context comprises the juxtaposition of (a) being as presence emergent as a derivation from the history of metaphysics and (b) the ensemble of abstract realities–unity, sameness, difference–drawn from that very same history and also necessary as foundational elements in order to articulate deconstruction as a theoretical gesture.

The purpose of being as presence within Derrida's thought is not only to stand as a single, all-embracing characteristic (with various modes of application and individual systematic differences) accurately describing the entire western tradition of metaphysical reflection but also to suggest seminally that something significant has been missing, overlooked or distorted throughout this history, even up to the point where Derrida's texts complete the closure of that history. However, the array of concepts functioning as the foundation of deconstruction as an articulated position is itself drawn from the history of metaphysics. The realities referred to by these concepts are therefore necessarily affected by being as presence since it is axiomatic for Derrida that being as presence pertains to the entire history of metaphysics.

The concerted attention to being as presence as a structured response to the history of metaphysics must therefore be taken in conjunction with the tacit dependence on a series of general terms, all cardinal concepts instantiating being as presence. But it now becomes evident that this confluence produces an arresting methodological tension, indeed a tension which upon analysis becomes logical incompatibility of massive proportion. Derrida cannot consistently hold that the history of metaphysics is inherently repressive and then extract and exploit elements derived from precisely that repressiveness in order to justify through conceptual configuration a Heraclitean skepticism issuing from his philosophical approach to the structure of language. For if being as presence, although necessary to the history of metaphysics, encapsulates a closed and therefore in some respects a deficient system, how can Derrida appeal to basic characteristics of that very system in order to ground tacitly the pivotal stances he takes on the structure of meaning and reference in the sphere of language? Derrida's texts concerning the fundamental elements of deconstruction rest on the same kind of foundational substructure supporting the advancement of his own positions regarding language as the targeted and traditional foundationalism permeating the history of metaphysics which these very positions purport to deconstruct. But it appears logically incompatible (a) to affirm that a certain commonality proper to the nature of being once identified and brought into the open should be overcome and then (b) to embrace, even if implicitly, this same commonality in order to justify a related series of skeptical conclusions concerning the structure of language.

Bluntly stated: Derrida critically evaluates the history of metaphysics in its entirety, then uses aspects of that very history for his own substantive ends with respect to formulating positions concerning the structure–or, perhaps, the inherent lack of structure–of language itself. Can Derrida have it both ways?

Until this fundamental incompatibility as well as the other issues raised in this chapter are addressed and resolved, I maintain that the edifice of deconstruction rests on an unstable foundation at best; consequently, any results tendered in its name become commensurately suspect if these results are said to be inspired by deconstructionist gestures rather than through independent inquiry (with an implicit or indeed explicit reliance on a set of in many respects traditional metaphysical and logical concepts and principles).

8.6.3 Being as Presence and Deconstruction: The Question of Priority

The logical proportionality of the two problems just presented warrants a brief comment. Even if the first, internal inconsistency could be resolved, the second, external inconsistency remains in full force, with an implication which affects the fundamental cogency of Derrida's entire project. Assume that (a) the required formal terms necessary to formulate deconstruction as a coherent ensemble of interpretive gestures are identified and then (b) subjected to deconstructive examination. If this program of analysis can establish the import of unity, sameness and difference in such a way that each of these terms retains sufficient structural integrity to display a measure of formal stability, the implication would be that deconstruction, directing itself against its own inherent formal characteristics, would not necessarily become self-destructive in this very process.

True enough, this kind of result seems to run counter to the general tenor of what deconstruction does, since the application of whatever devices of deconstruction are deemed pertinent in this matter would not only establish stability but would in fact be mandated to seek such stability as a necessary condition for deconstruction's very existence in language as well as for the maintenance of that existence throughout its duration as a viable philosophical vehicle of inquiry. From this perspective then, deconstruction requires foundationalism in order to be in a position to undercut foundationalism. However, that reservation to one side, it remains the case that the joint appeals to revitalized and restated senses of unity, sameness and difference would continue under the aegis of being as presence–unless, of course, it is possible to reach a realm of insight and language for the articulation of these insights which can function in the requisite senses somehow apart from being as presence and its fluidly ongoing epochal legacy.

A final thought in this context: In a typically taut, epigrammatic fragment, Heraclitus is credited with writing that "the path up and down is one and the same."[106] Deconstruction preserves this spatial contrast but with divergent results–break down the sacrosanct concepts and realities that have controlled the history of metaphysics in order to open up what has been concealed or distorted throughout that history by the uncompromising sameness of being as presence. But this contrast is only apparent (as, indeed, it is for Heraclitus), since embarking on the way down eventually leads the same practitioner of reflection to the way up–until, if the seeker continues the cyclical journey with sufficient energy and perseverance, the way up ineluctably slides into, yet again, the way down....

The way down for the deconstructionist penetrates into the depths of metaphysics but does so through gestures and techniques resting atop a series of fundamental realities; as a result, going deconstructively "down" into the concealed core of metaphysics only leads the explorer back "up" again into the region where the same fundamental realities appear out in the open in the sense that the realities implicitly present during the journey down now have emerged as they have been explicitly thematized by the epochal metaphysicians of the past. Thus Heraclitus' "one and the same" names the uniform content of the pertinent principles marking *both* deconstruction itself *as well as* deconstruction's intended object. The circular motion marked in Fragment 60 may appear repetitively uninformative but if that motion is situated hermeneutically along the most fundamental level of deconstructive activity and the foundational realities presupposed as necessary for that activity, this fragment becomes the source of an illuminating perspective on the necessarily interpenetrating structure of these two discursive phenomena.

8.7 Deconstruction, Presence and the Play of Consequences

The pair of inconsistences described above emerge from a contrasting internal/external interpretive base. The implications drawn raise serious questions about the cogency of deconstruction both by itself and, in an even more pronounced sense, when the

106 Heraclitus, Fragment 60. See Kirk and Raven 1957, 189.

canons of deconstruction are conjoined with Derrida's insistence that we read the history of metaphysics as theme and variations on being as presence. But additional problems arise if we consider separately the elements serving as the primary focal points throughout the critical treatment of Derrida advanced in this essay and apply these perspectives to contexts juxtaposing deconstruction and being as presence.

8.7.1 The Ontologizing of Language

We argued in Chapter 4 that the ontologizing of language rests on a joint denial of difference and overextension of sameness in the relation between language as such and extra-linguistic domains. Even if stringently qualified, the dictum "There is nothing outside the text" leads inevitably to many intractable difficulties. Thus it seems evident that if the import Derrida assigns to language, especially writing, were implemented within the contexts of practicality defining the ordinary lives of people, indeed even those who spend time philosophizing about reality and language, the upshot from all involved in coping with the implications from this principle would be vigorous testimony calling attention to Derrida's excessive theoretical zeal. A passage from the conclusion of the *Cratylus* reinforces this concern: "Whether there is this eternal nature in things, or whether the truth is what Heraclitus and his followers and many others say, is a question hard to determine, and no man of sense will like to put himself or the education of his mind in the power of names" (*Cratylus*, 440c–Jowett's translation). Most if not all advocates of Derrida would quickly reject the initial moment in this thesis–that there is an "eternal nature in things"–redolent with foundationalist conviction as this metaphysical pronouncement appears to be. However, the passage also maintains that the other alternative derived from Heraclitus is equally difficult to substantiate if, as Socrates claims, we put the education of our minds "in the power of names," that is, names understood as all the words used to identify things in our experience of these things, whether that experience be mundane in its practical scope or grandly philosophical in purpose.

Derrida was quoted as admiring the metaphysical sayings of Heraclitus.[107]If, however, connections between words and things are as intimate as Derrida would make them, then the argument of the *Cratylus* gives reasons to question whether, as just noted, either Heraclitus or Plato could justify a systematic metaphysics if the only evidence was based on the "power of names." If a fundamental difference is granted between language and whatever language can refer to, then whatever measure of power the philosopher ascribes to language in relation to extra-linguistic reality will

107 David Wood 1988, xi.

always be marked by distance of some sort between the two sectors of existence. Many of Derrida's positions–e.g., on the continually shifting boundaries of contexts and similarly mobile identities of supposed "transcendental signifieds"–certainly lend themselves to a Heraclitean worldview, assuming that anyone philosophizing according to those core positions would want to make a theoretical (and, of course, historically conditioned) pronouncement in that regard. But if metaphysics pitched in such global terms is set aside, we may repeat the point that even if we grant the stipulated sense of the word "writing" as Derrida argued that sense in *Of Grammatology*, it remains unclear what has been gained by this move with respect to winning insight into, speaking in a broadly generalized way, the reality of the objects so signified.

What is clear are consequences derived from the denial of unity and stability when this denial is considered with respect to the import of *différance* and the status of a context. For Derrida, the unity of a word with divergent meanings justifies the conclusion that contexts as a certain kind of reality cannot display an equivalent unity. It might be thought that divergently constituted contexts relative to the use of the same word in both contexts entail that such words exhibit divergent meanings–that is, that a word of this sort should be receptive to this divergence because the difference in context will, given the complex metaphysical surroundings which define any context, mandate a difference in meaning. But Derrida reverses the priority–terminological unity controls (in fact, denies) contextual unity in any and every case (with exceptions indicated by Derrida–see Chapter 9), including the resultant stability such uniformity would establish for determining fixed reference and meaning. Thus, the evolution of a word in terms of a sequence of diversified meaning throughout that word's history dictates what can and cannot be relevant and real as far as contexts within which that word may appear. For Derrida, the "power of names" is of such proportion that it overwhelms and negates all the affects the elements characterizing a given context have in determining the meaning of words–and without some determinate measure of stability in meaning, it is not obvious how words can exercise any semblance of "power" over anything extra-linguistic. As we have demonstrated, such a reversal has serious and arguably unsustainable consequences.

Furthermore, Derrida's denial of the *moment* as a feasible element in metaphysical thinking depends in part on the ontologizing of language–he wants to move away from the present, the moment, insofar as the present serves as focal point, if not the paradigmatic constituent of being as presence. There will be no "moment" as a matter of metaphysical fact if an occurrence of language, whether spoken or written and ostensibly taking place "right now," must be deconstructively linked to all other occurrences, past and future, in order to ensure the possibility that the given occurrence can become and is meaningful.[108] Is there a way to save the conceptual integrity of "the

---

108 Derrida denies in various contexts the philosophical viability of the notion of "the moment": see WD, 14, 263; MP, 58; OG, 159; SP, 61, 62-3; PS, 6; N, 23. See also DFL, 250n9.

moment"–and thereby insert a wedge against Derrida's ontologizing of language–so that it receives due attention from the perspectives of both metaphysics as such and the relevance of the moment to language? I argue in Chapter 9 that the answer is in the affirmative, a response Derrida has invited in part by reason of his vigorous if not vitriolic response to John Searle in the matter of reading and interpreting John Austin's theoretical treatment of speech acts.

8.7.2 Particularity, Signification, Concepts

The collective force of *différance* as a gathering and summation of statements defining the history of metaphysics, working in tandem with being as presence as a phenomenon circumscribing that entire history, renders it theoretically impossible (on pain of inconsistency) for Derrida to propound a substantive account of particularity. Furthermore, the same problem arises when deconstruction and its emphasis on flux and change denies the existence of a stable transcendental signified. Note the correlative consequences for Derrida's position: (a) the inability to enunciate a fixed stance concerning the metaphysical character of individual entities mirrors (b) the denial that a stable transcendental signified can be secured through philosophical means. It follows from Derrida's promulgation and adherence to deconstruction that neither a particular as such nor a cognitive correlate to a particular can be recognized and articulated as components of a philosophical system.

Consider the reversed direction of movement characterizing this implication–the inability to formulate particularity from within the theoretical domain defined by *différance* as well as the omnipresence of being as presence moves from (a) the totality, the history of metaphysics, to (b) a given part of that totality–i.e., the traditional interest metaphysics has in stating properties of particulars. By contrast, the denial of the transcendental signified originates from within experience and extends, necessarily for Derrida, outward to the limits of experience as contoured by perception and imagination working in concert with one another, both avenues of experience collaborating to support the rejection of a limited cognitive object which can stand as transcendentally stable. Although the directions driving the implication are reversed, the point to be made is identical–either way, Derrida loses even the possibility of any theoretical hold on particularity, whether as such (in the manner of metaphysics in its standard practice when dealing with the theoretical structure of extra-mental entities) or as the object of linguistic and cognitive concern (in the guise of what Derrida has designated as the transcendental signified).

The fundamental gap in Derrida's metaphysics of individual entities parallels the imposed insistence that particularity cannot be secured as it plays out, or would have played out, in linguistic contexts regarding the conjunction between referential object and its counterpart as transcendental signified. The implications are considerable. This joint instability where neither obtain, i.e., particularity in extra-linguistic regions or the existence of a transcendental signified in the cognitive realm, renders it difficult if not impossible to envision theoretically the formation of concepts. If it is a characteristic of concepts to "pick out" certain features of intended objects (the content of these objects understood very broadly) and then to "hold them together" for purposes of examination and utility, how can such cognitional stability be consistently approached and deployed given that the conditions for achieving such stability cannot coherently coexist with the deconstructive drive for investigative variableness and indexical flow?

Stating facts about the world depends on language and its ability to connect via extra-linguistic media to specific referential realities, then move in the opposite direction, as it were, in order to initiate a conceptual response to that referentiality. Words framed as propositions go "out" to refer to determinate extra-linguistic reality, then this entire relation goes "in" and funds a pertinent cognitive evocation and realized identity. Derrida's deconstruction denies both components; how then can concepts be formed?

If the pursuit of philosophy requires concepts, then a critical evaluation of this effective denial that concepts can be fashioned and function in discourse is to ask: to what extent does unity of an entity as object of a potential fact depend on unity of a word or words referring to or describing that entity? Whether an entity will bifurcate itself either from within or due to external provocation does not seem to involve anything regarding language we employ when in the presence of this entity. The *Cratylus* is again pertinent–whatever entities do with respect to their own structure is independent of what words referring to those entities may do, especially given the unbounded vagaries of discourse. The deconstructionist must show that putative dimensions of difference within words when referring to entities somehow reflect diversifications proper to the entities themselves. In brief, the advocate of Derrida must do metaphysics. It remains a question whether undeveloped reliance on gestures taken from deconstruction alone is sufficient to accomplish this end. In the face of such pursuit, philosophy would become either a very different type of activity from what it has been historically or, at the other extreme, it would dissolve its apparent possibilities aimed at explaining realities and issues in the world into expressions of discourse approximating what is now named rhetoric (on which more below).

8.7.3 Opposition

Heraclitus and Derrida both appeal in different yet commensurately basic ways to opposition but it is perhaps worth wondering whether either thinker reflected on the structure of opposition as such–for in its traditional formulations, the concept of opposition is inherently complex, presupposing as elements the secondary notions of difference, plurality, generic identity and relationality. And yet opposition taken as a concept also exhibits underlying unity in several distinct senses–unity of each of the opposites considered as such as well as the *relation between* the two opposites which must be ordered to preserve any emphasis in importance of interest emerging from either opposite relative to the other. Opposition as a uniformly ordered concept displays a palpable sameness of function establishing it as a bedrock organizational instrument for millennia of philosophical (and, indeed, non-philosophical) thought. But as such, opposition must function structurally with unvarying stability; if it does not, then as noted earlier any consequence deconstruction might infer from a given instance of opposition cannot be asserted as problematic since this apparent difficulty derives from an intended structure lacking sufficient weight and integrity to remain standing long enough to engender such elements of thoughtful and critical response.

Deconstruction assumes that opposition does indeed exhibit this kind of structural regularity and stability. The uniform approach deconstruction takes to opposition renders this conceptual configuration to be a gambit of singular penetration, a locus of especially fertile interpretive possibilities. Indeed, the advocate of deconstruction may appeal to substantive and decidedly variegated results generated by its application to instances of opposition–e.g., Derrida's discoveries as described in *Of Grammatology* of realms of latent significance in Rousseau's texts. One might object, however, that commensurately "jumbled" results can be achieved in assessing contexts defined by opposition without resorting to deconstruction in order to achieve these results. But caution is necessary in pursuing these quests.

Concepts readily deployed along a span of opposition often serve as proximate source of one of the more common logical mistakes. Almost any comprehensive logic textbook will discuss informal fallacies bearing such names as "black and white thinking" with the implication that a careful student of phenomena will appreciate that reality as such, i.e., the complete continuum underlying conceptual end points characterized formally as "black" and "white," contains much that is "grey," existing between the extremes set by this kind of starkly drawn opposition. Assume that one of deconstruction's cardinal strategies is to reverse the privileged opposite and then to reinvestigate issues governed by this reformulated instance of opposition. It is not clear whether a thorough reexamination of a given instance of opposition would not, at some point in the survey of what the pair of opposed concepts had been taken to mean and what had been inferred from that pair in its original and sustained deployment, have discovered the same kind of reoriented perspective on the phenomenon (or phenomena) under scrutiny. In sum, if deconstruction is replaced by an elementary course in informal logical fallacies, there is no reason to believe that the same analytically directed results will not accrue.

This suggested approach to the potentially problematic construal of opposition considers the conceptual structure of opposition strictly in terms of individual instances which may but need not assume problematic proportions; it does not propose to construe opposition as such, wherever pairs of concepts may appear as so related throughout the history of metaphysics, as a monolithic notional block which guarantees distortion of its elements regardless what those elements might be.[109] A Wittgensteinian motto drawn from standard practice in *Philosophical Investigations* may be appropriate here: the strategy embodied in the above approach adopts an attitude of "Look and see" just what exactly has been embedded in each of the opposed elements taken individually and with respect to one another within the deployment of the relation of opposition. Even if upon investigation it should happen that one opposite dominates its counterpart, it would appear premature to assume that the fact of domination as such automatically invalidates this priority; again, look and see why the dominance has been the case, then proceed with pertinent implications and revisions as required.

Proponents of deconstruction will respond that its treatment of opposition is considerably more vigorous and expansive than that provided by the rudimentary guidelines presented in elementary logic and its recognition of the tendency for opposition to generate informal fallacies. Thus the exploration of a reversed direction of dominance is only one of and indeed not the most radical option available to a deconstructive treatment of opposition. Derrida is clear when he contends that the ultimate goal of deconstruction with respect to an individual instance of opposition–if not to the employment of opposition in general regardless of the content of the opposed elements–is to reject the range of reality covered by the scope of the opposites taken as a comprehensive unity and to replace that range with a completely new and apparently innovative identification and description of "what is there." Thus the very notion of opposition itself is to be deconstructed, then displaced entirely; the aim is not merely to engender a revision limited to reversing the direction of dominance displayed by the opposed elements as originally formulated.

If, however, deconstruction's objective is to recast the matter of concern completely outside the scope of opposition, then the feasibility of this speculative gambit depends on negation, complementary classes, and articulation of totality with respect to negation and class diversification. The deconstructionist faces an approximate overhaul of another series of fundamental metaphysical considerations especially as these components play into the restated scope of philosophical relevance once a creative resolution of the original skewed oppositional setting has been secured.

A traditionalist in this area of philosophical methodology and analysis might respond variously to this more commanding statement of purpose: first, whether the dominant opposite remains in control throughout some segment of the history of metaphysics is an empirical matter, that is, the continued dominance depends on the

---

109 Aristotle details various different senses of opposition in Chapter 10 of Book V, the "Philosophical Lexicon" of the *Metaphysics*, 1018^{a} 20-25. For examples of Derrida's concerted interest in opposites, see SP, 134, 148-9; MP, 63, 151-2, 325, 329-30; WD, 229, 244, 282-3; P, 41, 90; LI, 34, 67, 68-9, 93; Derrida 1992, 51. For discussion of the centrality of opposition for Derrida, see Harvey 113-4, 116; Stocker, 43-4; Rorty 1989.

career of the instance of opposition through the passage of time and circumstance. Furthermore, if deconstruction could re-examine the relation of opposition with respect to a given instance of dominance, there is no reason to believe that only deconstruction has the means to achieve this interpretive approach. Even if the deconstructionist insisted on excising the given instance of opposition as a prelude to replacing its scope with a revamped account of the pertinent matter covered by that instance, it still seems open to an independent investigator to secure the same results without embracing any of the occasionally arcane strategies embodied by canonic deconstruction.[110] After all, if the referential features in Derrida's notion of protowriting do not exist as intended by that formulation, then opposition as such, a deployment of entities as reflected in a certain turn of language, becomes at most a purely theoretical arrangement of concepts with respect to their objects. So construed, opposition exists relative to epistemological and linguistically-oriented patterns of arrangement but it does not follow that the stipulated patterns have an equivalent hold on the extra-linguistic realities these patterns identify. "Opposition" is a powerful name in the annals of human cognition; the question is whether entities deemed "opposites" have been accurately and usefully categorized by the mere imposition of that name.

Opposition as a concept with its own identity is constituted by factors all of which unite in gathering and configuring diverse realities. This identity must remain the same in order to establish opposition as a fixed element capable of allowing its instances to become the subject of deconstructive analysis. But such structural configuration and constancy elevate opposition to a level of foundational significance so that it too, opposition as such, should also become subject to deconstruction. The advocate of deconstruction must surely recognize the range of complex and powerful assumptions that fund the notion of opposition and make them just as much an object of inquiry–formal abstractions underlying the framework contouring the specific content of instances of oppositions–as those instances themselves.

Furthermore, if the theoretical justification for the metaphysical foundation of deconstruction turns out to be unstable, what impetus justifies the factor of reversibility in the context of opposition? Why should deconstruction's attempt to dislodge and invert the relationship of priority pertaining to members in a given instance of opposition be granted the requisite theoretical force to accomplish this end? In general, why should opposition as carrier of stipulated relations and, implicitly, of potential insight into its concealments be taken as a "given" in its appearances throughout the history of metaphysics? The question is not whether an opposite not currently privileged should become privileged after all, a frequently

---

110 For an exposition and critical assessment of several attempts to codify or normalize the practices of deconstruction, see DFL, 161-8.

sought end in deconstructive application; rather, more fundamentally, it is whether opposition should be considered as privileged in terms of its status as one of the primary targets of deconstructive treatment.

8.7.4 Logic, Self-reference and the Rise of the Rhetorical

We demonstrated earlier in this chapter the logical incompatibility–pitched at an especially high theoretical level–permeating Derrida's thought, i.e., the apparent inconsistency emerging from conjoint criticism and espousal of the same metaphysical elements. However, this logical crux is a problem only if the student of Derrida cares to observe the canonic rules of correct critical thinking. If there is no concern for these rules, then one is free to play philosophically but to do so in a predominantly pejorative sense, that is, to create a variety of word constructs and present them to the philosophical world (and anyone outside that world who cares) as self-justifying ventures. These constructs when fashioned as exercises obeying few if any formal rules for rightness will succeed or fail with respect to staying power on the philosophical scene depending on degrees of popularity granted by audiences with similar interests and an equivalent lack of concern for what might appear to be mere logical niceties and groundless adherence to purely formal configurations. One may wonder, however, whether any of these hybrid creations will find a viable and justified place within the extra-deconstructive domain of philosophical inquiry.

The extent to which deconstruction depends on a metaphysical base itself subject to sustained and serious if not destructive criticisms is the extent to which deconstruction loses its relevance as a doctrine capable of being embraced and applied by the student of either (or both) philosophical or literary texts. In fact, the logical inconsistency resulting from the confrontation between being as presence and metaphysical considerations underlying the promulgation of deconstruction reduces the philosophical enterprise, to the extent that the practice of philosophy presupposes integration with the project of deconstruction, to species of rhetoric. The modes of language produced may engage certain audiences in terms of an emotive response or perhaps, with sufficient modification, even as a point of departure for additional reflection. But regardless how provocative this discourse may become, such language by itself will not be enjoined to satisfy seemingly necessary requirements for formal rightness or adequacy with regard to either pressing practical demands in the extra-philosophical domain of public life or to the concerns of that large sector of the philosophical community which does not subscribe to Derrida in particular or deconstruction in general. Philosophers (e.g., Plato and Aristotle) have analyzed rhetoric as an object worth serious and sustained discursive discussion. At least by tradition, however, rhetoric and philosophy in its many diverse modes differ in their ends and are not equivalent forms of language. Presumably this separation should be preserved to benefit the integrity and purposes of both endeavors.

8.7.5 The Primacy of Difference

Earlier in this chapter, we detailed a series of contexts in which a purely formal difference is necessary to engender the possibility of formulating pivotal features of Derrida's program, especially regarding deconstruction as an ensemble of distinct yet correlated strategies. This section now enumerates implications derived from a denial of the priority of difference–a denial which, as seen on several occasions, Derrida makes explicit.

8.7.5.1 *Difference and the Possibility of Coining "différance"*

There is and must be a fundamental distinction–in a word, difference–between (a) the import of the technical term *différance* standing as one and only one element within deconstruction as an aggregation of functionally unified strategies and (b) difference insofar as it characterizes the distinction between any one word and any other word in a given natural language. For if such a distinction were lacking, then *différance* becomes indistinguishable from any other word in terms of its lack of necessarily precise referential capacity when it is used propositionally, a consequence required by the application of canonic deconstruction. *Différance* itself cannot fall under the conforming sway of *différance* because then its structural comprehensiveness, covering as it does the entire history of metaphysics, is threatened with compromise either in the present or at some point in the future. In short, *différance* would provide conditions for the deconstruction of "*différance*" itself. To put the point in scholastic terms–*différance* in act would eliminate even the possibility that *différance* could be potentially and coherently formulated. If "*différance*" exercises force and direction within Derrida's thought, there must first be a difference between the possibility that *différance* could exist and all discourse other than *différance*, especially when the discourse in question constitutes the history of metaphysics.

8.7.5.2 *Difference and the Possibility of Discerning "Being as Presence"*

"Being as Presence" Difference is necessary in order to separate (a) being as presence in its role of governance with respect to the history of metaphysics and (b) the aggregated techniques comprising deconstruction. Without difference functioning in this context, being as presence could not have been formulated (as something different from the total history of metaphysics). Furthermore, "closure" of the history of metaphysics presupposes the difference between reading the history of metaphysics, circumscribed as it has been by being as presence, and recognizing this unified totality from an "other," different philosophical perspective. Thus difference must obtain between the history of metaphysics insofar as closure pertains to that history as a uniform progression of texts and deconstruction as an interpretive approach which is and must be in some measure *outside* that history in order for an observer to appreciate that it can be construed as, in point of philosophical fact, "closed."

8.7.5.3 *Difference and the Possibility of Formulating Deconstruction*

If deconstruction is self-referential, then Derrida's presentation of that position in his own texts must be kept separate from the implications of that very position. However, this separation entails a series of assumptions which, individually and collectively, assume foundationalist status. The most fundamental of these assumptions is, again, that difference must precede *différance*.

This priority is necessary in order to establish the possibility that deconstruction as a diagnostic instrument for detecting metaphysical excesses can be posited in the first place and remain safeguarded from its own leveling activity, much less to be implemented as a uniform array of techniques aimed at potentially revelatory analysis of other texts, whether literary or philosophical. Anything substantive and deployed in a format intended to be other than the history of metaphysics must differ from (a) that history as one organically whole sequence of thought as well as (b) every single instance of thought within that sequence. If difference does possess this degree of independent and autonomous stability, difference will formally block the implications of deconstruction, assuming the relevant mechanisms were proposed to apply to its own bedrock conceptual elements. It follows then that the difference marking the distinction between deconstruction with its full complement of specialized vocabulary and everything written in the language of metaphysics which that position is intended to deconstruct cannot itself be an instance of *différance*. Without this difference firmly in place, deconstruction as such–a unified position with a limited number of elements as well as a determinate function–becomes just another moment in the history of metaphysics, just another extension of being as presence.

In sum, difference must be kept separate and also recognized as privileged, a "greatest kind," in order to preserve the integrity of deconstruction as a set of potentially realizable theoretical gestures. Even if it were possible according to some configuration of deconstructive techniques to salvage and preserve every necessary general term in order to ground the possibility that deconstruction can be articulated as a theory, at least one of these terms, difference, must be embraced metaphysically in order to preserve distinctions within the formulation of deconstruction as an ensemble of words intended to realize ends marked by discernibly theoretical characteristics. Difference in these contexts becomes equivalent to sheer otherness in a sense stated by Aristotle and anticipated by Plato. It is ironic that a position concerning an especially basic element of logic and advanced by the principal progenitors of the grand sweep of western metaphysics must remain necessarily in place in order that a thinker, living at the culmination of this epoch and announcing its "closure," should be capable of formulating the theoretical stances intended to make that very announcement!

8.7.5.4 *Difference and the Possibility of Preserving Logical Rightness*

The advocate of deconstruction must accept the privileged status of difference in order to formulate the principles of deconstruction. The question becomes whether deconstruction when applied according to its own canons will countenance endorsing the traditional, formal principles that establish and control logical rightness.

Consider the principle of contradiction. As we have seen, Derrida once encapsulated formal logic as, "itself an interpretation."[111] If this evaluation is not ironic and is duly pressed, would it follow that the principle of contradiction is also "an interpretation" of consequences following the proffering of a set of propositions characterized formally in certain ways? For example, two positions derived from the history of metaphysics are, when juxtaposed, logically divergent to such an extent that the resulting consequences cannot be held in the mind as significant. The two positions are, in fact, contradictory. To assert that the proposed juxtaposition exists "outside" the domain of contradiction and all other logical principles is a doctrinal gambit which cannot succeed. This gesture of defense only locates the conjunction in a conceptual context impossible to assimilate or even to approach for purposes of thoughtful investigation, at least for anyone aware of the relevant structural feature of formal logic–from a contradiction anything follows.[112]

If, however, the application of deconstruction and the regulative function of contradiction are indeed compatible with one another, it will be necessary in order to preserve the efficacy of the principle of contradiction to endorse the unique status of its components when embodied in this principle. Thus all formal elements constituting the principle of contradiction must remain intact as purely formal elements existing independently of the sway of *différance* and its effects on metaphysical notions. The advocate of deconstruction must both presuppose and impose an application of difference which must remain the same as itself and which also cannot be affected by any investigative sortie deconstruction might make in its direction. Deconstruction must therefore endorse difference as a foundational and constant element inherent in the formulation of the principle of contradiction–an instance of precisely the kind of element deconstruction intends to explode–because difference is necessary in this purely regulative context.

---

111 Cf. Ch1n19. For Christopher Norris 2012, 79, Derrida considers it "well beyond the pale of rational acceptability" to renounce the principle of contradiction. Norris frequently emphasizes Derrida's dependence on classical logic in order to enable deconstruction to arrive at legitimate results–Norris 2012, 83, 128, 131, 132, 138, 142-3, 147, 148. If this dependency is as intimate as Norris claims, then the necessity to deconstruct the formal principles which underlie the mechanics of classical logic becomes even more vital. Norris does not consider possible repercussions insofar as his firm endorsement of contradiction and related concepts and axioms of classical logic pertains to the practice of deconstructive analysis of general terms.

112 For discussion, including the proof regarding logically necessary implications from asserting or allowing a contradiction, see DFL 246-9n6.

The problem for deconstruction is determining how to distinguish between the formal realities underlying the possibility of formulating the theoretical gestures comprising deconstruction insofar as they derive from the history of metaphysics, from which it would follow that these realities are permeated by being as presence and insofar as they, functioning as formal necessities with respect to the preservation of formal logic and its regulatory functions, are themselves impervious to deconstructive analysis. In fact, if certain "formal" realities command this kind of existence when they animate concepts and principles of logic, it requires an argument–not merely a simple denial–to show that they cannot also extend this independence into substantive metaphysical dimensions and patterns.

In view of the above contexts and their associated implications, we must grant that difference functions as a primitive, a greatest kind. In this respect, difference must be more fundamental than *différance* as a necessary condition for preserving all the distinctions itemized above. Derrida's subsumption of difference under *différance* contributes to the weakening of difference regarding its necessity in preserving deconstruction as well as *différance* itself from self-referential effects–that is, inherent dislocation and, ultimately, complete disintegration.

8.8 Deconstruction and "Making a Difference"

Early in Part II of *Process and Reality*, Alfred North Whitehead refers to the crucial importance of the "endeavor to interpret experience in accordance with the overpowering deliverance of common sense...."[113] Just prior to this claim, Whitehead had mentioned almost as if it were self-evident the difficulty modern philosophy faced and faces of "describing the world in terms of subject and predicate, substance and quality, particular and universal," since these traditional instances of opposed pairs do "violence to that immediate experience which we express in our actions, our hopes, our sympathies, our purposes...."[114] The rejection of such seemingly fundamental oppositions (many others fall away as well for Whitehead) situates Whitehead's process philosophy side by side with deconstruction in not only questioning the legitimacy of these opposed approaches to understanding reality, regardless how ingrained they may seem to be in the philosophical mind, but forthrightly replacing them with alternatives.

Common sense says that we live in the moment. Derrida often inveighs against appeals both to "common sense" and to the "moment" but at this juncture of the essay,

113 Whitehead, 50. Derrida frequently dismisses the philosophical relevance of "common sense" and "ordinary life." See, e.g., WD, 307n3; LI, 91; Derrida 1992, 42. (But cf. Ch10n134 for a contrasting view from Christopher Norris).
114 Whitehead, 49.

I submit that we read and internalize Whitehead's guidance. Thus, the interaction of *this* agent with *this* object and *this* person at *this* instant at *this* place and at *this* time collectively embody an "overpowering" set of data. Our experience is continually characterized–or "delivered" in Whitehead's terminology–through the intersection of these multiple phases of reality. Of course, the moment does not happen by itself, a purely temporal snapshot. Moments occur within complex metaphysical frameworks and, as Whitehead so spiritedly observed, the immediacy of our lives as lived–that is to say, our lives!–transpires in such settings.

Derrida too lived in the moment and vividly displayed this mode of temporal existentiality through some of his most passionate theoretical writing. The moment in question involved a particularly sharp critical reaction by a particular contemporary philosopher of note to a particular position, one of pivotal importance in Derrida's canon of principles. Derrida reacted in a certain way to this criticism and his "actions," in writing, received extensive justification in order to deliver to the readers of this writing a firm and fixed grasp of Derrida's "purposes." Derrida had had "hopes" that fellow professional seekers of wisdom would have greeted these purposes with greater "sympathies." But at least in this one instance, Derrida's audience was not enthralled by what he had written. The reader will have observed that all the quoted words earlier in this paragraph are from Whitehead's perceptive list of basic human experiences; all find exemplification in the texts from Derrida to be considered in Chapter 9.

This appeal to the relevance and power of the moment will be situated in Chapter 9 within a broader context of Derridean proportion. As we saw in Chapter 6, Derrida challenged the history of western metaphysics with a densely developed concept, what he referred to as "trace." We offered in that chapter an introductory account of the scope of trace deployed against the backdrop of the more familiar (and, presumably, more riddled with the effects of being as presence) notion of negation. In order to extend the earlier account, we will add in the next chapter the breadth inherent in the concept of possibility, in particular the latent possibilities in Derrida's notion of trace. This line of inquiry will demonstrate that other considerations must also come into "play"–a term functioning partially in Derrida's quasi-technical usage and partly according to its customary idiomatic sense–in order to establish the possibility that Derrida's vigorous response to the challenges of the moment are not inconsistent with his, Derrida's, canonic versions of the conceptual foundations of deconstruction. For example, is there a way to address the questionable status of particularity from within Derrida's own texts, especially with regard to the fact that those texts espouse the gestures of deconstruction? This question becomes relevant if not crucial given the demands of the moment. If a response to a reaction occurs in the moment, it is a particular agent who experiences that reaction; furthermore, what has provoked that reaction is also circumscribed within a set of particulars designated with decided metaphysical focus and contour. Without a vibrant sense of richly articulated particularity, the moment is merely an undifferentiated flash, or at best a flutter of evanescence.

The ultimate wedge concept Derrida exploited while he lived in a philosophical moment of high drama, a moment aimed at restoring the credibility of his position, is *difference*. Not "difference" as a derivative of *différance* but difference as underlying an approach to a barrage of criticisms which for Derrida were blatantly unjust. In a context contoured by often virulent dispute, we must appreciate the fundamental and seemingly "dry" dimension of difference in Derrida's writing. Although individuals will idiomatically say (in English) that they strive to "make a difference" in life, considering difference from the perspective of fundamental metaphysical considerations allows us to realize that we do not "make" a difference at all, nor do we locate difference under something else (i.e., *différance*) as a phase in a program of thought intended to achieve a speculative goal. Rather we, with Plato, acknowledge difference as a "greatest kind," a formal element necessary as a precondition separating the rush of reality into singulars and plurals, and ultimately for establishing at least a semblance of order in discourse about what so exists.

Is there room in Derrida for such recognition? Chapter 9 addresses and answers this question–in a word, yes. The discussion therein will embrace provocative and fertile claims and positions drawn from Derrida's work on being as presence as well as deconstruction understood as a unified series of techniques for reading texts.

9 The Deconstruction of Deconstruction: Prelude to a Metaphysics

It may seem that deconstruction emerging from Derrida's texts as a schematic approach to evaluating philosophical theory has collapsed from unrecognized reliance on its own residual metaphysics. Derrida's concerted regard for and attention to the history of philosophy and its effect on contemporary thought would not then have resulted in a position with sufficient stability to reveal untapped possibilities resident in that very history. But this summary and ultimately dismissive judgment is in a way premature. Even if the criticisms aimed at Derrida are accepted, the elements collectively standing as "deconstruction" hold out possibilities for additional inquiry and potential insight into the experience and articulation of traditional metaphysics. Chapter 9 addresses these attendant possibilities.

The pivotal component is the notion named "trace." In Chapter 6, we outlined and discussed Derrida's complex and convoluted treatment of trace functioning as a key component in signification. But now we shall see that Derrida's notion of trace can become an access to ways of looking at and thinking about "reality," a code word for the subject of metaphysics. This access engages what should be countenanced from western metaphysics but then routes that engagement through a style of discourse which, at least by intent, reduces residual effects of having emerged from the terminology of metaphysics. The resulting language embraces what looks to be fresh and revelatory means of expression, or at least optimizes chances of eventually accessing such material. Although on the surface it may appear difficult, perhaps even impossible, to transcend the outermost limits of being as presence, we approach Derrida's disposition of trace with a high degree of optimism regarding its potential for producing insight at this level. It may in the end remain viable to "do" metaphysics in something like the ways it has been done for the last three millennia.

The notion of *possibility* identified in the terminology of classical metaphysics stands as both an underlying context and ongoing theme throughout this chapter. In general, a possibility may–or may not–be actualized. The reader should keep this broad tentativeness in view as the chapter derives an argument based on texts from deconstruction which leads gradually to positive results in metaphysical matters. In one respect, being as presence is and has been a continually closed and closing door; but in another respect, finalizing one set of possible outcomes found in and defined by the history of metaphysics may engender another set of entrances to new destinations.

9.1 Derrida's Truth

The scene is the arena of contemporary philosophy. That fundamental differences obtain between analytical and continental philosophy is a view commonly held by many practitioners of each mode of expression as they respectively pursue the love of wisdom. One of the most dramatic and intense examples of this difference emerged from exchanges between Jacques Derrida and John Searle on the topic of speech acts, especially as this concept was formulated and developed by John Austin. Derrida critically evaluated Austin on speech acts, Searle critiqued Derrida's assessment with considerable energy and also an apparent lack of good will. For Derrida, Searle's reply was not just laced with ill humor, it was also fundamentally wrong. Derrida said so, bluntly and unambiguously:

> **Passage 9A** The answer is simple enough: this definition of the deconstructionist is *false* (that's right: false, not true) and feeble; it supposes a bad (that's right: bad, not good) and feeble reading of numerous texts, first of all mine, which therefore must finally be read or reread. Then perhaps it will be understood that the value of truth (and all those values associated with it) is never contested or destroyed in my writings, but only reinscribed in more powerful, larger, more stratified contexts (LI, 146–italics in text).

Philosophers often disagree with evaluations of their positions, whether negative reactions appear in print or are "live" at meetings, so in one respect Derrida's blunt (if not apparently overwrought) rejoinder is not of special interest. If it were just a matter of intellectual biography (in what may be one of its more indecorous moments), then this episode is interesting to the extent that public idiosyncrasies of authors are interesting; however, if only this were at stake, little if anything illuminating for those philosophically inclined would remain from this fractious interchange. What is striking from a more dispassionate perspective is that Derrida's response to Searle appears to be flatly inconsistent with Derrida's canonic positions as they emerge from a plateau for interpreting texts based on deconstructionist principles.

For Derrida, Searle's reading of Derrida's text is simply "false." But if an articulated reading of a text is false, then it is reasonable to expect that another reading of the same text will be "true." Yet on deconstructionist principles, how is a true reading possible? Derrida's insistence on rightness in this case may have been only a momentary aberration (motivated by pique and aggrieved ego), a temporary deferral of dispassionate deconstructionist approaches to reading.[115]

115 In the Afterword to LI, written ten years after Derrida's original reply to Searle, Derrida admits that this response "was not devoid of aggressivity" (LI, 113). However, he also insists that this mode of expression was consonant with the "violence" characterizing the context surrounding his work and Searle's critique of it. Derrida says in the Afterword that he will return to "straightforward" discussion so that his observations "will thereby gain in clarity" (LI, 114). Whether Derrida's "aggressivity" is read as a manifest instance of polemic depends on one's taste in academic writing. But the more

But if this appeal to rightness is indeed seriously intended, we ask: first, what conditions must obtain to enable the possibility of such a reading; second, are these conditions, if enunciated with sufficient clarity and comprehensiveness, compatible with deconstruction as an aggregation of interpretive principles aimed at evoking determinate ends?

9.2 "As we speak...."

By way of preamble to an exploration of these joint concerns, let us review what is true of many linguistic contexts, intractably and stubbornly so, in ways which philosophers must acknowledge and incorporate into their reflection on the issues at hand. Consider as paradigmatic Dewey's example, asking someone for something. This event happens *now*, not a few minutes ago, not later on. The event happens *here*, not there or anywhere else.The reader of this essay is doing so here and now, surrounded by a determinable number of entities or people, perhaps both.

Generally speaking, an utterance such as, e.g., a request for a drink of water relates a speaker existing in this here and now to an auditor also existing in this here and now. The resultant meaning of linguistic sequence voiced within an environment circumscribed by such temporal and contextual constraints is straightforward, stable and unambiguous for all concerned participants. But following Derrida and deconstruction with its undifferentiated and comprehensively unified ensemble of interpretive strategies fully in place–a unity tacitly spanning all contexts and all dimensions of temporality–the theoretical possibility of securing such a meaning is necessarily out of bounds. However, the primary reason for this loss of stability has nothing to do with considerations relative to language proper. As argued in Chapter 8, the deconstructively motivated conclusion derives its theoretical force from reliance on metaphysics plain and simple–in fact, a continually ongoing rush of contexts and temporality, with this reality spreading in all possible directions from the confines of a single moment or definite duration and the equally determinate limits of a context with a necessary measure of givenness and limit. But this rampant Heracliteanism apparently does not extend to at least some of Derrida's texts as well as to the readings which he himself produced and imposed on those texts. This fact must give us pause.

important philosophical point is how any sense of "clarity" is possible in Derrida's own writing given the deconstructive approach to linguistic structure exemplified in Derrida's theoretical work and the interpretive assumption that the statement of deconstructionist theory in Derrida's texts is itself fully self-referential.

9.3 Deconstruction and the Margins of Self-reference

We assume that Derrida's reaction to Searle is intended seriously and also that this seriousness can be justified philosophically. But how?

The cogency of Derrida's written reactions to what others say in texts about his texts presupposes the existence and reliability of a series of realities and principles based on those realities. First, the text in question must be identical for both the critic of that text and its defender. Furthermore, this text must remain identical during both (a) the process of criticism advanced by a reader of a text as well as (b) the defense raised against that criticism on behalf of the text's apparent integrity. If the text is moving and unstable in its import, then it follows that the critic and defender may not be referring to the same ordered series of words as these words produce unity of signification.

In addition, the defender must understand the text in a certain way, an understanding existing apart from the myriad possible ways in which the text could be taken by other readers. This separation of textual givenness and an understanding of that givenness is necessary in order to establish a condition for the *possibility of rightness* from the interpretive perspective of the defender; without this distance between text and understanding, the defender's claim of rightness against alternate readings rests solely on what the defender believes to be the case.

Finally, the process of reading and defense occupies a certain sequence of time. This sequence need be only of the moment if the defender recognizes instantly that the reading offered by the critic is false. Thus in his response to Searle, Derrida implicitly acknowledges the potentially dominating presence of the present moment as well as the fact that a speaker, dwelling in the moment, exists in relation to diverse streams of reality regarding the reading of texts, with each stream enjoying its own measure of self-identity and overall stability. Even if the defense possesses duration in its realization and development, this segment of extended time has a unity and self-identity which runs against Derrida's denial of the existence of these logical elements, whether these elements emerge in moments or in stretches of duration.

These assumptions and their underlying principles undercut the theoretical feasibility of Derrida's established position regarding the primacy and value of deconstruction, a position asserted over many years and in many texts and on many topics. How then can Derrida as the originator and primary protagonist of this position proclaim that Searle's reading is wrong and that his, Derrida's, reading is correct? At issue is not whether Derrida's response to Searle adequately confronts and refutes Searle's critique; the relevant point concerns the very possibility that such a response can be mounted at all, quite apart from its success, whether partial or complete or, indeed, its failure as a defense of Derrida's position. How then to proceed in the quest to determine the cogency of Derrida against Searle?

Derrida often expressed keen interest in the "margins" of texts.[116] Thus an important anthology of essays on various authors bears the name *Margins*. But all texts examined in this volume (and related deconstructive analyses) are by other authors. Recall that Derrida wrote that texts advancing deconstructionist ideas and ideals are subject to their own dictates and strategies.[117] Is it possible to deconstruct the margins of Derrida's own texts? Yes. We will assume that Derrida's texts, written gestures sharing in discourse controlled by being as presence, can themselves be subjected to deconstructionist strategies. If so, then the application of this straightforward instance of consistency (not to mention the authority of Derrida himself!) with the apparently universal scope of deconstructive efficacy engenders the interpretive legitimacy of examining places in Derrida's writings which appear to marginalize certain concepts, to locate them on the fringes of the main avenues of texts in which those marginalizations occur. Related examples of this self-reflexive venture follow in subsequent sections of this chapter.

9.4 The Possibility of Stability

To show that one reading of a text is wrong–Derrida's position against Searle–does not imply that another reading of that same text is necessarily correct, i.e., true. In fact, multiple wrong readings do not elevate a remaining reading as preferable, much less correct, since ways to misconstrue a text are limitless. However, Derrida's response to Searle includes both that (a) Searle is wrong about Derrida's text and (b) Derrida's own elucidation of that text is correct. What conditions must obtain for Derrida to state that a certain reading of his own work is right and that another reading laid out in print is wrong? We ask this question not merely to garner support for the legitimacy of one author's view of that individual's written work but, more extensively (and, ultimately, of greater use philosophically), to elicit the possibility of exhuming a source of stability in general, not merely stability as applied to the written word.

We begin with a matter of principle: as noted in Chapter 8, the tactics of deconstruction frequently aim at texts embodying instances of opposites. The strategy is to upend any priority which, given traditions of philosophical inquiry, has accrued for one opposite with respect to its counterpart. At this juncture, the relevant opposition is motion vs. rest, or stability rather than rest if we may introduce a word with fewer mechanical overtones. Deconstruction emphasizes motion, indeed unceasing motion, in a number of respects (signification, meaning, referential signified, context, formality of logical principles); yet Derrida's response to Searle presupposes that stability must be at least countenanced if not fully endorsed

116 See, e.g., MP, xxiii; LI, 34, 44, 70, 118; WD, 32n3. See also DFL, 239-40n6.

117 See Ch8n105.

regarding the notion of interpretive correctness. The question then becomes: what are the necessary conditions for the kind of stability which will ground the possibility of determining the correct reading of a text?

A belief in stability is a cardinal conviction of western metaphysics, or so a number of epochal philosophers have thought. As a metaphysical property of broad import, stability can exist in various guises and degrees. Therefore, the extent to which stability becomes feasible for a thinker approaching the world from a deconstructionist perspective is the extent to which the properties of stability can emerge–or, perhaps more accurately, be allowed to emerge–from consideration of the history of metaphysics. This interpretive prerequisite will incisively direct our attention to Derrida on *closure* and its function with respect to that history. This initial phase of inquiry involves heeding consequences which follow from Derrida's approach to the history of metaphysics as a totality which has in fact been closed.

The investigative context for this sought-for stability is complex in that it must be sought from the only available source which, given the fabric of Derrida's entire system, could provide a schematic to accomplish this end. Indeed, this schematic rests on the need to span *both* dimensions of a speculative whole composed of elements which are in some respects antithetical to one another. To anticipate: at the culmination of this chapter, we argue that the most that can be secured at a point marked by the intersection of the history of metaphysics and Derrida's speculative appraisal of that history is a series of *provisional categories*. These categories (I develop a sample set in Chapter 10) provide a foundation for various metaphysical necessities. One such necessity includes articulating assumptions which must be in play so that Jacques Derrida can write what he wrote about what John Searle wrote concerning what Derrida had written about John Austin. The stylistic modality of the last sentence is significant in terms of what it implies, that is, that the convoluted character of the interrelationships involved in this dispute must be carefully preserved–in a word, stable–in order to appreciate who said what about whom, as well as the philosophical significance of what all these exchanges might mean.

9.5 Presence, Closure, Trace

We have been emphasizing throughout this essay the importance of *différance* both as a fundamental formative agent in the history of metaphysics and in its structural considerations pertaining to deconstruction as an organized response to that history. The following passage underlines the function of *différance* with respect to the unity of the history of metaphysics but it also concludes with a useful signpost for the concerns of this chapter:

Passage 9B Reflection on this last determination [i.e., Heidegger's fundamental distinction between Being and beings] will lead us to consider differance as the *strategic* note or connection–relatively or provisionally *privileged*–which indicates the closure of presence, together with the closure of the conceptual order and denomination, a closure that is effected in the functioning of traces (SP, 130-1–italics in text).

The "provisional privilege" enjoyed by the term "*différance*" is displayed in at least two senses. First, it is unique with respect to the lexicon of philosophy and therefore it cannot be located under a more general or generic sortal concept or arrangement. The uniqueness of *différance* implies that it cannot be redefined in more derivative and explanatory concepts. Second, *différance* denotes the "closure" of western metaphysics. In being "provisionally" privileged in this respect, *différance* may as a unique term be of finite duration as applicable to philosophical inquiry. Thus "provisionally" does not indicate that *différance* could be replaced by a superior notion while the history of metaphysics is itself ongoing but that its explanatory efficacy is circumscribed by the outer limits of that very history. While the history of metaphysics is still with us–or, more accurately, while the practical and theoretical existence of human beings remains within the boundaries of the history of metaphysics–*différance* is as fixed and unquestionable in its functions as human cognition, with Jacques Derrida its primary protagonist, can secure by force of inquiry.

What does Derrida mean by "closure"? Closure is deployed to encompass a set of comprehensive totality conditions covering both what has actually been advanced by metaphysicians throughout its history along with everything which could possibly be posited by future metaphysicians. Even if metaphysics appears to assume new forms (Whitehead's process metaphysics or the more current ventures of analytic philosophers doing "metametaphysics"), Derrida's "closure" indicates that any theoretical items of apparent novelty will not touch anything in the fundamental matters metaphysics has developed throughout its history.We may then tentatively characterize "closure" as referring to *a range of determinate positions which collectively exhaust certain possibilities* insofar as these positions constitute the history of metaphysics. This definition is clearly limited in its denotative force by the notion of "possibilities."

It is important to understand that closure is a conjunction of closures–first, "the closure of presence," second, "the closure of the conceptual order and denomination," suggesting that closure pertains to every systematic metaphysics with respect to their configurations of being as well as to the systems of concepts ordering that content. Presence is inevitably present throughout the history of metaphysics; furthermore, the ensemble of concepts comprising each metaphysical system as well as the conjunction of all such systems–regardless of apparent terminological divergencies–constitute the totality of a set. This set taken both individually and collectively is declared to be closed.

At the conclusion of **Passage 9B**, we are told that closure achieves its effect through the notion of "trace." Traces (in Chapter 6 we assumed that trace admits of plurality) penetrate the entirety of presence's informing of the history of metaphysics. But we see later that traces do not exhaust their function merely by establishing, as it were, the outer limits of metaphysics. Traces in some respects continue beyond the limits of presence and systematic metaphysics. Although the argument of Chapter 8 offers compelling reasons to block the attempt to elicit principles of stability directly from Derrida's canonic texts, given his assumptions and the general lines of his deconstructive analyses, the most promising arena to pursue something approaching stability derives from Derrida's speculations concerning traces. We will discuss this notion as a tentative source for providing at least potential principles of stability which will in turn initially ground the assumptions Derrida must have had in mind as he formulated his rejection of Searle's blunt critical response.

We may wonder, based on speculative possibilities to be pursued shortly, whether closure will include on the other side of metaphysics (again, speaking imagistically) a recognition of the need to establish the formal impregnability of contradiction. If so, crucial consequences follow. For example, this need presupposes sufficient self-identity of propositional reference and meaning covering either an instant or some duration of time in order for the principle of contradiction (understood in its material rather than formal mode) to underlie and order the cogency of discourse, whether written or spoken. If, collectively viewed, closure's preservation of self-identity, difference and the principle of contradiction is accepted, then considerations derived from the fundamental character of these logical elements can be applied *within* the ambit of the history of metaphysics to the foundational character of being as presence. And if these principles are also allowed the possibility of formally regulating the rightness of discourse *beyond* the limits of the history of metaphysics, then the elements of these principles–e.g., identity and negation taken as a primitive–sanction the additional possibility that any purported novel approaches to the standard issues of metaphysical thought will continue to be expressed (and open to critical analysis) in light of these "logical" principles.

Of course, the issue antecedent to these logic-based concerns is whether anything at all can be asserted after the closure of metaphysics and, if so, what type of utterances or inscriptions will emerge. This question is discussed below.

9.6 Approaching the Possibility of "Metaphysics"

The closure of the metaphysical epoch is not the end of philosophizing with purpose and speculative intent; form, taken as the exemplar of metaphysics from Plato through Heidegger, has not won the day in all possible philosophical respects. Here is Derrida (the much lengthier continuation of the same passage is discussed in the next section):

Passage 9C It might then be thought that the sense of *being* has been limited by the imposition of *form*–which in its most overt function, and ever since the origin of philosophy, would, with the authority of the *is*, have assigned to the sense of being the closure of presence, the form-of-presence, presence-in-form, or form-presence (SP, 127n14–italics in text).

We assume with Derrida that the history of metaphysics is the history of being as presence. Heidegger recognized this history and the epochal significance of presence. But according to Derrida, Heidegger remained within its constrictive sphere of influence, especially with his later, more speculative and assertive works dealing with being as a type of temporality.[118] By contrast, Derrida has embraced and absorbed this history, fully marking its closure. But Derrida insists that "the imposition of form" in conjunction with "the authority of the *is*" should not be seen as equivalent to an essential limitation of "the sense of being." In one respect, being is closed off by the history of metaphysics but in another respect this closing off is precisely what opens being and its interpretive specifications to other possibilities.

This implication reinforces the crucial importance of achieving the highest degree of clarity possible in elaborating the meaning and scope of "presence." In **Passage 9C**, the connection between being–what being "is"–and presence is as tight as Derrida can make it with the language currently available. Therefore, if another approach to the ultimate referent (or referents) of being is possible to secure, realizing this possibility must be as carefully described with respect to fostering future philosophizing as the attempts (argued for in this essay) directed at stating the meaning of Derrida's archetype phrase, "being as presence." To signal the potential arrival of a non-presence account of being, we will use quote marks–"metaphysics"–understood as an appropriate name for the post-deconstructionist rendering of what since the time of Aristotle has been simply called *metaphysics*.

For Derrida, trace has set the limits of metaphysics, has permeated if not defined metaphysics proper, the actualized textual history of metaphysics as a collection of positions all marked by being as presence. But again, that is not the conclusion of the tale traces can tell. Derrida's development of the notion of "trace" suggests that the relevant question is not *whether* traces also transcend the limits of metaphysics but *how* they achieve this end.

118 For exposition and analysis of temporality as integrated in Heidegger's notion of "presence," see White 1985, "The Structure of Being as Presence (*Anwesen*)," Ch8, 175-205. Heidegger is an epochal figure in Derrida's reading of the history of metaphysics. See the essay "*Ousia* and *Grammē*: Note on a Note from *Being and Time*" in MP, especially MP, 33n6; WD, 97, 138, 140.

9.7 Trace and the Elements of "Metaphysics"

In **Passage 9B**, Derrida briefly surrounds prerequisites for asserting the possibility of a true or correct meaning of a text with a convoluted historical context embracing the entirety of western metaphysics but containing elements coincident with and yet in a sense transcending that context. This historical complexus revolves around the notion of trace.

The following extended passage focuses on trace in connection with elements which constitute the core of Derrida's position and also intersect with traditional formulations of basic concepts in the history of metaphysics. This passage is noteworthy for its length and concentrated attention to abstract properties related to trace. Derrida relegated this discussion to a footnote (beginning with the text of **Passage 9C)**. In standard scholarly protocol, footnotes are places for secondary source references, parallel passages from primary sources, and supplemental comments as addenda to an argument developed in the "main" text. But in this case, Derrida's footnote serves as a pivotal focus for conceptual data useful if not necessary for understanding how traces function with respect to the history of metaphysics.[119] The entire passage is now quoted, then divided into sections for more accessible analysis:

> **Passage 9D** Form (presence, evidence) would not be the final recourse, the last instance, to which every possible sign would refer–the arche or the telos.... In a sense–or a non-sense–that metaphysics would have excluded from its field, while nonetheless being secretly and incessantly related to it, the form would already and in itself be the *trace* (*ichnos*) of a certain non-presence, the vestige of the formless, announcing and recalling its other to the whole of metaphysics–as Plotinus perhaps said. The trace would not be the mixture or passage between form and the amorphous, between presence and absence, etc., but that which, in escaping this opposition, renders it possible because of its irreducible excess. Then the closure of metaphysics...would not move *around* the homogeneous and continuous field of metaphysics. The closure of metaphysics would crack the structure and history of this field, by *organically* inscribing and systematically *articulating* from within the traces of the *before*, the *after*, and the *outside* of metaphysics. In this way we are offered an infinite and infinitely surprising reading of this structure and history. An irreducible rupture and excess may always occur within a given epoch, at a certain point in its text (for example in the "Platonic" fabric of "Neo-Platonism") and, no doubt, already in Plato's text (SP,127-8n14–all italics in text).

9.7.1 Form, Sign, Possibility

The parenthesis–"(presence, evidence)"–added immediately after the initial appearance of "form" is important regarding the precise meaning of being as presence. The following inferences are suggested: (i) "presence" and "evidence"

119 For representative passages illustrating the use of trace in a variety of contexts in Derrida's earlier texts, see: WD, 68, 95, 98, 101, 161, 175, 247, 265, 295; OG, 73; MP, 65-6, 313.

are two, perhaps the only two, instances of form; (ii) presence and evidence differ from one another insofar as both are related to form (although the parenthesis offers no obvious way to describe this difference); (iii) "evidence" pertains to form but it would be premature to conclude that "presence," as in "being as presence," does not also have its own modes of evidence. Derrida specifies form rather than being as presence as the central element in this approach to trace. Furthermore, if the history of metaphysics is strictly defined and limited by form as the sole object of signs, then Derrida asserts that form is epochally determined. As a result, "outside" (more on this and other spatial qualifiers below) the history of metaphysics, signs can refer to something other than form and thus to something other than the totality of the history of metaphysics which has been, so to speak, informed by form.

Although semiology as an autonomous discipline considers signage as more than a purely linguistic phenomenon, we will assume that the reference to "every possible sign" pertains to possibility solely within the sphere of language (although this sphere may have to be considerably widened in denotative scope). This dimension of possibility maintains the transcendental perspective marking so much of Derrida's terminology and theoretical concerns during his discussions of the foundational elements of language. In context, the denial that "every possible sign" must refer to form reaffirms the canonic position that *during* the epoch of metaphysics, "every possible sign" *does in fact* refer to form. But this denial also opens up the more fundamental possibility that "after" the closure of presence, after the end of the history of metaphysics, another type of relation between language as such and everything other than language may indeed be receptive to expression and articulation.

9.7.2 Presence, Non-sense, *Différance*

The "non-sense," i.e., the "non-presence" Derrida has posited is at one and the same time (a) excluded from the field of metaphysics and (b) secretly and incessantly related to that field. It seems fitting then to ask: if the exclusion cited in (a) is total and complete in all respects, how can the subject of this exclusion (b) be "incessantly related" to the field of metaphysics in any way except as sheer indeterminate otherness?

It will be crucial therefore to determine or at least approximate how "non-" in "non-presence" should be understood. Derrida explicitly uses "non-" as equivalent to "other" ("...recalling its other..."). However, the connection between whatever falls under this "other" and the history of metaphysics should not be interpreted as merely another instance of *différance*. Thus difference, not *différance*, must separate non-presence, whatever its content may be, from presence. The reason: if this neutral, purely formal type of separation is not observed at this point, then the distinction between presence and non-presence becomes subject to infinite regress. For if the interplay between presence and non-presence is itself a product of *différance*, then there is no difference between (a) the conjunction of presence and non-presence even when pitched in this aboriginal setting and (b) any two instances of systematic metaphysics occurring *within* metaphysics as the history of presence. This supposition, if it were the case, would in turn entail sameness common to both presence and non-presence. But with this consequence, Derrida loses the factor of exclusion regarding the content of non-presence, i.e., non-presence would share with presence the kind of sameness that we examined so closely in the interpretations and arguments constituting Part I of this essay.

Derrida apparently wants to separate this "other," whatever this non-presence may be as a totality, as widely as possible from all differences defining the sequence of positions constituting the history of metaphysics. It would then follow that *différance* functions only *within the epoch of metaphysics*; thus if dimensions of being are accessible other than those generated within the history of metaphysics, *différance* will not factor in their generation or relations to one another (assuming the generations admit plurality). We shall see that this conclusion has seminal consequences.

9.7.3 "Non-presence" and Negation

We confront the "certain non-presence" which exists only as a "vestige of the formless" and the associated fact that trace "announces" and recalls "its other to the whole of metaphysics."

It will be instructive to compare the range of negation, the "non"-factor developed in the passage discussed in Chapter 6, with the parallel negational factor in the passage at hand. In the analysis of **Passage 6C** in Chapter 6, the context suggests that Derrida's use of "not" relative to trace should be limited primarily to temporal considerations. The passage specifies temporal considerations essential to trace by juxtaposing the efficacy of trace next to "what is called" past, present, and future. This specification would be appropriate if the primary emphasis in Derrida's exposition at this juncture is the role trace plays in actualizing language, whether spoken or written, given the temporality built into the *verbal* dimensions of language. However, limiting the scope of non-presence in its metaphysical completeness strictly to temporal considerations must be justified for two distinct but related reasons: first, because the traditional referential range of non- as a class complement is exhaustive and completely indeterminate; second, given this unlimited indeterminacy, any particular or determinate partitioning of the comprehensiveness covered by the complement class must be identified, described and justified with respect to the limitless range of other possible components.

But surely such specificity will be difficult to secure. The certain non-presence announces and recalls "its other to the whole of metaphysics...." We take the whole of metaphysics to refer to the entire set of positions advanced in the name of describing and analyzing being as presence. "Non-presence" then refers to everything other than presence, at least if, again, "non-" indicates a complementary class. If so, then if whatever falls under trace can indeed be realized, its content will be completely other than the "the whole" of metaphysics, i.e., this content will be in all respects irreducible to any moment in the history of metaphysics. Thus the content of every rupture or irruption (assuming a plurality of such episodes) will be different from *the content of the entirety of that history.*

Derrida fills this gap of nothingness, of sheer otherness, by identifying "the vestige of the formless" as "already and in itself" existing as "the trace" relative to form now understood as a synonym for presence. If we emphasize the terms "already and in itself" along with "vestige," the picture that emerges is of a parallel stream of reality that exists "in" the mode of being named form but resident only as a "trace," i.e., something "secretly" related to being as presence *qua* form. Thus the connection between the two distinct modes of reality is not evident at any one point within the history of metaphysics given the pressure being as presence exerts in permeating that history. But Derrida insists that a "certain non-presence" is nonetheless "incessantly" related to canonic metaphysics, that is, at no point during the entirety of the history of metaphysics is trace as the "vestige of the formless" *not* in attendance–albeit secretly–to that history.

This "certain non-presence" is and has been established at least indirectly by the very concept and reality of form as it has dominated and permeated the entire history of metaphysics. All differences between Plato and Aristotle and by extension between and among each and every instance of systematic metaphysics belong to metaphysics in this global sense, in the sense encompassing the totality of western metaphysics. This complete set of differences, duly and doctrinally elicited by Derrida's notion of *différance*, clearly constitutes the "presence" class. But now, the import of a "certain non-presence" has been extended to include the possibility of something over and above the actualization of the history of metaphysics, a possibility which can be accessed or revealed by trace once trace is allowed to transcend the limits of the closure of metaphysics when being exists as presence.

Derrida puns on "non-sense" by introducing a feature of his thought which may appear to readers thoroughly imbued with the history of metaphysics as "nonsense." For Derrida, "non-sense" is not "nonsense," although the latter term may register the initial reaction of those who take seriously the millennial traditions of metaphysics when they confront the revelations bestowed in and through traces. These individuals may well feel uneasy if not thoroughly disoriented when they behold the "monstrous" possibilities which apparently arise once deconstruction has conscientiously "shaken" the language of traditional metaphysics and epistemology, possibilities doubtless striking such a viewer as "nonsensical" by comparison with accepted metaphysical doctrine. But an appropriately sympathetic response to Derrida's project is to approach such impending "non-sense" as the source of what is not only meaningful in some essential respect but also crucially meaningful given the limiting confines of western metaphysics as it has realized itself under the overly restrictive reins of being as presence.

Trace then becomes the necessary gateway to actualizing the dimension of non-presence, i.e., "non-sense," in relation to being as presence. It seems legitimate therefore to "translate" this evanescent, negatively characterized presence into the language of possibility in the sense that at any given point along the history of metaphysics it is and remains theoretically possible for "non-presence" to have emerged or to emerge as an "irruption." Thus "certain non-presence" functions as an ongoing and apparently unlimited potentiality to complement the totality of instances of systematic metaphysics defining the western tradition. Only when this historical epoch was in fact "closed" could this systematic dimension be recognized for what it was, is and presumably will be. Only when Jacques Derrida "saw the light," as it were, was this realm of reality recognized as *systematically and epochally present to* the history of metaphysics and embodied within the potentialities resident in the domain of traces.

9.7.4 Opposition and Possibility

At one point, Derrida wrote that "trace" and "*différance*" can apparently function interchangeably.[120] Such terminological fluidity is consonant with deconstruction's preference for lack of rigid technical language. But **Passage 9D** locates at least part of the import of trace *outside the limits of metaphysics proper*. If so, then since *différance* has engendered the history of metaphysics to be what it is, the straightforward substitution of "trace" for "*différance*" will fail to preserve the difference between (a) what could result from traces which is new and important to metaphysics when characterized solely by its canonic texts and (b) what has already happened *within* the history of metaphysics as a set of established (not to mention well-studied and discussed) texts.

If non-presence refers to a region of reality completely other than everything encompassed by being as presence, then Derrida intends a counterpart totality, something as referentially encompassing as that controlled by being as presence. Traces bridge the divide differentiating these two sectors of reality; and form, as defined and characterized by the history of metaphysics, constitutes the "trace" between that history and the theoretical proposed otherness. But at this juncture Derrida makes an important move. This sense of trace should not be construed as a "mixture or passage"–in more neutral terminology, as a relation–between form as the cardinal reality of the history of metaphysics and the indeterminate "other" which, Derrida insists, nonetheless exists. A mixture or passage would presuppose an opposition of stark immediacy and separation between form as presence and non-

120 See **Passage 1M**. According to the argument in this chapter (developed and illustrated in Chapter 10), Derrida wrote on trace in ways which suggest, if not imply, that traces convey measures of potentiality for future thought which are not resident in the creative operations of *différance* as such.

presence. Rather, Derrida posits that trace *renders this opposition possible* because of "its irreducible excess."

The perspective Derrida takes at this speculative juncture is decidedly transcendental; he describes what must obtain to ground the possibility that opposition of this generality can emerge and become viable if not definitive for an entire epoch. But if this single source makes opposites possible or, more precisely evoked, what are construed as opposites from the perspective of a vantage point within one of those contrasting realities, does it follow that this source, taken by itself as a distinctive and foundational unity, must have structure characterized by uniformity in at least one respect? For if two streams of realization emerge from the same source, then it seems plausible to infer that at some point the two streams will share at least one characteristic (perhaps more than one). This assertion is based on the assumption that one and the same origin confers some measure of sameness on whatever emerges from that origin, i.e., that the sequences Derrida describes are in some respects analogous to an organic process.

But caution is crucial. Derrida presumably skirts the necessity of this inference by appealing to the "irreducible excess" of trace. Trace can beget an "other" relative to the history of metaphysics but the depths of this irreducible excess are hardly plumbed by that history, long and spectacularly creative as it has been (or, more accurately from Derrida's standpoint, "was," assuming that the history of metaphysics is now closed). However, even if the irreducible excess is not exhausted by form as defining the history of metaphysics and non-presence as accessible through traces, it does not follow that these two streams, properly described as "opposites," might not share at least one (possibly more than one) discernible characteristic. We will return to this supposition below.

9.7.5 The Matter of Excess

For Derrida, trace escapes the opposition between "presence and absence;" in fact, trace reaches a level of fundamentality such that it renders this opposition possible due to, as we have seen, "its irreducible excess." Trace is not a "mixture or passage," i.e., not a *relation between* presence and absence such that one could derive what it accessed when it was in contact with presence and then correlate the substance of that eventuality with what it then came in contact with when it had entered the realm of presence. Trace "escapes" this opposition and allows what is other than the history of metaphysics to appear. It appears justified then to infer that Derrida's use of negation in "non-presence" (and cognates) *replicates* the totality-dimension of being as presence. If so, then an isomorphism obtains between being as presence and the range of possibility Derrida builds into negation. Derrida's "not" includes a multi-pronged dimension of potentiality or possibility since whatever falls under this "not" will, when it has been expressed, counterbalance the excesses of western metaphysics. The question is, however, whether it is possible to say or even merely to intimate any of the content of whatever falls under the class of "non-presence," a question which in turn presupposes some sort of interpretive access to the inexpressible source itself.

This source, we are assured, exists beyond the confines of terminology restricted to presence and absence since it is ultimately that from which both presence and absence originate. What then can be said–if anything–about this source? Reflection on a reality of such fundamental character perhaps reminds us of the reaction immediately after Socrates tells the company at *Republic* VI that "the good itself is not essence but still transcends essence in dignity and surpassing power."[121] At this moment of high philosophical drama, Plato has Glaucon exclaim: "Heaven save us, hyperbole can no further go" (509c–Shorey's translation). What Derrida has in mind is doubtless far distant from Plato's *to agathon*, the Good, although a parallel exists in that the power of the Good transcends essence–presumably all the Forms insofar as Forms make all things to be essentially what they are–whereas Derrida's source also transcends the opposition between presence and absence, an especially primordial instance of opposition since its combined scope has characterized and defined the entirety of western metaphysics. In sum, to assume that "presence" and "absence" are jointly exhaustive is apparently a mistake; if Derrida's pronouncements are correct, something exists which transcends this singular, metaphysically-driven and neatly-stated dichotomy.

How then should "irreducible excess" be understood relative to traditional metaphysical concepts? Observe, first, that with this excess in place, Derrida maintains that the closure of metaphysics should not be envisioned as a surrounding aura with the totality of metaphysics on the "inside," as it were. Closure is not a process culminating in a conclusion, or a winding-up. Closure of metaphysics does not mean the end of metaphysics in a sense determinable by historical dating or even by a finite number of existing texts. Closure refers to what can be "organically inscribed and systematically articulated" from what has been said or written *within* the history of metaphysics.

It follows, second, that Derrida is not appealing to possibility *qua* possibility, that is, to an aspect of something existing which has the potential or capacity to be realized but which may for a variety of reasons not be realized. Indeed, his usage of possibility invokes actualization of this possibility as long as we take an appropriate perspective on the sense of possibility in play. Thus the appeal to "incessantly" means that at any given point it is possible that, as noted earlier, circumstances will be aligned in such a way as to enable irruption of content actualizing and expressing the parallel realm of "metaphysics." Derrida's position depends for its cogency on

---

121 Derrida variously used this passage on the Good in the *Republic* as an emblem of the foundational dimension of the history of metaphysics. See: WD, 56-7, 85, 98, 153; MP, 266, 268.

the real possibility of potentialities resident in the accounts comprising the history of metaphysics which, if these potentialities were to realize themselves–irrupt, in Derrida's vocabulary–would apparently challenge if not nullify the hegemony of presence and form as the overarching principle both guiding and underlying the history of metaphysics.

We should not lose sight of the considerable range, depth and speculative dimensions of Derrida's proposals. What sanctions the assertion of this heretofore hidden source? Where does it originate? Or must it simply be accepted as a most fundamental and decidedly foundational theoretical given? Finally, why should such a massive potentiality be considered even marginally present to that history? The extent to which this source is, in Derrida's ontological vision, itself hyperbolic remains to be analyzed. A final consideration: even if we grant as a logical possibility that such content lies dormant in the "irreducible excess" of trace, how will this content be expressed given that any and all language currently available to thoughtful application will necessarily evince being as presence insofar as language itself, whether technical or colloquial, is epochally determined by being as presence? Apparently we must understand language in the broadest sense, a system of "expressive gestures," to be able to appreciate the realization of traces when they do occur.

9.7.6 Non-presence: Otherness and the Transcendental Dimension

Earlier in **Passage 9D**, Derrida insisted that "non-presence" had announced "its other to the whole of metaphysics..." Later, Derrida speaks of reconciling what trace had "cracked" from the history of metaphysics as derived from "systematically" inscribing "the before, the after, and the outside of metaphysics."

We might construe trace as a route of exploration which approaches metaphysical thought from the "outside" of the totality of that thought–"outside" referring to modes of possible intersection with reality which would predate the spare language of the pre-Socratics and postdate Heideggerian fundamental ontology and *Denken* (as a name for the last phases of Heidegger's more "poetized" reflections[122])–and yet penetrates the totality of western metaphysics.

Derrida has provided directional markers for this quest by taking advantage of the linguistic presentation of opposition as an ordering rubric in order to formulate the contrasts advancing his argument–e.g., the contrast between all pertinent dimensions of "sense" animating being as presence and a certain "non-sense." Note also the introduction of images driven by straightforward spatial opposition,

122 For commentary on this dimension of Heidegger's thought, see White 1978, Chapters 7 and 9.

i.e., appeals to the "before" and "after" of the history of metaphysics as well as to what is "outside" that history, a spatial sector presupposing that this history is marked, in some sense, by an "inside." If we attempt to think beyond the spatial imagery of Derrida's terminology, we may wonder whether, if before/after/outside are intended to apply to a unified and organically whole set, it will follow that traces must engage each member of the set? What would "before" and "after" metaphysics refer to? Although Derrida has cushioned the resident sharpness of these terms with a series of qualifying gestures, I submit that deconstructive concern to break down opposites should also be in play in these contexts. The fact that Derrida contours this passage in terms of a series of opposites indicates that even at this seminal moment in transcribing his anticipations for future thought, his own texts remain resolutely within the framework of metaphysical representation.

In that vein, it is important to appreciate the full-fledged transcendental aspects of what is being proposed. The appeals to the totality-notions discussed previously indicate that Derrida stipulates a set of limit-conditions for realizing a given possibility in order to achieve a given end. His thinking has taken a position outside the scope of an entire tradition of philosophical thought, measured the fundamental character of that thought, then expressed an authoritatively stated dimension of significance which will rectify a skewed commonality defining that thought.

This level of high abstraction and transcendental concern is reinforced with Derrida's use of "systematically." Is the intended sense of this qualifier equal or even similar to system as found in the history of metaphysics? Does system in this sense presuppose a uniform configuration characterized by a part-whole relation? Is there then an implicit reliance on principles, assuming that "systematically" will exhibit itself according to order in some sense? Will this ordering dimension itself depend on the concepts and principles of "logic"? If so, then the rules of formal order must either (a) be determined in a manner which is difficult to conceive of while we continue to reside in the epoch of metaphysics or (b) inevitably draw on the traditional rules of logical formality, implying that in this respect the elements of metaphysics continue their dominance even when trace has disclosed entirely new dimensions from its irreducible excess. This last inference also suggests that if system in this context means anything like what it means for a variety of instances of western metaphysics, then what trace will reveal from the "excess" may not be all that different from what the history of metaphysics has already made available.

Finally, we take "within the traces" to denote that traces are not just limit-setters to an entire epoch of millennial duration; they are also principles of movement within the epoch itself. In fact, the discussion of trace in Chapter 6 detailed ways in which Derrida construed trace in its integral role in the transmission of linguistic meaning, whether in contexts of ordinary or philosophical character. Trace has a double function: (a) as theoretical element that accounts for the possibility of signification insofar as that concept necessarily functions in philosophical discourse as well as areas of common life involving the practical efficacy of language. And also (b) as source of potential escape or liberation from defining–and confining–characteristics of being as presence in terms of the vantage point of thought which has sufficiently grasped implications of the structure of trace. The presence of trace thus exceeds the limits of presence even as it funds being as presence; rather, trace itself is what, in conjunction with that history–and therefore transcending the metaphysical notion of opposition–enables and informs both that history and what is other than that history.[123]

9.8 Trace and Irruption: The Factor of Surprise

In the final phase of the passage under scrutiny, Derrida refers to "an infinite and infinitely surprising reading of this structure and history." He elaborates: "An irreducible rupture and excess may always occur within a given epoch, at a certain point in its text (for example in the "Platonic" fabric of "Neo-Platonism") and, no doubt, already in Plato's text." The notions of surprise and rupture or irruption may connect to the fundamental reality of the irreducible excess as evinced by the appearance of traces.

9.8.1 History and Source

The "its" of "its text" apparently refers to the totality of systems constituting a "given epoch," although the passage does not indicate what determines the limits of one epoch as opposed to any other epoch. Furthermore, the reference to "structure and history" appears, in context, to be about the history of metaphysics in its entirety as an organic whole, from its beginnings to its stipulated "closure" with Heidegger. If so, then the entire history of metaphysics constitutes an epoch, differentiated in turn by a number of "given epochs." The appeal to "epoch" is significant because the distinctive nature of a given irruption will depend at least in part upon whatever textual limits constitute the boundaries of a given epoch. The "irreducible rupture and excess" resulting in what Derrida refers to as "infinite" realizations must mean that whatever "irruptions" occur within a given epoch will always and necessarily emanate from the same source of otherness. In addition, future expressions regardless of their novelty, eccentricity, or even monstrosity will arise within the ambit of this excess given that

123 This point is stated abstractly and, at this juncture in the essay, more as a programmatic possibility. But I submit that trace so read stands as an entry for articulating the dichotomous differences Derrida deems crucial. This approach to the notion of trace is developed later in Chapter 9, the section "Deconstruction and the Possibility of Categorization."

all such realizations are merely the next advances in the "infinite" range of resident traces as they emerge from this excess.

9.8.2 Epoch and Sequence

The "given epoch" refers to the period between Plato and Plotinus. In general, the sequence of positions defining this epoch is fixed and invariant–thus Aristotle comes after Plato and however the epoch of western metaphysics is construed and whatever irruptions may emerge from that epoch, Aristotle always comes after Plato. This is an empirical matter decided irrevocably by the accepted history of pertinent texts. The sequential order of the figures comprising the history of metaphysics will, as invariant, doubtless affect the range of possible irruptions. Presumably then any specific doctrinal advance which occurs during the history of metaphysics does not automatically quality as an irruption since the content of that advance is directly limited to and conditioned by the position of its antecedent, e.g., Aristotle on form compared to Plato on form. What justifies an irruption's occurrence must therefore be identified.

9.8.3 The Determination of Surprises

There were irruptions in the texts of Plotinus and "doubtless" also in Plato's texts. Note how the two examples differ: if Neo-Platonism means adaptation of Plato by Plotinus, then the irruption is a relation between later and earlier philosophers; but an irruption is claimed for Plato himself as emergent within "Plato's own texts." How can a student of these two thinkers determine such epiphanies? Thus how would we distinguish between an irruption and a straightforward alteration of or advance from a previous position? When Aristotle famously maintains that forms exist only in particulars and do not exist independently of particulars, there is no reason to believe that this sequence constitutes an irruption in the intended sense; it could be merely Aristotle's reaction to problematic implications in the Platonic position concerning the independent existence of Forms. The same considerations would obtain for any two-position sequence throughout the history of metaphysics as well as any multiple-position sequence (e.g., the progressive change in the constitution of knowledge from Locke to Hume).

Two questions: how do we recognize an irruption when it pertains to a sequence of positions? And second, how do we detect an irruption when it pertains to an alteration of doctrine *within* the texts of just one philosopher (e.g., Plato)? It is surely necessary to distinguish between an advance from a previous position, whether within the personal history of a given philosopher or between a sequence of philosophers, and a position which would constitute a "surprise" based on a relation between something novel and something already stated. A review of the structure of *différance* underlines the importance of this question.

9.8.4 Surprise and *Différance*

It may seem odd to analyze "surprise" as a structural component. However, "surprise" is now a technical term; attention must therefore be paid to identifying its determinable elements.

It is axiomatic that an irruption may occur at any moment along the history of metaphysics. But by definition of trace, an irruption will emerge formulated in terms other than *any* terms resident in the entire history of metaphysics, not just any one or several positions as appearing within that history. According to this interpretation, irruptions are not marginally incisive deployments of existing texts; rather, they are fundamentally new positions given the content of what has preceded them. From this perspective, an irruption must differ from any given sequence of positions extant during the history of metaphysics as that history has been handed down over millennia.

Derrida uses Plotinus to illustrate the realization of a surprise, with this irruption apparently having an epochal effect on all of Plotinus if not the western tradition as a whole. Derrida mentions the likelihood of, but does not cite, an example of a correlative surprise in the dialogues of Plato. I suggest a possible example in *Philebus*. The metaphysics advanced in the *Philebus* comes after the metaphysics of, say, the *Republic*. There are surprising features in *Philebus*, i.e., the elevation of a reality (the "unlimited"), often interpreted as the Platonic version of matter, which as one of Socrates' four "new tools" is apparently just as essential to the structure of overall reality as the Forms (see *Philebus* 23bff). The assertion of a principle of matter would surely have an epochal effect on all Platonic dialogues in reorienting the focus of metaphysical inquiry with respect to particulars, since presumably they would then no longer be, as claimed in the *Republic*, a straightforward blend of being and non-being (478d). If a reading reveals a potent insight into a certain figure with repercussions that affect not just the understanding of that figure in isolation but also the status of that figure within the entire epoch, this kind of insight can be aptly described as surprising.

The argument then is that surprise in the relevant sense illustrates a reading which presents a novel perspective on a particular figure within a given historical epoch, a perspective with repercussions pertaining to the entire epoch as a unity. If such textual consideration is *not* what Derrida had in mind with the appeal to "surprise," then we need additional clarification in order to recognize when a surprise in the matter of the classic texts of western metaphysics is upon us. In the usual sense of the term, "surprise" is a relational concept, that is, an event is surprising when it appears (a) with respect to something other than itself and (b) in a way which runs outside boundaries that control patterns or sequences of normalcy. The second factor in this account of surprise must be clarified in more general terms so that the concept of surprise establishes a measure of legitimacy within the theoretical context under scrutiny.

9.8.5 The Limits of Surprises

An appropriate reception of a trace offers "an infinite and infinitely surprising reading of this structure and history." Two distinct but conjoined properties are at hand–first, the reading in question is "infinite," second, this reading is "infinitely surprising." But why "*infinitely* surprising"? I suggest that "surprising" is qualified by "infinitely" on condition that the content of the reading in question cannot be reduced to any single moment within western metaphysics when viewed *as a totality*. For if an irruption were identical to *any* existing position, it would not only become arguably indistinguishable from that position by being subsumed under the all-embracing control of being as presence but also merely duplicate a position already existent within the arena of metaphysics. Thus whatever this language evokes will not be readily connected with any antecedent located within the history of metaphysics, whether in part or in whole. Is it possible to discern an example of a surprise at such an "infinite" level of significance?

9.8.6 The Content of Surprises–A Proposal

Here are two possible surprises of very different order and with decidedly different implications. If "infinite surprise" encompasses a surprise which is measureless in its implications (keeping in view the description of trace as announcing itself to the "whole" of metaphysics), then Derrida's own recognition of the epochal nature of the history of metaphysics as well as the diagnosis of being as presence characterizing this epoch might be a pertinent instance. This reading reveals a cardinal apprehension covering the entire history of a given epoch (e.g., the history of western metaphysics). If so, then Derrida's theoretical position could refer to himself and to deconstruction as an irruption occurring within the epoch of modernity. Or, more concisely yet, Derrida's announcement of the existence of irruptions is itself an irruption. Thus the apparatus of deconstruction and all texts dealing with the exposure of being as presence can be considered an instance of an infinitely surprising reading of the entire tradition.

The second possibility or "surprise" as it pertains or, more cautiously, could pertain to Derrida's own texts would be the re-emergence of precisely those elements occurring and indeed defining the history of metaphysics, the very elements which deconstruction has taken upon itself to shake, uproot, and expose as unduly repressive. An appropriately convoluted summary of this possibility: the announcement of this surprise is itself surprising if the point of the surprise is the realization that a proponent of deconstruction should never have been surprised by this surprise! Indeed, these examples could be renamed “meta-surprises” in that they are derived from “outside” Derrida’s texts. They illustrate the possible application of surprise when that notion applies to Derrida himself and to implications from his texts.

The transcendental dimensions of this position again emerges; regardless of the content of an irruption, it will represent an “infinite” surprise. We may also have cause to wonder whether this kind of reading occurs continually, regardless of the specific character of the irruption, and merely awaits an appropriate mouthpiece. Or will this currently concealed otherness irrupt on its own, sporadically, if properly enticed or enabled?

9.8.7 Trace, Surprise and the Possibility of Assimilation

The set of basic questions and problems enumerated in this section of Chapter 9 culminates in the question of *assimilation*. The appeal to “the whole of metaphysics” implies that whatever emerges as an irruption will be so distinct, so different–in a traditional sense of divergency or sheer otherness and, as such, not reducible to the activity of *différance*–that both its deep structure and surface sheen may be too jarring to assimilate readily into western philosophical sensibilities. But assimilation is required at some point; otherwise, the philosophical adventure in its historical entirety will be marked by irreconcilable stops and starts, a zigzag pattern without endowing the possibility of collecting and combining insights and visions as they emerge from the various realized instances of irruption in tandem with the traditional messages of the established history of metaphysics from the pre-Socratics to Heidegger and his successors. How then to secure such assimilation?

9.8.8 The Accessibility of Irruptions

We consider in more detail irruption as a factor in the function of trace providing a passage for the emergence of what has been hidden throughout that history. This discussion continues to elicit the transcendental dimensions inherent in the notion of trace.

An “irruption” typically conveys volcanic suddenness. If this image continues to be appropriate in a philosophical setting, then the articulation of those hidden elements would be sudden, perhaps dramatic in its presentational force and in its effect on those concerned with the existence and emergence of such thoughtfully-laden phenomena. Furthermore, the fact that trace derives from the absolute otherness of the source common to the origins of western metaphysics will presumably establish its uniqueness of articulation and originality of content. Derrida's position implies that it is possible for language to irrupt at any moment within any given closed system. But even if irruptions in fact occur, how will the language of irruption, whatever the configurations or unusual variations it may display, be recognized as such and also concurrently escape the epochally driven effects of being as presence?

Productive deconstruction promises novelty when applied to traditional texts. But any statement of metaphysical proportion intended to move outside the closure of metaphysics would doubtless be conditioned to some extent by what has occurred *within* the textual confines of that epochally determined whole. However, if western metaphysics has missed something important or even if it has merely overemphasized elements which, measured by deconstructive appropriateness, require moderation, modification or revision, then it is not evident how any of these objectives could ever be formulated and expressed in language given that language itself has also been necessarily contoured or, more accurately from a strict deconstructionist perspective, distorted by metaphysical considerations.

The following implication must also be considered. Any claims made about reality currently concealed by the oppressive history of metaphysics must presumably be *justified* in some way as indeed novel and incisive, even if this justification is purely rhetorical and not argumentative, that is, not controlled by the putative principles that have traditionally orchestrated the structure of right thinking as such thinking has appeared within metaphysics' history. If this justification at some juncture appeals to evidence, understanding evidence in the broadest possible sense, then as argued in Part I we wonder whether such evidence must still be "present to" anyone receptive to the new insights into reality vouchsafed by deconstructive means. As a result, the argument Derrida brings to bear against the controlling and repressive history of metaphysics with respect to its reliance on being as presence appears to apply with commensurate force to the results of any attempt to use deconstruction for purposes of saying or writing something "new" and heretofore unseen and repressed by that same history.

But if so, then by parallel reasoning the novelty of traces will be compromised by the very fact of such experience. Therefore, if an irruption were to occur, its articulation would presumably be subject to the same strictures marking the total output of the history of metaphysics. If the reply to this criticism is that "present to" trades on ambiguities in this phrase for its critical force, then this response reinforces one of the central theses of this essay, that Derrida's notion of presence is unwaveringly ambivalent, at least in the various texts we have examined, and therefore that it requires considerable clarification and precising before presence can do the kind of speculative and critical work Derrida asks of it, especially as a summary encapsulation of the entire history of metaphysics.

We suspect then that deconstructively inspired reformulations of being as presence as emergent through the seminal agency of traces would currently be such that no account of being could be offered without that account also falling under the notional constraints of being as presence, thereby reducing that projected account to just another moment in the continuing sequence of such moments as they defined the history of metaphysics. This conclusion seems difficult to avoid. We should nonetheless hold open the possibility of envisioning this (indeed any) such irruption so that its statement will be expressible in a manner completely other than that characterized by the kind of discourse implementing all the positions comprising the history of metaphysics. How should a proponent of Derrida's position proceed in order to maximize the fulfillment of this hope?

9.9 Deconstruction and the Possibility of Categorization

It may seem surprising if a position (or positions) of stability became visible from sources combining the determinate entries governed by trace and the process elements inherent in irruption. However, implications and presuppositions dwelling in Derrida's own texts as well as in assertions explicitly made through those texts–all this material bearing unmistakable signs of transcendental strategy–support the possibility that we can secure such stability. If so, then the theoretically posited "surprise" in question should not, as the idiom has it, take us completely by surprise.

The original assumption underlying the argument in this chapter was that Derrida must accept a measure of stability in order to formulate a set and resolute response to Searle's critique of Derrida on Austin. In fact, it may be argued that Derrida's texts on characteristics pertaining to the history of metaphysics coupled with his calm confidence that a given reading of one of his own texts is not right but wrong yield the possibility of categorizing moments derived from that history and, by extension, visualized through the notion of trace. This investigation will lead to developing a system of categories–perhaps only provisionally accurate at this point in an informed response to Derrida's position–which collectively guarantee the possibility of the requisite kind of stability. We may read such a yield as an irruption proper or at least as a result similar to an irruption in Derrida's technical sense in that its content will be surprising, even if its effect on a target audience is not "infinitely" surprising.[124]

---

124 This dimension of possibility is developed in Chapter 10 (and may be taken to address Norris' point, that deconstruction is aimed at "discovering truths not accessible to current modes of discourse"–see Norris 2012, 95; also 96). The setting fits the sense of irruption which Derrida once referred to as the "thought of Being" (WD, 82). For representative observations on possibility, see MP, 319, 324, 325, 327, 328; Derrida 2001, 108-9, 111; N, 23.

Derrida has asserted axiomatically that form is presence. But we now have learned, given Derrida's conception of negation and the directional power inherent in trace, that words can refer in ways other than those invoking their intended referents as instances of being as presence. The telos of discourse need not necessarily be form. Thus evidence for interacting with "being" can assume another dimension, another "metaphysical" guise. This possibility is one metaphysics "would have excluded from its field." The problem is to interpret the relation between the history of metaphysics and what trace allows to emerge from negation as non-presence–something which for Derrida stands as an infinitely surprising storehouse of possibilities. Can this exclusion become discernible depending on the care with which a student of "being" reviews and analyzes (in ways which incorporate the strategies of deconstruction) the history of metaphysics? Derrida's notion of irruption establishes this possibility at least if this notion collaborates with the insights ascribed to what can appear through a trace. In fact, the "irruption" of the proposed provisional categories will accomplish two ends, in contexts initially marked by explicit particularity as well as broad comprehensiveness.

First, these categories justify Derrida's conviction that right reading of his work exists. However, the immediate implication is that if one author can do this, then any author can achieve the same result as long as the requisite conditions are met. It would, after all, certainly seem arbitrary and self-serving to grant special privilege with regard to reading, language, and interpretation, that is, to exempt Jacques Derrida's texts from the dismantling procedures of canonic deconstruction, when there is considerable reason to believe that these procedures apply uniformly to *all texts* regardless of the identity of their authors. Second, For Derrida, all philosophical discourse has been affected by *différance* as modes of language have begun to exist within the bounded course of western metaphysics. If so, then it would be premature to expect a sharp and decisive split between traditional discourse and that which expresses, or reveals, something of the "irreducible excess" which ultimately grounded both the history of western metaphysics and closure running side-by-side with that history, yet unrelated to that history in anything resembling cause-effect linkage or oppositional bonding.

Recall again Derrida's appeal in *Of Grammatology* to shun traditional categories, literary and philosophical, and to be ready to confront a "monstrous" account of fundamental considerations in philosophy. Also assume that the philosophical monstrosity Derrida refers to can be and is realized, and that its articulation exhibits a tolerably smooth succession between past metaphysics and future thought. This transitional discourse will be transformational in intent and focus but its mode of presentation will also exhibit at least the trappings of language. This language (for lack of a better word, since it may not be like language in traditional modes) has emerged, says Derrida, from an "infinite" source. As a result, there may be many variations, embellishments and advances on whatever first appeared to the contemporary student of philosophical discourse as "monstrous." This sequence of novelties will also, presumably, have its own history as an extension of thought beyond the present bounds of philosophical propriety.[125]

The argument of this chapter has contended that what we should look for from Derrida's texts are ways to coordinate our experience and the resultant philosophical vocabulary deemed suitable for its expression so that it is possible to situate the desired attributes of stability. But Derrida insists that these attributes originate from a source necessarily encompassing the history of metaphysics, since that very history and the currently undetermined and "unstated" dimensions of being–what "trace" will reveal–originate from *the same source* with, in consequence, what appear to be an infinite range of options for future thinking. Therefore, it will be a necessary condition of any attempt to encapsulate the undetermined or currently "missing" aspects of being that any and all of these indeterminate ontological shadings must also be accessible to and contoured by considerations of categorical strength which have been and continue to be appropriately shaped by the history of metaphysics. This condition will not be sufficient, however, since the content of what remains unthought and unsaid may affect, whether in a fundamental way or only marginally, the structure of such categorizations. Yet it seems clear that some elements of their structure must remain, otherwise Derrida could not have maintained that both (a) the history of metaphysics and (b) whatever traces will disclose "other" than the content of that history are in fact related with the degree of intimacy which he so confidently asserted.

But now an important implication may be drawn. The closure of metaphysics does not imply that the history of metaphysics, from Heraclitus to Heidegger, will simply vanish, its content and effects swallowed up, fully digested and then transformed by the infinitely surprising new visions produced by properly attuned (and deconstructively experienced) thinkers. If this history continues into the present and beyond, then the result is a pair of histories, with the former, that of metaphysics in the traditional sense, coupled with its sequel as opened by deconstructive practices and gestures. Therefore, *both* histories must be capable of resonating with one another for the speculative consideration of the student of philosophy on those occasions when the discipline recognizes that it has generated a complete cycle and then will regenerate itself in new and, if Derrida's surmise is correct, irruptive incarnations.

What categorical possibilities remain viable to execute these demands? Given the emphasis on novelty that emerges from canonic deconstruction, it would be "surprising," although perhaps not "infinitely" surprising, if the very categorical structures and perspectives which had appeared emblematic of the stable qualities

125 Derrida challenged the traditions of orthodox philosophical exposition with the distinctive text formats employed in *Glas* and *Signsponge*. But will such orthographic novelty affect the substance of philosophizing in responding to and shaping the lifeworld? Does the visual field in which a text is presented somehow bear on the meaning of that text?

of being and, as such, had been dominant throughout the history of metaphysics were in fact necessary to guarantee continuity with that history as well as to establish feasible contributions to the philosophical enterprise. We suggested this possibility in the previous section of this chapter. These specifications would turn the element of surprise against deconstruction's purported inspiration of surprise, since deconstruction would resurrect from metaphysics and elevate as necessary for the ongoing application of deconstructive inquiry precisely what in the canonic texts of the history of metaphysics it apparently aimed at subverting and reversing and exposing as concealed. In the same vein: will this residual surprise, as it might be termed, also encompass a fully-fledged dimension of formality for the purpose of "logically" ordering any forms of the prospective novel discourse, regardless of the content of this novelty? If so, then whatever necessity installed the formality of logical concepts and principles as essential elements in western rationality will remain just as necessary during any irruption of deconstructive insight emergent from the now closed system of western metaphysics.

In order then to accommodate future philosophizing in all its multiform guises, categorization must be available to cover coupling the content of metaphysics as first realized with novel (albeit monstrous) revelations of being in its post-metaphysical emergence. It follows that the categorization specified must be sufficiently supple and inclusive to encompass ways of thinking based on traditional modes of analysis and the content which that analysis produced during almost three millennia of philosophical labor. The ultimate goal is to recoup relevant features or perspectives lost within or concealed by the cumulative weight of a repressive history of metaphysics and then, at least given the realm of possible conceptualization, to integrate these elements within a set of basal concepts or categories as these vehicles of organization have been more traditionally conceived. Whatever irruption does indeed emerge from the infinitely surprising source of negation, i.e., non-presence, will have to pass through the tainted terminology of the prism of presence; therefore, the best that can be done now, at this moment in history, is to prepare the way with *provisional categories* in service to the range of philosophical novelties awaiting their determination.

These categories will stand as intermediate parameters in service to potential transition *from* discourse controlled by being as presence and the stultifying history of western metaphysics *to* articulation of what has heretofore remained unsaid in the "irreducible excess" funding both being as presence and all that is and could be other than being as presence. Such a large-scale possibility controls this essay's conclusion. The categories developed in Chapter 10 are provisional possibilities in the formulation of aggregating notions which allow a student to view both historical sequences–metaphysics proper as well as the tracing of post-metaphysical experience and thought–along comparative lines.

We assume then that tension between the history of metaphysics as an organic but closed unity (cf. **Passage 9D**) and the revelatory openness provided by deconstructive attention to texts will be eased by formulations and systematic codifications aiming toward and at least partially informed by traditional metaphysics. This approach suggests that any substantive creation engendered by assemblages of deconstruction may result in an account not all that different from some moment (or moments) appearing within the history of metaphysics. Whether this possibility is viable awaits analysis and thus, at the moment, is indeterminate, a book printed on pages of potentiality but yet to be cracked open.

9.10 Summary: Closure and Surprise–the Code of Metaphysics

We have advanced the deconstruction of deconstruction from two directions and in two different respects. *Internal* pressure emerges from the fact that Derrida's position threatens to self-destruct in the process of its very articulation unless certain concepts and realities are imported from standard sources in the history of metaphysics and remain functional throughout all Derrida's texts. By contrast, *external* pressure exerts a positive theoretical force; if *différance* establishes closure to the history of metaphysics, then the student of metaphysics must be prepared to experience and if possible to inscribe an "irruption" from that history as an event revealing elements about reality which metaphysics–presumably all moments in its history–has either overlooked or suppressed.

The apparent logical impasse generated by the internal pressure just detailed may be effectively blunted, however, if the external pressure applied by considerations pertaining to deconstruction is allowed to serve as impetus to metaphysical speculation. Potentially fertile categorical discriminations then begin to emerge. We approach this tentative and perhaps tantalizing domain by asking a question with its own sharply contoured contextual focus: what conditions must be presupposed in order for Derrida to contest Searle's reading of the former's critique of John Austin? This is, of course, a narrowly circumscribed situation but if a certain kind of irruption transpires from, as Derrida would have it, the negational dimensions correlated with or at least tangent to being as presence, this question can be answered. In fact, with this answer in hand, it will be possible to address a number of other concerns which have arisen from the various lines of criticism already introduced and developed in Chapters 8 and 9.

In Chapter 10, we conclude by establishing an interpretive framework suitable for describing a set of residual and restorative categories. These categories possess a certain defined structure. When applied to philosophical concerns, they address individual and restricted matters–such as the dispute between Derrida and Searle. However, they are also sufficiently supple to accommodate and facilitate much broader and more inclusive issues. The latter include modulating the transitional period between the closure of metaphysics and articulation in whatever linguistic form that articulation may take of the new modes of philosophizing Derrida advised us to anticipate. These modes will result ultimately from the concerted incisiveness of the deconstructive approach to the structure of language and all that has been written in language, especially the language of philosophy advanced throughout the western tradition.

The essay thus concludes on a speculative note directly inspired by the essay's extended interpretive and critical approaches to Derrida's texts.

10 Toward a Deconstructed Metaphysics

The improvised deconstruction of Derrida's texts advanced in Chapter 9 has revealed an ensemble of concepts–totality, closure, trace, irruption–forming a reservoir of resources which collectively may yield insight into fundamental matters in metaphysics. These endorsements should be connected to Derrida's sustained interest in transcendental or at least quasi-transcendental lines of thought. Indeed, his own critical approaches to metaphysics and language rest on a thoroughly transcendental foundation embodied in the retinue of realities discussed in Chapters 8 and 9. It may be suggested then that the lure of the transcendental animates so much of Derrida's work because he divined at some level that the transcendental was already present within the formulation and development of all the putatively deconstructive results secured throughout his writing career. In his work on James Joyce, for example, Derrida may be read as attempting to bring into the open a discourse which would in some way evoke the transcendental dimension grounding his earlier critical reflections on metaphysics and language–including, of course, the selfsame essays on Joyce.[126] In the concluding chapter of this essay, we develop a speculative but reasoned reading intended to unveil at least presentiments of such discourse. This account is in fact consonant with the thrust of Derrida's earlier period of work on fundamental issues animating the history of philosophy, in particular the phases of philosophy devoted to metaphysical considerations.

10.1 Introduction

A passage toward the beginning of *Of Grammatology* registers Derrida's studied hesitancy concerning the place of his own thinking with respect to the epoch of metaphysics and ways of responding to that epoch:

> **Passage 10A** This would perhaps mean that one does not leave the epoch whose closure one can outline. The movements of belonging or not belonging to the epoch are too subtle, the illusions in that regard are too easy, for us to make a definite judgment (OG, 12).

Once Derrida accounted for "closure" of the epoch of metaphysics, the argument has been that he neither could nor did he intend to leave that epoch, at least not entirely. We recall, however, Derrida's appeal to the reader at the outset of *Of Grammatology* to prepare to embrace a philosophical monstrosity. How then do we proceed if the goal

---

126 For discussion of Derrida's Joyce essays from various metaphysical perspectives, see DFL, especially Chs 1, 2 and 5. See also n131 below.

of philosophical inquiry is to follow directions presented by deconstructive theory and its application? Will this goal be a middle way between respecting the demands of closure and embracing the destabilizing novelty of the monstrous?

One might wonder whether this quest is worth pursuing (assuming that, if the answer is affirmative, it can in fact be pursued at all). Here is one reason to justify this quest. According to Derrida, metaphysics has a history which is meaningful *as history* at least in part insofar as it has continuously if not progressively concealed something under the protracted conceptual veneer enveloping the doctrinal positions studied by philosophers when they read and reflect on philosophical texts. Does this unknown sector have any relevance to understanding, grandiosely put, the nature of reality or perhaps to clarifying the ways we guide our behavior by what philosophers have commonly construed and represented as morality? If not, then even if we grant that the entire history of metaphysics has concealed extended dimensions of significance, unearthing these unknown strata would not seem to matter. But assume that the unknowns embedded in the history of metaphysics are serious in their import. If so, are they then open to some form of revealment by human beings when they think and engage in the public sphere of language? Although the "serious-not serious" dichotomy just introduced could readily be construed as one of the instances of opposition which deconstruction intends to turn on its head or, more drastically, to dissolve into a play of fine shadings, we preserve it here in order to give point to the urgency to follow through on the promises Derrida inserted within his development of deconstruction's principles and gestures.

Assume then that Derrida and deconstruction are important if not essential for the philosophical enterprise and that advocates of this mode of thinking continue applying strategies of deconstruction with all available precision and relevance to basic human concerns. How should such an investigation begin?

For Derrida, an irruption can occur at any moment within the textual confines of any position resident in the historical sweep of metaphysical systems. Imagine then a philosopher sensitive to deconstructive concerns, at least moderately well-versed in literature, and conversant with both the history of metaphysics as well as the guiding canons of logical concepts and principles. That individual will confront the following crucial question, deployed in linked phases: if deconstructing Derrida's deconstructionism reveals that Derrida's acceptance of referentiality as well as the actuality of right answers in the reading of texts points toward an implicit–but necessary–metaphysical position, then can this position be uttered and brought into the open? How will this discursive assemblage once stated in language (however provisional this statement may be) then compare with any particular moment (or moments) already resident in the history of western metaphysics? If our hybrid philosopher is also sufficiently courageous to venture into currently unspoken and unwritten realms of thought, he or she will be in position to augment Derrida's influence throughout the world of contemporary philosophy by connecting and extending the core principles of deconstruction to selected and pivotal moments of western metaphysics. Seriously embracing and attempting to resolve these concerns will, I think, eventually reinforce rather than nullify the practicality of Derrida's approach to substantive philosophical matters.

Consider this example: one could initiate metaphysical inquiry in line with the spirit if not the letter of deconstructive activity by arguing that any theoretical approach toward the effective agency of referentiality would appear to presuppose *particularity*. This conclusion carries plausibility in that the referring function of language cannot be always indeterminate, merely aiming the word beyond its tangible mark and toward an undifferentiated field of experience which cannot in principle be localized and individualized with respect to content. Words in propositions do, although they need not always, refer to things outside of language proper. And in some cases, these things exist as particular entities. Therefore, the category of particularity is a vital element in the theoretical formation of referentiality.

This terse inaugural analysis emphasizing the necessity of particularity may or may not be consonant with a settled sense of deconstructionist principles, but it represents a line of thought retaining a measure of possibility even if the immediate context is dominated by an astringently skeptical reading of those principles. Is there a way to present a position with such content so that it strikes a middle course between slavish dependence on trace elements and disarmed embrace of the monstrous? The prospect of witnessing such a daunting event is surely to be looked at with a uniquely philosophical sense of anticipation. The following pages represent a tentative exploration moving in that direction.

10.2 Speculative Preamble

This section is called "speculative" since it emerges against a backdrop of Derrida's radical questioning. Although everything said in this section is straightforward, perhaps even perfunctory, what appears straightforward to a traditional philosophical sensibility may not in fact be so apparent when experienced within Derrida's world of inquiry. Therefore, the reader is asked to entertain this account, a preamble with a decided theoretical agenda, as a joint statement combining respect for implications derivable from the notion of trace as well as an anticipatory appreciation of the monstrous.

Entities and language about entities differ from one another, and the difference is radical. Thus entities could exist and continue to exist without any language whatsoever. In such a world, metaphysics understood as an account of reality depicted in general terms would not exist even if entities did exist since metaphysics presupposes a student of reality as well as the capacity to express linguistically this student's experience of reality. Assume also that language is always in some respect self-reflexive, that is, that it reflects some of its own abstract and formalized characteristics in process of referring to something outside itself. If so, then any metaphysics will also be in some respect philosophy of language since metaphysics is necessarily linguistic in character. It will therefore not be possible to separate in all respects the purely referential function of metaphysical language–the drive to encapsulate something essential, important, perhaps even interesting about extra-linguistic reality–from the fact that it is and necessarily remains language, with a host of inherent formal, historical and social features, textures and conventions.

Many human affairs involve events of language, a good number of which are successful in their outcome. If so, then it becomes necessary to account theoretically for these results. As a minimal condition for describing such success, consider a state of affairs we may call *equilibrium*. Equilibrium names a set of realities as well as relations between realities and users of language about these realities so constituted that success can–not always, of course, but can–transpire in realizing intentions by human protagonists of language.[127] To account for this equilibrium, we must sketch, however briefly, an account of extra-linguistic reality–what language is about as a field of objects (understanding "objects" in a very wide sense) and as stated in general terms. We must also emphasize features in this account which will render extra-linguistic reality amenable to human affairs when instances of this dimension of reality, what exists by itself and apart from any linguistic connection, become factors in events marked by linguistic success.

Derrida's approach to deconstruction suggests that it is philosophically naive as well as potentially harmful if not destructive to the point of inquiry to pursue metaphysics but to divorce such accounts of reality, however they may be couched and developed, from the capacity of entities, whatever is analyzed and described, to exist in such a way that they can become susceptible to language. Entities never exist "as such" if "as such" implies that a metaphysics can instantly enter into and elaborate the "inner" workings of reality–e.g., something like Aristotelian substance–without taking into account what conditions must be present in order for language to propose *anything at all*, Aristotelian or otherwise, about those inner workings. It follows then that even a provisional attempt at metaphysics will render the abstract structure of a given category, i.e., a vehicle intended to reveal something about reality, so that it necessarily includes a direct correlate with some element or elements in the actualization of language.

---

127 Consider this scenario as a premise: Daniel Dennett proposes the following minimal conditions for a successful response from a subject in a specialized, controlled context–scientific experiments pertaining to any aspect of "consciousness." The subject must express a "speech act" which involves "the sincere, reliable expression by a single, unified subject of that very subject's beliefs and opinions," 76. Presumably Derrida would caution Dennett about all its resident assumptions. However, this type of explanation can surely be matched by all instances of realized speech acts in the lifeworld marked by the same contextual circumstances. To question the reliability of this position, whether in sciences involving human participants or in the everyday matters of the lifeworld, is to countenance a virulent skepticism that would, if acted upon, undercut the possibility of these areas of science and radically alter the stability necessary for productive human action. For Christopher Norris, Derrida's position never goes to these extremes–Norris 2012, 81, 84, 98. See also n134 below.

Now an axiomatic principle for the argument in this chapter: the primary characteristic enveloping all philosophically-determined elements of language is that meaning and reference are secured with sufficient stability to guarantee success in those activities of the lifeworld in which language is or can become an integral component. An attempt to state at least some of the elements establishing this stability appears below.

10.3 Deconstructive Metaphysics: Structural Considerations

If deconstruction is self-referential then, as argued in Chapter 8, Derrida's presentation of that position must be kept separate from the implications of that position. This separation entails a series of assumptions based on logical elements which, as foundational for the very possibility of formulating deconstruction in language, deconstruction cannot address and affect.[128] Nevertheless, if pressure applied by factors pertaining to deconstruction is allowed to serve as an impetus to metaphysical speculation, a series of potentially fertile conceptual decisions–categories, in a word–begins to emerge.

The exercise to follow will not develop anything like a comprehensive theoretical account of the respective categorical options to be considered. Rather, I offer a more modest presentation which introduces considerations following the revamped or transcendental approach to deconstruction as Part II of this essay developed that approach. These categorical considerations–which I will name "zones of specificity"–establish *unity* and *stability* as ultimate metaphysical data, directly responding to Derrida's preference for fragmentation and flux. As the essay has shown at length, Derrida's version of deconstruction implies serious and perhaps unanswerable difficulties regarding a philosophical treatment of language. This self-destructive consequence remains unavoidable unless all his accounts in this regard become subordinate to the kind of foundational perspectives presented below.

The substantive terms serving as the foundation for development in the following discussion, development inspired by and observant of deconstructive considerations, will carry with them a patina of conceptual and discursive configurations imbued with shadings of various moments which have occurred throughout the history of metaphysics. These auras will doubtless have some bearing on at least the initial direction of reformulation, what we are claiming here is a potential contribution to philosophy in general insofar as deconstruction has intersected with the venerable philosophical institution of metaphysics. The primary goal in the following accounts is to sketch provisional descriptions and adumbrated lines of argument balancing

128 The material mode articulation of logical concepts and principles is replete with metaphysical elements. It may be possible then for an adroit exercise in deconstruction to "shake," clarify and in some respects deepen the "formal" character of these logical components. See below, i.e., Assumption A in the body of the chapter.

the apparent novelty embodied in deconstructive analyses and the seemingly solid pillars of stability typically presupposed by traditional metaphysical reasoning. The history of metaphysics may not have accorded all these designations due categorical status; however, if their introduction in this essay is given at least a hearing in terms of possible relevance, then we have made effective use of the critical impetus of deconstruction as an entry into what has been suppressed or overlooked in the history of reflection on the nature and structure of reality.

This unique wedge function sanctions the need for "transcendental categories," i.e., zones of specificity. These zones display obvious metaphysical pedigree but here all include elements combining traditional extra-linguistic structure and provenance with the presence–a word carefully chosen in this context–of linguistic considerations.

The categorical designations employed, the names of the zones, may appear prosaic, lacking the *brio* which a student of Derrida might assume would be embodied in language naming these designations if such discourse were engendered from a Derridean setting. Such language, coined in this spirit, would strike a reader accustomed to the bland samenesses of the history of western thought, in particular its epochally defining metaphysical moments, as evocatively novel and vivid. Consider the designed ingenuity in Derrada's usage of "trace," an elegantly terse and incisive coinage doing serious work during highly abstract thinking. That Derrida intended something important and, within limits, stable and unmovable by using a word such as "trace" cannot be denied, as shown in the previous chapter. But the word also carries ambiguity and tenuousness whenever it appears in Derrida; the reader must conscientiously impose relevant limits on its scope yet concurrently allow the word to be applied loosely and with wavering senses of direction. By contrast, no apology is made here for the prosaic terminology the reader will soon experience. Derrida's challenge is sufficiently penetrating to require reestablishing metaphysics at a rooted, even foundational status. The zones of specificity are intended to accomplish or at least to approximate something of that status. In this regard, the explicit content and description of each zone is of far greater import philosophically than the nomenclature chosen to designate it.

10.3.1 Assumption A: The Tentative Status of "Formal" Properties

We assume that self-identity, difference, unity, and limit–formal properties termed, following (and adapting) Platonic terminology, greatest ordering kinds–can be appealed to at appropriate moments to establish the requisite conceptual and theoretical stability in the descriptions of the respective zones of specificity. This list of ordering kinds is illustrative rather than complete. The claim is that at least these realities must be in play to establish basic intelligibility concerning both linguistic and extra-linguistic orders of reality. To deny one, some or all of these formal concepts generates extensive and debilitating consequences, although now is not the place to demonstrate these effects.[129] The formal properties in question as well as logical axioms based on these properties are therefore not tentative as conditions to establish and preserve the possibility of significant discourse; in this respect they are necessary. They are tentative in a sense indicated by Alfred North Whitehead, that is, in their relation to the most fundamental principles of a metaphysics and its coordinating vision.[130]

Derrida's statements about being as presence depend implicitly for their intelligibility on unity, self-identity and difference taken as pure otherness. The substantive conclusions asserted in Derrida's critique of traditional elements of language also presuppose the formal quality of unity and the relevance of difference. But if deconstruction is to have a positive effect with the broadest possible reach in philosophical investigations, acknowledging the necessity of these formal properties is not sufficient. New language in service to philosophical ends disclosed by deconstructive investigation must presumably be sought, coined, and developed.[131] Will developing this language be controlled by rules which limit what can and cannot be said as dictated by canons of formal rightness? If so, then the structure of these canons will assume massive proportion with respect to their foundational albeit formal control of future discourse. But if these formal canons do not apply–perhaps an outcome emergent from Derrida's hint of philosophical monstrosity!–then the burden shifts to what imagination envisions will occupy the attention of future philosophers and, importantly, how they will express the results of such reflection.

Derrida's insistent questioning concerning the status of general terms prevalent throughout the history of metaphysics activates and energizes the need to inquire into what may be hidden when philosophers appeal to "formality" as the grounding condition for the ready application of such realities. We recall Derrida's own crucial distinction between sameness and identity as separate factors in the functioning of *différance*. It is hardly a stretch therefore to approach the domain of concerns at hand by adding equivalently general realities to the intersection of sameness and identity, then to situate all such "formal" concepts as subjects for future analysis, whether that analysis derives its direction and substance from deconstructive gestures or is contoured in some other way. This is a project of considerable depth and difficulty; its importance as an outgrowth of reflection on Derrida is certainly

129 Relevant sections of DFL concentrate on self-identity: 71-2, 181-5, 187-9; also on contradiction: 145-9, 161-8, 192-6, 202, 246-9n6.

130 Consider Whitehead's studied hesitancy concerning metaphysical concepts and claims, 8: "Metaphysical categories are not dogmatic statements of the obvious; they are tentative formulations of the ultimate generalities." Recall also Whitehead's appeal to imagination–see Ch7n88. The imagination is pivotal at this point in Ch. 10.

131 T. S. Eliot's lines (the third motto of this essay) adumbrate with vivid poetic clarity the transitional character of the philosophical discourse Derrida's position invites us to seek.

acknowledged, but I hope it will be obvious, given this essay's scope, why the project cannot be attempted here.

10.3.2 Assumption B: The Arena of Language

Since our concern in this chapter is primarily metaphysical, certain assumptions will be in force regarding basic concepts in the traditional philosophical approach to language.

10.3.2.1 *Substantive*

A concept such as reference is closely studied in contemporary philosophy and various options are available as far as articulating theoretically compelling accounts of its structure.[132] But that reference obtains is generally accepted and will be so here as well. We also assume that the connection between meaning and reference as well as the effective and successful actualization of linguistic events all exist as justified theoretical constructs. The question at hand concerns the most efficient and accommodating way to frame their existence within categorical dimensions which also perform metaphysical ends in addition to registering consequences applicable to language as a human phenomenon.

10.3.2.2 *Formative*

We take as givens the processes by which language evolves and the various orthographic or aural forms it may assume. Some connection may obtain between (a) the ways in which the formation of purely linguistic elements occur and (b) the referential objects of language–for example, connections depending on cultural differences. However, these process elements will here remain subservient to the identification and structure of the zones of specificity. The present context emphasizes metaphysical considerations in relation to language once language has been established as an empirical and social given, when language has become firmly settled as an ineradicable component in the lifeworld and prior to any reflective or philosophical attempt to analyze what it is and does.

---

132 For discussion, see Henry Staten 1989. And for a useful compendium of viable theories of reference, see Aloysius P. Martinich 1984.

10.4 Zones of Specificity: Applicable Contexts

A point by way of contrast: in Chapter 3, we introduced several terms of high generality–e.g., relationality and intersectionality–concerning transcendental expansion of Derridean deconstruction. These concepts were *ad hoc* devices providing a more stable foundation for approaching being as presence and also language as the vehicle for the utterance of presence. The zones of specificity introduced and discussed below encompass logical prerequisites (i.e., the formal necessities discussed above) as well as the explicit speculative content revolving around Derrida's notion of future philosophizing as "monstrous." In these respects, the zones are intended to be fully comprehensive as far as covering both end points of this continuum–the formal and the monstrous–and thus are applicable to whatever direction philosophy might pursue once it has been duly informed by deconstructive methodologies and goals.

It will be useful to indicate at the outset that the zones of specificity will be directly relevant to the following contexts, arranged in terms of breadth of concern:

10.4.1 Derrida vs. Searle

The most immediately obvious setting for the introduction of the zones of specificity is the Derrida/Searle exchange. If Searle misread Derrida, then Derrida could mount a defense against this misreading if all the zones of specificity were in place. This deployment does not guarantee, of course, that Derrida's defense adequately met Searle's criticisms (assuming these criticisms were indeed warranted). However, the point is that Derrida's canonic approach to language via deconstructive techniques does not allow him even the possibility of stating a defense, much less being so confident in its reticulated force that he could maintain, as he does, that Searle is wrong in what he attributed to Derrida's text. Derrida could indeed defend the position he advanced in his written work but only in conjunction with the zones of specificity circumscribing the structure and direction of that defense.

10.4.2 Deconstruction as Unified Program

Derrida vs. Searle is only one event in philosophical communication, dispute and discussion. The theoretical boundaries circumscribing the movement known as deconstruction are much more extensive and significant. Thus the extent to which stability characterizes referentiality and truth as elements resident in the deconstructive agenda is the extent to which it is essential to secure a determinate and commensurately stable metaphysics in order to ground the possibility that referentiality and truth can exist. In this vein, we recall (**Passage 9A**) Derrida's insistence that he never abandoned the concept of truth. Additional zones of specificity–context is a crucial case in point–are also necessary if deconstruction applied to given texts can produce positive and coherent results.

10.4.3 Accesses to a Retroactive Appreciation of "Being as Presence"

For Derrida, being as presence names the essential flow of the history of metaphysics up to and including its closure. Now we assume, as Derrida surely would, that the past–the historical past defined by the history of metaphysics–never ceases to exist and thus that it continues to exert pressure into and through future manifestations of presence. Derrida's notion of *différance* supports if not entails this assumption. If so, then the movement from the present into the future wherein lies possible rectificatory metaphysical language must be so configured to take into account this still-living dimension of history. Regardless of the radical results produced by deconstruction and whether or not these results derived their provenance solely from deconstructive applications, the philosophical investigator cannot forget about the history of the philosopher's discipline.

The residues surrounding concepts and realities drawn from the epoch of being as presence will be included, necessarily, in articulated instances of content appearing within the zones of specificity. However, these residual elements will facilitate access to those dimensions of being as presence which will play a part–even if that part be of only cameo duration–in determining what the history of metaphysics has either suppressed or overlooked. The zones of specificity preserve relevant materials within their stated limits for situating and pursuing "present" elements which may be reincorporated or at least reconsidered when moving from past philosophizing into future modes of thought.

10.4.4 Transitional Delimitations–the Past and "Future" of Metaphysics

Consider the moment or duration of Derrida's own entrance into and contributions to the history of thought. As such, he represents a transitional phase in this history, gathering elements from the past, consolidating them with his own preferences and notions, then aiming the results into the future for others to embrace and explore. The zones of specificity taken as unities are fixed conceptual limits moving along this continuum. In this respect, they function as gathering places for the work required to implement Derrida's juxtaposition of the history of metaphysics, inherently repressive and concealing, with the future of thought as a locale more open to stating what the earlier epoch of philosophy had not seen, said and written.

This transitional function ascribed to zones of specificity is illustrated by Derrida himself. At key points, Derrida situated much of his technical language according to process or transitional specifications: "quasi-transcendental," for example, which appears often in the Joyce essays,[133] and *différance* as itself provisional or, to use Derrida's own words, "relatively or provisionally" privileged (**Passage 9B**). And finally, we cite the variable ways Derrida referred to deconstruction, sometimes as not subject to formulation as a methodology, sometimes as at least partially defined by determinate structure.[134] Amidst a number of convergent or "quasi-" streams of thought, the zones of specificity exemplify stable contexts for continuing the work of transitional thinking as the epoch of metaphysics flows into the subsequent epoch of philosophical thinking defined in part by the inspirations and insight of deconstruction.

10.4.5 The Heuristic Dimension

Level 3 above emphasizes the ongoing reality of the past as it emerges into the present through the commonality of sameness generated by being as presence. The heuristic dimension of the zones of specificity combines the past and channels it toward the future in exciting a new, more percipient awareness of what the history of metaphysics has terminologically rendered as reality. This novelty may bear only tinges of resemblance to past thinking, but determining the scope of the novelty may be inspired by reflecting on being as presence as a way into the future as a sort of wellspring. Even a lingering residence in the various zones will, or should, contribute to the possibility of discovering or at least moving toward what has been concealed or suppressed throughout the history of metaphysics.

---

133 For discussion of Derrida's emphasis on "quasi-transcendental" in the Joyce essays, see DFL, 32-3. See also the interview where Derrida notes his use of this term, cited in Ch3n34.

134 As already noted (Ch8n97), Derrida fluctuated in describing the status of deconstruction with respect to methodological structure and stability. However, the extent to which Derrida countenances that some measure of structure applies to deconstruction is the extent to which articulating those structural features constitutes a "metaphilosophy," vocabulary and syntax apart from deconstruction itself which are relevant to talking "about" deconstruction and doing so without undue (if any) distortion. For Norris 2012,149, "Derrida never goes so far as his post-structuralist disciples would wish in renouncing the distinction between object-language and metalanguage...." But cf. MP, 259, where Derrida wonders: "Can a metaphilosophy be dreamed of?" See MP, 219 for additional skepticism regarding the possibility of metaphysics; also LI, 120. Regarding the notion of a metalanguage: P, 111-2, n44; PS, 13. Other passages indicate at least a provisional receptivity to the possibility of new language related to conceptuality–see WD, 19; 28, 240, 272, 273, 284; Derrida 1987, 255; Derrida 1992, 73-4. See also the Derrida citations in Ch3n43.

10.4.6 Analysis and the Monstrous....

Unless extra-linguistic reality or language as it has evolved over millennia (or both!) become drastically altered, the zones of specificity will orient the thinker toward "reality" through residence in the modes of receptivity circumscribed within each zone. At that point, it is up to the insight, creativity and sense of rigor of an individual thinker to express what has been experienced in ways approximating or verging toward the limits established by each zone. How "monstrous" these subsequent accounts will appear can be determined only by juxtaposition and comparison with some, perhaps many, of the existing accounts of reality found throughout the history of metaphysics. Of course, at this point in the history of philosophical thought we should hold open the possibility that a reality determined by predominantly deconstructive means may well be very monstrous indeed. This supposition is tendered by heeding what Derrida himself said and being prepared for...anything.

10.5 Zones of Specificity: Structural Elements

The following zones of specificity are sufficiently broad-based to assist in locating what has been hidden or suppressed within and throughout the history of metaphysics. "Specificity" refers to the fact that each of the zones brings out a certain feature or range of features of reality. "Reality" in this context conveys only that whatever exists can become the object of philosophical attention and analysis. "Zone" implies that each of the designated areas displays properties of higher-order abstraction. Thus each zone is self-identical in a way that precludes that zone becoming something other than itself. Also, each zone is one zone; although some measure of overlap in terms of the entities subsumed and referred to under any one instance of a given zone may affect any other instance of a different zone, each zone is characterized with limits which belong only to all instances of that zone. This limit with respect to the structure of any zone does not, however, affect the correlative fact that a given entity can and will fall under more than one zone of specificity.

Although the zones of specificity display a resident openness to novelty and reconfiguration, certain boundaries apply to each zone as related to all other zones. The necessity for such limits is one of the pivotal lessons we have learned from this essay and its critical approach to Derrida and deconstruction, given Derrida's emphasis on fluidity and instability as foundational realities. Each zone can nonetheless be expanded *as a zone*, that is, as a principle of organization which includes heuristic elements linked to designated details in accordance with the uniform vision of a given metaphysical system. The zones introduce limited areas of basic concern and relevance to metaphysical issues from various perspectives–some focus on entities as such, others elicit relational aspects between and among entities. The set of six is illustrative rather than exhaustive. However, I intend to identify and briefly describe zones fundamental to both a classical approach to the history of metaphysics as well as to the demands and interests of the lifeworld.

An important corollary: each of the six designations is not a category closed off to variations *within* the boundaries so designated. For example, under the zone of specificity named Particularity, individual entities can be configured according to a wide spectrum of possible anterior concepts and divisions. The history of metaphysics already attests to the rich actualization of this possibility. Furthermore, as Derrida and deconstruction have made us aware, we will maintain a circumambient and ongoing openness to new and perhaps unexpected ways of experiencing particulars as well as the resultant need to voice and inscribe those ways. As such, these articulations embody complex modes of interaction which may currently be unavailable or unrecognized by interested philosophical investigators.

Deconstructionist-inspired attempts at anything like what has been traditionally referred to as metaphysics will, to the extent that they depend on terminology borrowing from the history of that tradition, be affected–if not tainted–precisely by that unique association. If the claim is that any more concrete development of a position within each zone cannot avoid the presence of being as presence, so to speak, then either some aspect of being as presence is necessarily a component of a metaphysical scheme or the direction that Derrida intended any results of deconstructionally inspired speculation to go is beyond the currently available repertoire of terminology to exhibit.

Finally, the deployment of zones of specificity in terms of their order of succession is important from a narrative standpoint even if it may not be crucial from a logical perspective. The following set of zones is arranged according to an urgency based on Derrida's persistent and consistent skepticism regarding what can be said about concepts traditionally ascribed to the fact of language. Thus the first zone is the most important in establishing a metaphysical system which will stand as a response to theoretical constructs which deconstruction has seen fit to dislodge, with subsequent zones arranged to complement and perhaps approximate a systematic treatment of the relevant issues and elements pertaining to language as a human activity. Various logical dependencies between the zones are pointed out as they become evident and relevant but here again in this brief and introductory account I have made no attempt to canvas all possible logical entailments or dependencies as these factors may structure the set of six zones.

10.5.1 Particularity

The first zone of specificity posits existing particular entities–this stone, tree, animal, house, human being. To begin this way guarantees existence of entities invested with greater or lesser number of properties and capable of being talked and written about, of existing as something which can become of linguistic interest and concern.

These entities, and this zone, can therefore represent an unabashed and straightforward realism. The entities in question exist whether or not anyone ever takes notice of them, whether or not anyone ever talks or thinks about them. But if these entities do not exist, language does not exist. Note also that entities endowed with this kind of existential certainty are also compatible with being construed metaphysically according to idealist approaches. The condition for this possibility is that conscious awareness exists in relation to entities as a prerequisite for entities being recognized as engaged with something other than themselves, but without the correlative assertion that entities *require* such awareness either (a) to exist as entities or (b) to display connectedness with other entities.

Each entity is one thing, and in this respect each entity is unique from every other entity. The unity of each entity necessarily includes its self-identity, this in order to guarantee that $entity_1$ is not identical to $entity_2$. Thus self-identity would obtain even if only one particular entity existed as the sum total of reality. Entities are subject to processes, including change, but how they are so subject must remain unanalyzed here. Furthermore, regardless how the specific nature or character of a particular entity is represented in a given metaphysical system, a particular will be *one* particular however its internal character is then described by the metaphysics in question.

The insistence on unity and self-identity safeguards the mode of existence of particulars in order to guarantee that language as referential will have a determinate object, not just a neutral "something" existing in a purely indeterminate "other" than the fact of language itself. Thus whatever the relation may be between "transcendental signified" and particular as referent, the latter mode of existence–a given particular insofar as it has become the object of linguistic interest–guarantees the significance of language at least in the sense that discourse is assured it can aim accurately and uniquely at something other than itself.

Pending clarification of the relation just mentioned, between transcendental signified and referent, the advocate of Derrida can continue to deny that the former can be specified. And in a related context, once this problematic (according to canonic Derrida) distinction is in full view, there will continue to be a variable status given to the entity as referent, that is, insofar as an entity becomes a participant in a theory about language. As noted earlier, current philosophy of language accepts the existence of entities as the object of reference but disagrees concerning the most expeditious way to account for reference, that is, as part of a given theory of reference. The assumption here is that it is possible to develop an account of reference within a more inclusive theory of language without necessarily coming into conflict with the precise sense of Derrida's notion of the transcendental signified.

Finally, self-identity, the logical guarantor of stability necessary to a particular, maintains a reality which does not undergo alteration in terms of its predicational possibilities, in particular (given the context at hand) its applicability to a distinct particular. This position concerning self-identity and associated formal realities was adopted as one of the axiomatic assumptions (stated above) which, as such, remains independent of specification of the relevant zones.

10.5.2 Plurality

That many particulars exist is an empirical truism, but nominating Plurality as the second zone of reference ensures that a necessary metaphysical condition establishing the possibility of discourse about entities has been satisfied. Even if only one entity constituted reality as such, this singleton will generate a linguistic event only if it is sufficiently complex both to inaugurate language and to be aware of itself *as* complex (even if the complexity involved only the ability to speak and no other type of individuation–see below, the zone of specificity identified as Contrast).

A plurality by itself does not guarantee the existence of discourse since plurality as such is satisfied by, e.g., two stones–a world consisting of only two things and nothing else. Indeed, there could be many entities and many types of entities without the possibility that even at a superficial level of analysis, discourse about entities becomes realizable. Furthermore, although difference subsists under Particularity (if a particular is complex in any respect, then its constituent elements differ from one another *within* the particular as a unity), difference now also pertains to particulars insofar as they constitute a plurality, i.e., they differ from one another as unities (and, presumably, in other respects as well). An additional feature: sheer plurality also serves to ground the possibility that each particular may have an inherent disposition to exist either as solitary or to combine or be combined with other particulars to form associations of various sorts. This capacity will clearly be relevant to the notion of Context (see below).

10.5.3 Contrast

In order to realize a given instance of language, there must be, very broadly stated, a contrast between the kind of particular capable of discourse and the object of that discourse, a contrast allowing two dissimilar entities to interact. Even if the only two entities existing were human beings, one human being (and a potential user of language) would have to be completely distinct from the other human being in a spatio-temporal sense, surely a most striking contrast (especially given the impoverished character of such a possible world!). As noted under Plurality, however, it is logically possible for one entity to discourse about itself at least in the sense that this entity would be sufficiently aware of the contrast between its capacity to utter meaningful sounds and what these sounds intended to evoke about that entity.

Difference is present in both Particularity and Plurality, but more dramatically and pressingly so when difference emerges within the zone of Contrast, given the intricate functional distinctions between the language user and the particular entity referred to by that user or the particular aspect of that entity (should the entity under scrutiny be a singleton). It should also be noted that with Contrast and Plurality both in play, the possibility arises that language can have multiple referents and multiple distinct unities as referents.

Whether a qualitative distinction obtains between (a) the difference marking two distinct entities existing and construed at the level of Plurality and (b) the difference between an entity referred to in discourse and the speaker of that discourse is a question worth close examination. Derrida's position on the unavailability of the transcendental signified suggests that the answer to this question is in the affirmative–there are thus *kinds* of difference–but, again, the point remains moot pending clarification of the relation between transcendental signification and reference.

10.5.4 Evocability

It should not be assumed that language as written or verbal activity can intersect with something other than itself so as to enunciate or evoke some sort of difference. As a result, the metaphysics encompassing the nature of a particular will necessarily include a zone of linguistic specificity–Evocability. This zone represents the capacity of particular entities to enter into the activity of discourse so that some form of illumination is produced on the part of the agent of discourse, the speaker attending in some sense to an object marked by particularity. This illumination may concern the speaker in relation to the object, the object itself, or both speaker and object via the linguistic relation established between them. It is therefore a necessary feature of every particular to exist in such a way that it comes "alive" within an event of spoken or written discourse as a condition for the possibility that this particular can indeed become a component in realizing language. It may be observed that whenever this kind of event occurs, such appearance presupposes a context (see below, the zone of specificity identified as Context).

Evocability thus invites inquiry into regions of experiential media which constitute necessary conditions for implementing language–e.g., perception and imagination. The resulting connections between structural features of these media in relation to language would comprise a legacy emergent from Derrida's fluid if ultimately skeptical treatment of linguistic concepts proper.

10.5.5 Simultaneity

In *Of Grammatology*'s initial and primarily theoretical section, Derrida asserts: "Simultaneity coordinates two absolute presents, two points or instances of presence, and it remains a linear concept." We discussed (Chapter 1) the various aspects of ambivalence in this claim. For now, first consider Derrida's conclusion, that simultaneity "remains a linear concept." If we imagine two entities in close proximity to one another in time, simultaneity may be depicted conceptually as a line joining these two entities. In fact, from a strictly temporal perspective, "close proximity" is a misnomer; the two entities are identical with one another in the sense that they both occupy "the same" time. But since two entities are present and are not reducible to one another, the two entities are distinct. Therefore, it is feasible to have a line–one line–of temporality connecting both entities.

However, we argued earlier that, second, what Derrida appeals to in this passage as an "absolute present" is an empty abstraction. There can be no absolute present without an entity existing in such a way as to generate or become involved in a temporal present. Now if the present of one entity is identical to the present of another, different entity, then the two entities are simultaneous. Thus simultaneity presupposes a temporal field necessarily capable of differentiation, i.e., difference must obtain within this field in order to generate the possibility that the same time could apply to *distinct* entities. But this limited sameness need not extend to other entities. For if all entities are simultaneous with all entities, then the predication of temporal attributes becomes otiose. Furthermore, if two entities resident in the lifeworld exist simultaneously–in order to be connected linearly by this kind of temporality–these two entities do not exist in solitude. Rather, they exist within an environment encompassing other entities as well as other dimensions, both temporal and spatial. Therefore, a more accurate geometrical model would be not to simultaneity represented by a line, as Derrida contends, but rather by a *sphere.* Two entities are simultaneous insofar as they coexist within a sphere which includes time explicitly but also (a) space and (b) a plurality of entities, a minimal plurality in the sense of two entities related to one another in such a way as to engender considerations defined by temporality.

Simultaneity is an especially diversified zone of specificity and we may for the sake of illustration demarcate some of its additional structural implications. However, the following list does not exhaustively describe the structure of Simultaneity as a zone of specificity.

Simultaneity underlies the temporal relationship between a user of language and any object immediately referred to by that speaker. Consider requesting a glass of water. Simultaneity is a necessary condition for a speaker to request a glass of water, i.e., to become related to the glass of water referentially. The glass need not be immediately at hand (the individual so asked may have to get it), but unless an entity as immediately related to by language exists simultaneously with that language, the connection between language and a referent of that language could not obtain.

Simultaneity apparently presupposes Plurality since it seems intuitively credible to deny that any one being is simultaneous with itself on the grounds that the temporality so appealed to is vacuous and without effectively determined content; this justification, if accepted, implies in turn that time requires some sort of fundamental complexity as a source of necessary differentiation–e.g., the standard temporal demarcations of past, present, future.

Consider as a corollary the relation between Simultaneity and the concept of the momentary. Derrida denied the existence of the moment[135] and also, by extension, any philosophical conclusions dependent on such temporality. But allegiance to the reality of simultaneity invites scrutiny concerning the apparent necessity of the moment. This matter cannot be pursued here except to note the possibility that moment is the broader concept–thus, all instances of simultaneity occur at a moment but that all moments include simultaneity does not follow. (A singleton could exist in a moment but this moment does not exhibit simultaneity since the singleton has nothing with which to be simultaneous.)

Furthermore, spatio-temporal determinations are not elements separable from the metaphysical configuration of particulars. Rather, a particular exists necessarily with certain spatio-temporal aspects. Furthermore, these aspects are subject to continual alteration and diversification depending on what does or does not happen to a given particular.

Two (or more) entities "sharing" the same time means that they exist "in" time such that a designated moment becomes a property each of them possesses. During a linguistic event, simultaneity thus presupposes a perspective on something by an observer, whether explicitly or by implication. Simultaneity is therefore perspectival in that an observer must indicate a given slice of time such that only these beings coexist "within" that slice.

10.5.6 Context

The preservation of immediacy becomes a basic concern for all zones of specificity. What is simultaneous is immediate but the converse does not follow since immediacy can also exhibit motion while mobile objects are moving through time. Thus immediacy appears as ground in terms of spatio-temporal characteristics situating emergent particular entities, co-existence of a plurality of entities, any and all contrasting properties of or perspectives on these entities, and the capacity to receive and codify data from all these elements into linguistic expressions employed by speakers for widely diverse purposes.

What gives unity to these strands of immediacy is Context. In fact, a context becomes a link in one of Derrida's "chains," a technical term Derrida introduced in order to account for the generation of meaning. Only if words are related to an ensemble of entities and relations can they consolidate in such a way that one link, that is, a given word, can emerge as an element or a member in a chain. Context becomes an environmental zone, as it were, in that the variegated elements constitute a unity

135 Derrida frequently denied the reality of the moment–see Ch8n108.

which must be present in order for discourse to become referential and meaningful. A context displays discernibly relevant limits. Within these limits, a linguistic event can and does occur. Although the unity of a context is flexible, the limits of a given context are determinable as necessary metaphysical conditions for the possibility of referentiality and meaning emerging from *within* those limits.

Two corollaries: Derrida's deconstructionist approach shows that contexts or, if the materials constituting a context are taken more broadly, event, must be included in any metaphysics. This conclusion follows if a given metaphysical configuration, a grouping of entities and discriminations of these groupings, also includes the capacity to articulate any perspective on that configuration. Such articulation is realized through discourse so that any reaction to any entities whatsoever will constitute an event. Of course, the concept of an event can certainly include instances characterized by entities and relations between entities other than those involving discourse. This possibility when actualized necessitates distinguishing between *types* of events (i.e., linguistic and non-linguistic) but it in no way negates the necessity to include event as a basic component of any such metaphysical system.[136]

The second corollary: if context necessarily includes linguistic activity, then (a) all contexts will include at least one human agent whether explicitly or indirectly (e.g., as a natural setting produced by imagination) and (b) all contexts are events but not all events are contexts. This logical dependency follows since, as noted, entities could coexist symbiotically with one another without an observer reacting to this entitative nexus by use of language of any sort.

Six zones of specificity. Their developments, although brief, speculative and provisional, are justified by reasoning. Furthermore, it may be emphasized that these sketches outline sets of broadly stated parameters which sanction the development of a considerable number of concrete theoretical options within each zone. These options await attention and analysis, depending on the goal of metaphysical inquiry. Furthermore, the linguistic components included in each of the six developments reflect the influence of Derrida and the attempt, inherent in the very project of promulgating such zones of specificity, to appreciate the incisive relevance of deconstruction as a philosophical position with a measure of theoretical force. This force includes an

136 The lack of punctuation in the essay title "Signature Event Context" conveys to the reader that Derrida intends to show the inherent malleability and indeterminacy of what can seemingly be identified with at least a measure of rigor as an "event." See MP, 309-30; LI, 1-25. For extended discussion on this possibility, see Michael Marder, 2009. The "ontology of events" is a recognized area of inquiry among analytic philosophers. For a very different approach to a metaphysical interpretation of an "atomic event," see Cynthia Macdonald 2005, 181-215. Christopher Norris 2012, 84, insists that deconstruction is intimately connected to "a great range of real-world objects and events" as well as "various modes of interpersonal discourse" involving "speakers' (and authors') intentions." Norris' strong and concerted realist imprint on Derrida's writing would seem to require detailed argumentation in the face of texts which appear to lean in another direction concerning the metaphysical underpinnings of such a possibility.

undeniable dimension of empirically driven factors. Although Derrida's texts may, at times, suggest a covert idealism, the zones of specificity are as noted earlier sufficiently broad-based and ampliative to accommodate any systematic metaphysical position. Such liberality is not conceptual looseness but rather respectfully reflects possibilities inherent in the movement of and implications derived from Derrida's thought.

10.6 Conclusion

This essay has culminated in an ensemble of metacategorical descriptions defined by considerable conceptual latitude. These zones of specificity address aspects of Derrida's basic overall position–being as presence and the application of being as presence to the formation of deconstruction–which, as the essay has argued throughout, are problematic by virtue of their studied if not avowedly skeptical indeterminacy. Although the inspiration to generate and describe these zones, however rudimentary these descriptions may be, results from an ultimately positive reading of Derrida, I envisage two probable critical responses from those followers of Derrida's work who remain in spirited sympathy with his overall project. These responses may be broadly characterized and are quite distinct.

First response: rejecting the entire metacategorical schematic deployed under the rubric of "zones of specificity" since the very endeavor to secure categories of any stripe results from a fundamental misreading of Derrida.[137]

I submit that the most effective justification for this rejection would be review of all passages from Derrida's works introduced in this essay (as well as supplementary passages if the critic deems them warranted) and a demonstration, by argument, that the sequence of passages examined as well as the interpretive directions animating the entire essay are either unfounded, premature, or, at different moments, both unfounded *and* premature. If such a critical rejoinder were comprehensive and sufficiently justified, it would be entirely possible that the need to address an alternate approach to traditional metaphysical concerns as attempted in this chapter would not arise.

Second response: appreciation for the relevance and feasibility of attempting secure and clearly delimited metacategorization but criticism, again by argument, directed toward either the entire slate of zones as a set or some one or several members of the six zones outlined above.

In sum, the response in its first alternative would be reasoned denial that categories of any sort are required–or, more fundamentally, that such categories are

---

137 Paola Marrati argues for the systematic impossibility of categorizing Derrida's thought–see Marrati, especially 196. The concluding sections of Chapter 10 represent such categorization, or at least an attempt at that project in the spirit of combining substantive possibilities resident in Derrida and common to classical metaphysics

even possible; the second alternative involves a comprehensive and entirely new metacategorical scheme or, less drastically, selective revisions and modifications of the existing approach arranged as a set of interrelated zones of specificity.

To those adopting the first approach, the question becomes: will post-Derrida philosophy be anything like what it has been for almost three millennia; if so, what sort of discourse and what discriminations should be in place in order to avoid, or to circumvent intelligently and with due cause, the excesses, pitfalls, and omissions marking the history of metaphysics since its origin? In short, how will the conversation of philosophy continue? Will it be Nietzschean as isolated dicta for the most part unseasoned by argument or will it take some other, non-discursive narrative form?

To those following the second approach, the question is more limited–which zone (or zones) of specificity should be modified (or replaced) and how should such modification, whether singular or plural, be pursued? Or if the entire approach in selecting this set of zones has been misguided, what rationale should serve as a preliminary guide for engendering a completely distinct set of categories? For those who elect this option and work to implement it, philosophy will proceed as it has since Aristotle. Early in his formative development, Aristotle enumerated a system of categories which for him better "fit" reality than counterpart elements advanced by his illustrious teacher. In a similar vein, the resulting post-Derrida disagreements will concern how to partition reality and how to justify those partitions. As a result, ongoing discussion in philosophy will address matters of moment concerning how best to understand and characterize reality rather than the intermediary issue of how language, whatever its structure (or lack of same) may be, buffers that understanding. For what it is worth, the former engagements are, for the author, more valuable and more informative than the more localized types of inquiry which typically appear under the latter heading.

***

This essay has taken tentative and provisional steps in a certain direction as far as critically responding to the early, foundational phase of Derrida's overall body of work. The analysis and attempted reconstruction of Derrida's notion of being as presence along with the extended amplification and application of that notion to linguistic considerations followed a circuitous but, at least by intent, organically coherent and connected route. As a result, those students of philosophy with an interest in Derrida should at least have acquired a greater awareness of some of the major implications and problems that arise from his theoretical work on presence and language.

The reader must decide whether to endorse the leading conclusions of this essay, to continue to read Derrida as an agent of provocation or to move elsewhere along the broad canvasses of available modes and styles of present or past philosophizing. This decision ranges over many options, one of which is, again, simply ignoring everything Derrida said and wrote. But even if one were to elect the path marked by a completely non-Derrida world view, we remain in debt to Derrida for raising issues about metaphysics as well as the relation between metaphysics and language, and discoursing about these matters in ways which provoke concerted thoughtful attention. This essay is one such response. The reader must determine whether Derrida's texts warrant additional critical and speculative energy for purposes of eliciting other interpretive possibilities as ways of continuing and advancing the adventure of philosophy.

Bibliography

Aristotle. (1941). *Metaphysics, Physics*. In *The Basic Works of Aristotle*. Richard McKeon. (Ed.). New York, NY: Random House.

Attridge, Derek (Ed.). (1992). *Acts of Literature*. London and NY: Routledge.

Augustine, St. (2006). *Confessions*. Indianapolis: Hackett.

Bacon, Francis. (1960). *The New Organon and Related Writings*. New York: Prentice-Hall.

Baker, Lynne Rudder. (2000). *Metaphysics of Everyday Life*. Cambridge: Cambridge University Press.

Baldwin, Thomas. (2008). "Presence, Truth and Authenticity." In S.Glendenning & R. Eaglestone. (Eds.), *Derrida's Legacies Literature and Philosophy* (pp. 107-17). New York: Routledge.

Bennington, Geoffrey. (2000). *Interrupting Derrida*. London: Routledge.

Berman, Art. (1988).*From the New Criticism to Deconstruction*. Urbana: University of Illinois Press.

Bernasconi, Robert. (1989). "Seeing Double: *Destruktion* and Deconstruction." In D. Michelfelder (Ed.), *Dialogue and Deconstruction The Gadamer-Derrida Encounter* (pp. 233-50). Albany: SUNY Press.

Brann, Eva T. H. (1991). *The World of the Imagination*. Lanham: Rowman & Littlefield.

Caputo, John D. (1989). "Gadamer's Closet Essentialism: A Derridean Critique." In D. Michelfelder (Ed.), *Dialogue and Deconstruction The Gadamer-Derrida Encounter* (pp. 258-64). Albany: SUNY Press.

________. (1993). "In Search of the Quasi-Transcendental: The Case of Derrida and Rorty." In G. Madison (Ed.). *Working through Derrida* (pp. 147-69). Evanston: Northwestern University Press.

________. (1997). *Deconstruction in a Nutshell: A Conversation with Jacques Derrida*. New York: Fordham University Press.

Chalmers, David, D. Manley, R. Wasserman (Eds.). (2009). *Metametaphysics New Essays on the Foundations of Ontology*. Oxford: Clarendon Press.

Churchland, Paul. (2013). *Matter and Consciousness*. 3rd. Ed. Cambridge: MIT Press.

Cohen, Tom (Ed.). (2001). *Jacques Derrida and the Humanities*. Cambridge: Cambridge University Press.

Cornell, Drucilla, M. Rosenfeld, D. G. Carlson (Eds.). (1997). *Deconstruction and the Possibility of Justice*. New York: Routledge.

Critchley, Simon. (1992). *The Ethics of Deconstruction Derrida & Levinas*. Oxford: Blackwell.

Culler, Jonathan. (1983). *On Deconstruction: Theory and Criticism*. London: Routledge.

Dasenbrock, Reed Way. (Ed.). (1989). *Redrawing the Lines: Analytic Philosophy, Deconstruction, and Literary Theory*. Minneapolis: University of Minnesota Press.

Daylight, Russell. (2011). *What if Derrida was wrong about Saussure?* Edinburgh: Edinburgh University Press.

Dennett, Daniel. (1991). *Explaining Consciousness*. Boston: Little, Brown and Company.

Derrida, Jacques. (1967a). *La Voix et le Phénomène*. Paris: Presses Universitaires de France.

________. (1967b). *De la Grammatologie*. Paris: Les Editions de Minuit

________. (1967c). *L'écriture et la Différence*. Paris: Éditions du Seuil

________. (1973). *Speech and Phenomena*. Translated by Allison, David B. Evanston: Northwestern University Press.

________. (1976). *Of Grammatology*. Translated by Spival, Gayatri Chakravorty. Baltimore: Johns Hopkins University Press.

________. (1981). *Positions*. Translated by Bass, Alan. Chicago: University of Chicago Press.

________. (1982). *Margins of Philosophy*. Translated by Bass, Alan. Chicago: University of Chicago Press.

________. (1987). "Some questions and responses." In N. Fabb & D. Attridge (Eds). *The Linguistics of Writing* (pp. 252-64). New York: Methuen.

________. (1988). *Limited Inc*. Translated by Weber, Samuel. Evanston: Northwestern University Press.

________. (1992)."This strange institution called literature." In D. Attridge (Ed.). (1992). *Acts of Literature* (pp. 33-75). London and NY: Routledge.

________. (2001)."As if it were possible, "within such limits"...". In M. Meyer (Ed.). (2001). *Questioning Derrida: With His Replies on Philosophy* (pp. 96-119). Aldershot: Ashgate.

________. (2002a) *Negotiations*. E. Rottenberg (Ed.). Stanford: Stanford University Press.

________. (2002b). *Who's Afraid of Philosophy*? Translated by Plug, Jan. Stanford: Stanford University Press.

________. (2004). "Deconstruction and the Other." In R. Kearney (Ed.). (2004). *Debates in Continental Philosophy* (pp. 139-56). New York: Fordham University Press.

________. (2007). *Psyche Inventions of the Other*, Vol. 1. P. Kamuf, P. & E. Rottenberg (Eds.). (2007). Stanford: Stanford University Press.

Descombes, Vincent. (2002). "Difference." In Z. Direk & L. Lawlor (Eds.). (2002). *Jacques Derrida Critical Assessments of Leading Philosophers*. (pp. 30-43, Vol III). Vol. II, III. London & New York: Routledge.

Dewey, John. (1958). *Experience and Nature*. 2d ed. New York: Dover.

Direk, Zeynep & Lawlor, Leonard. (Eds.). (2002). *Jacques Derrida Critical Assessments of Leading Philosophers*. Vols II, III. London & New York: Routledge.

Ellis, John M. (1989). *Against Deconstruction*. Princeton: Princeton University Press.

Evans, J. Claude. (1991). *Strategies of Deconstruction Derrida and the Myth of the Voice*. Minneapolis: University of Minnesota Press.

Fabb, Nigel and Attridge, Derek. (Eds). (1987). *The Linguistics of Writing*. New York: Methuen.

Farrell, Frank B. (1988). "Iterability and Meaning: The Searle-Derrida Debate." *Metaphilosophy*, 23, 53-64.

Fynsk, Christopher. (2001). "Derrida and Philosophy: Acts of Engagement." In T. Cohen (Ed.). (2001). *Jacques Derrida and the Humanities* (pp. 152-71). Cambridge: Cambridge University Press.

Gadamer, Hans-Georg. (1989). "Letter to Dahlmeyer." In D. Michelfelder & R. Palmer (Eds.). (1989). *Dialogue and Deconstruction The Gadamer-Derrida Encounter* (pp. 93-101). Albany: SUNY Press.

Gasché, R. (1986). *The Tain of the Mirror: Derrida and the Philosophy of Reflection*. Cambridge: Harvard University Press.

________. (1989). "Infrastructures and Systematicity." In J. Sallis (Ed.). (1989). *Deconstruction and Philosophy* (pp. 3-20). Chicago: University of Chicago Press.

Glendinning, Simon & Eaglestone, Robert. (Eds). (2008). *Derrida's Legacies Literature and Philosophy*. New York: Routledge.

Habermas, Jürgen. (1987). *The Philosophical Discourse of Modernity*. Translated by Lawrence, Frederick. Cambridge: MIT Press.

Haddad, Samir. (2013). *Derrida and the Inheritance of Democracy*. Bloomington: Indiana University Press.

Harvey, Irene. (1986). *Derrida and the Economy of Différance*. Bloomington: Indiana University Press.

Hill, Leslie. (2007). *The Cambridge Introduction to Jacques Derrida*. Cambridge: Cambridge University Press.

Hobson, Marian. (1998). *Jacques Derrida Opening Lines*. London: Routledge.

Hodge, Joanna. (2007). *Derrida on Time*. New York: Routledge.

Hume, David. (1993). *An Enquiry Concerning Human Understanding.* Steinberg, E. (Ed.). 2d ed. Indianapolis: Hackett.

Kearney, Richard. (Ed.). (2004).*Debates in Continental Philosophy.* New York: Fordham University Press.

Keller, Catherine, & Daniell, Anne. (Eds). (2002). *Process and Difference: Between Cosmological and Postructuralist Post Modernisms.* Albany: State University of New York Press.

Kirk, G.S. & Raven, J. E. (1957). *The Presocratic Philosophers.* Cambridge: University Press.

Lawlor, Leonard. (2002). *Derrida and Husserl: The Basic Problem of Phenomenology.* Bloomington: Indiana University Press.

Leibniz, G. W. (1991). *Discourse on Metaphysics and other Essays.* Translated by Garber, Daniel & Ariew, Roger. Indianapolis: Hackett.

Lucy, Niall. (1995). *Debating Derrida.* Melbourne: Melbourne University Press.

Macdonald, Cynthia. (2005). *Varieties of Things: Foundations of Contemporary Metaphysics.* Oxford: Blackwell.

Madison, Gary. (Ed.). (1993). *Working through Derrida.* Evanston: Northwestern University Press.

Marder, Michael. (2009). *The Event of the Thing Derrida's Post-deconstructive Realism.* Toronto: University of Toronto Press.

Marrati, Paola. (2005). *Genesis and Trace Derrida Reading Husserl and Heidegger.* Stanford: Stanford University Press.

Martinich, Aloysius P. (1984). *Communication and Reference.* Berlin: de Gruyter.

Meyer, Michel. (Ed.). (2001). *Questioning Derrida: With His Replies on Philosophy.* Aldershot: Ashgate.

Michelfelder, Diane P. & Palmer, Richard E. (Eds.). (1989). *Dialogue and Deconstruction The Gadamer-Derrida Encounter.* Albany: SUNY Press.

Miller, J. Hillis. (2002). "Derrida's Others." In Z. Direk & L. Lawlor (Eds.). (2002). *Jacques Derrida Critical Assessments of Leading Philosophers.* Vols II, III. (pp. 325-39). London & New York: Routledge.

Mitchell, W. J. T. (1985). *Against Theory Literary Studies and the New Pragmatism.* Chicago: University of Chicago Press.

Mouffe, Chantal. (Ed.). (1996). *Deconstruction and Pragmatism.* London: Routledge.

Naas, Michael. (2003). *Taking on the Tradition Jacques Derrida and the Legacies of Deconstruction.* Stanford: Stanford University Press.

________. (2008). "Derrida's America." In S. Glendinning & R. Eaglestone (Eds.). (2008). *Derrida's Legacies Literature and Philosophy* (pp. 118-37). New York: Routledge.

________. (2012). *Miracle and Machine: Jacques Derrida and the Two Sources of Religion, Science, and the Media.* Bronx, NY: Fordham University Press.

Norris, Christopher. (1983). *The Deconstructive Turn: Essays in the Rhetoric of Philosophy.* London: Methuen.

________. (1989). "'Philosophy is *Not* Just a Kind of Writing': Derrida and the Claim of Reason." In R. W. Dasenbrock (Ed.). *Redrawing the Lines: Analytic Philosophy, Deconstruction, and Literary Theory* (pp. 189-203). Minneapolis: University of Minnesota Press.

________. (1990). *What's Wrong with Postmodernism.* Baltimore: Johns Hopkins University Press.

________. (2001). "Deconstruction, Ontology, and Philosophy of Science: Derrida on Aristotle." In M. Meyer (Ed.). *Questioning Derrida: With His Replies on Philosophy* (pp. 39-65). Aldershot: Ashgate.

________. (2012). *Derrida, Badiou and the Formal Imperative.* London: Continuum.

Oxenhandler, Neil. (1989). "The Man with Shoes of Wind: The Derrida-Gadamer Encounter." In D. Michelfelder & R. Palmer (Eds.). *Dialogue and Deconstruction The Gadamer Derrida Encounter* (pp. 265-68). Albany: SUNY Press.

Perraja, Luis G. (1999). "Whitehead, Deconstruction, and Postmodernism." *Process Studies*, 28, 68-84. Reprinted in C. Keller & A. Daniell (Eds.). (2002). *Process and Difference: Between Cosmological and Postructuralist Post Modernisms* (pp. 73-90). Albany: State University of New York Press.

Plato. (1961). *The Collected Dialogues of Plato*. Hamilton, Edith & Cairns, Huntington. (Eds.). Princeton: Princeton University Press.

_______. (1991). *The Dialogues of Plato. Volume II. The Symposium*. Translation & commentary by Allen, R. E.. New Haven: Yale University Press.

Priest, Graham. (1994). "Derrida and Self-reference." *Australasian Journal of Philosophy*, 72, 103-11.

_______. (2006). *Doubt Truth To Be a Liar*. Oxford: Oxford University Press.

Rorty, Richard. (1982). "Philosophy as a Kind of Writing." In *Consequences of Pragmatism: Essays 1972-1980* (pp. 90-109). Minneapolis: University of Minnesota Press.

_______. (1985). "Philosophy without Principles." In W. J. T. Mitchell. *Against Theory Literary Studies and the New Pragmatism* (pp. 132-38). Chicago: University of Chicago Press.

_______. (1989). "Two Meanings of "Logocentrism": A Reply to Norris." In R. Dasenbrock (Ed). *Redrawing the Lines: Analytic Philosophy, Deconstruction, and Literary Theory* (pp. 204-16). Minneapolis: University of Minnesota Press.

_______. (1991). "Is Derrida a Transcendental Philosopher?"In *Essays on Heidegger and Others Philosophical Papers*, Vol. 2 (pp. 103-10). Cambridge: Cambridge University Press.

Sallis, John, (Ed). (1989). *Deconstruction and Philosophy*. Chicago: University of Chicago Press.

Shusterman, Richard. (1992). *Pragmatist Aesthetics*. Oxford: Blackwell.

Sokoloff, William M. (2005). "Between Justice and Legality: Derrida on Decision." *Political Research Quarterly*, 58, 341-52.

Staten, Henry. (1989). "The Secret Name of Cats: Deconstruction, Intentional Meaning, and the New Theory of Reference." In R. Dasenbrock (Ed). *Redrawing the Lines: Analytic Philosophy, Deconstruction, and Literary Theory* (pp. 27-48). Minneapolis: University of Minnesota Press.

Stocker, Barry. (2006). *Derrida on Deconstruction*. London: Routledge.

Wheeler, Samuel C. (2000). *Deconstruction as Analytic Philosophy*. Stanford: Stanford University Press.

White, David A. (1978). *Heidegger and the Language of Poetry*. Lincoln: University of Nebraska Press.

_______. (1985). *Logic and Ontology in Heidegger*. Columbus: Ohio State University Press.

_______. (1989). *Myth and Metaphysics in Plato's* Phaedo. Selinsgrove: Susquehenna University Press

_______. (1992). "On the Limits of Classical Reason: Derrida and Aristotle." *The Journal of the British Society for Phenomenology*, 23, 120-6.

_______. (1993). *Rhetoric and Reality in Plato's* Phaedrus. Albany: State University of New York Press.

_______. (2011). *Derrida on Formal Logic An Interpretive Essay.* Lanham: Lexington Books.

Whitehead, Alfred North. (1978). *Process and Reality*. New York: Free Press.

Wills, David. (2001). "Glossary." In T. Cohen (Ed.). *Jacques Derrida and the Humanities* (pp. 315-23). Cambridge: Cambridge University Press.

Wood, David & Bernasconi, Robert. (Eds.). (1988). *Derrida and Différance*. Evanston: Northwestern University Press.

Wortham, Simon Morgan. (2010). *The Derrida Dictionary*. London: Continuum.

Index

FSC
www.fsc.org
MIX
Papier | Fördert gute Waldnutzung
FSC® C083411

Kolibri 360 GmbH
Ferdinand-Jühlke-Straße 7
99095 Erfurt, Deutschland
produktsicherheit@kolibri360.de